Edge/Fog Computing Technologies for IoT Infrastructure

Editors

Taehong Kim
Seong-eun Yoo
Youngsoo Kim

MDPI • Basel • Beijing • Wuhan • Barcelona • Belgrade • Manchester • Tokyo • Cluj • Tianjin

Editors

Taehong Kim
Chungbuk National University
Cheongju
Korea, South

Seong-eun Yoo
Daegu University
Gyeongsan
Korea, South

Youngsoo Kim
Korea Institute for
Defense Analyses
Seoul
Korea, South

Editorial Office
MDPI
St. Alban-Anlage 66
4052 Basel, Switzerland

This is a reprint of articles from the Special Issue published online in the open access journal *Sensors* (ISSN 1424-8220) (available at: www.mdpi.com/journal/sensors/special_issues/Edge_Fog_IoT).

For citation purposes, cite each article independently as indicated on the article page online and as indicated below:

LastName, A.A.; LastName, B.B.; LastName, C.C. Article Title. *Journal Name* **Year**, *Volume Number*, Page Range.

ISBN 978-3-0365-1456-7 (Hbk)
ISBN 978-3-0365-1455-0 (PDF)

Contents

About the Editors

Taehong Kim

Taehong Kim received his B.S. degree in computer science from Ajou University, Korea, in 2005, and his M.S. degree in information and communication engineering from Korea Advanced Institute of Science and Technology (KAIST) in 2007. He received his Ph.D. degree in computer science from KAIST in 2012. He worked as a research staff member at the Samsung Advanced Institute of Technology (SAIT) and Samsung DMC R&D Center from 2012 to 2014. He also worked as a senior researcher at the Electronics and Telecommunications Research Institute (ETRI), Korea, from 2014 to 2016. Since 2016, he has been an associate professor with the School of Information and Communication Engineering, Chungbuk National University, Korea. He has been an associate editor of IEEE Access since 2020. His research interests include edge computing, SDN/NFV, the Internet of Things, and wireless sensor networks.

Seong-eun Yoo

Seong-eun Yoo received a BS degree in electronics and computer engineering from Hanyang University, Seoul, Korea, in 2003 and MS and PhD degrees in information and communications engineering from Korea Advanced Institute of Science and Technology (KAIST), Daejeon, Korea, in 2005 and 2010, respectively. Since September 2010, he has been a faculty member with the School of Computer and Communication Engineering, Daegu University, Gyeongsan, Korea. His research interests include real-time communication in wireless sensor networks and Internet of Things, and real-time embedded systems.

Youngsoo Kim

Youngsoo Kim is a researcher in the Battlefield Informatization Lab of the Military Development Research Center of the Korea National Defense Research Institute (KIDA). He received a bachelor's degree in computer science from the Republic of Korea Air Force Academy (KAFA) in 1994, a master's degree in computer science engineering at Sogang University in 2001, and a doctorate in computer science engineering, Korea Advanced Institute of Science and Technology (KAIST) in 2009. His research interests include artificial intelligence, IoT, cloud computing, and edge/fog computing.

sensors

MDPI

ditorial

:dge/Fog Computing Technologies for IoT Infrastructure

aehong Kim [1], Seong-eun Yoo [2] and Youngsoo Kim [3,*]

1 School of Information and Communication Engineering, Chungbuk National University, Chungbuk 28644, Korea; taehongkim@cbnu.ac.kr
2 School of Computer and Communication Engineering, Daegu University, Gyeongsan 38453, Korea; seyoo@daegu.ac.kr
3 Military Development Research Center, Korea Institute for Defense Analyses, Seoul 02455, Korea
* Correspondence: pineland@kida.re.kr; Tel.: +82-2-961-1378

itation: Kim, T.; Yoo, S.-e.; Kim, Y. :dge/Fog Computing Technologies or IoT Infrastructure. *Sensors* **2021**, *1*, 3001. https://doi.org/).3390/s21093001

eceived: 14 April 2021
.ccepted: 23 April 2021
ublished: 25 April 2021

The prevalence of smart devices and cloud computing has led to an explosion in the amount of data generated by IoT devices. Moreover, emerging IoT applications, such as augmented and virtual reality (AR/VR), intelligent transportation systems, and smart factories require ultra-low latency for data communication and processing. Fog/edge computing is a new computing paradigm where fully distributed fog/edge nodes located nearby end devices provide computing resources. By analyzing, filtering, and processing at local fog/edge resources instead of transferring tremendous data to the centralized cloud servers, fog/edge computing can reduce the processing delay and network traffic significantly. With these advantages, fog/edge computing is expected to be one of the key enabling technologies for building the IoT infrastructure.

Aiming to explore the recent research and development on fog/edge computing technologies for building an IoT infrastructure, this Special Issue collected several dozens of submissions and finally published 10 articles (one review and nine full-length articles) after the thorough review process. The selected articles cover diverse topics such as resource management, service provisioning, task offloading and scheduling, container orchestration, and security on edge/fog computing infrastructure, which can help to grasp the recent trends, as well as the state-of-the-art algorithms of the fog/edge computing technologies.

The review article "Resource Management Techniques for Cloud/Edge and Edge Computing: An Evaluation Framework and Classification" authored by Adriana Mijuskovic et al. [1] provides a comprehensive review on the resource management techniques applied for cloud, fog, and edge computing. They first classify various techniques into six classes, such as discovery, load-balancing, off-loading, deployment, QoS (quality of service), and energy management according to the goal and methodologies, and then analyze each algorithm from the viewpoint of the resource management types, such as resource allocation, workload balance, resource provisioning, and task scheduling.

The next article "Optimal Service Provisioning for the Scalable Fog/Edge Computing Environment" written by Jonghwa Choi and Sanghyun Ahn [2] proposes a service provisioning algorithm to optimally place service images for the service demands obtained from the prior time interval. They propose two heuristic algorithms such as MC-SP (maximal coverage service provisioning) and FC-SP (flexible coverage service provisioning) based on the logical fog network. The evaluation results show better performance than the on-demand resource provisioning mechanism in terms of the number of service image placements and the network cost per service request.

In the article [3], Md Delowar Hossain et al. propose a fuzzy decision-based task offloading management (FTOM) scheme in multi-tier MEC (multi-access edge computing) systems. The FTOM scheme selects an optimal target node to offload the task based on the server resource status and network condition, where it is designed to prefer the local or nearby servers by considering the latency sensitiveness of the tasks. The performance evaluations show that the proposed scheme outperforms the existing scheme in terms of the successful task offloading rate and the task completion time.

Another article, "Dynamically Controlling Offloading Thresholds in Fog Systems" authored by Faten Alenizi and Omer Rana [4], proposes an offloading scheme that can dynamically adjust the threshold to offload the tasks based on two algorithms such as the dynamic task scheduling (DTS) and dynamic energy control (DEC). The experimental results prove that the delay and the throughput can be improved by dynamically adjusting the ratio of tasks to be processed locally and the tasks to be offloaded based on the resource status of the local fog node and its neighboring nodes.

The article "Deep Reinforcement Learning-Based Task Scheduling in IoT Edge Computing" authored by Shuran Sheng et al. [5] formulates a Markov decision process (MDP) model for the resource allocation and task scheduling problem in the IoT edge computing environment, where computation-intensive tasks are scheduled and processed by an individual virtual machine with heterogeneous capacity. They apply deep reinforcement learning (DRL) to solve the MDP problem, and they demonstrate that the proposed algorithm achieves better performance in terms of task success ratio and the cumulative task satisfaction degree than the benchmark methods.

The next article "Efficient Implementation of NIST LWC ESTATE Algorithm Using OpenCL and Web Assembly for Secure Communication in Edge Computing Environment" written by Bosun Park and Seog Chung Seo [6] proposes methods to efficiently operate the ESTATE crypto algorithm using Web Assembly and OpenCL parallel processing for securing edge computing applications. The experimental evaluation shows that the proposed mechanism is five times faster than the C implementation by simultaneously encrypting data to be transmitted to multiple devices through OpenCL parallel processing.

In the article "Identification of IoT Actors" [7], Suada Hadzovic et al. focus on clarifying the identification of IoT actors used in one thousand IoT-related standards. Based on the five layers of computing paradigm such as cloud, fog, edge, mist, and dew computing, they define the IoT model by mapping diverse IoT actors as well as IoT components such as a thing, gateway, service, user, etc., which is expected to provide a clearer clarification of the blurred definition of IoT actors and their relationship.

The article "Hyper-Angle Exploitative Searching for Enabling Multi-Objective Optimization of Fog Computing" authored by Taj-Aldeen Naser Abdali et al. [8] formulates a novel fog computing optimization framework to achieve multiple objectives, such as time latency, energy consumption, energy distribution, renting cost, and reliability at the same time. They propose an HAES (hyper-angle exploitative searching) algorithm to prioritize solutions within the same rank. The evaluation results show that HAES outperforms the benchmark protocols in terms of various performance metrics, such as the hyper volume measure, the number of non-dominated solutions, generational distance measure, delta metric, and the set of coverage measure.

The last two articles focus on the Kubernetes container orchestration platform. The article "Horizontal Pod Autoscaling in Kubernetes for Elastic Container Orchestration" authored by Thanh-Tung Nguyen et al. [9] addresses that the Kubernetes provides diverse autoscaling mechanisms to support the high availability and scalability of the service. They investigate HPA (horizontal pod autoscaling) through diverse experiments. By including the comparison of the difference between Kubernetes resource metrics (KRM) and Prometheus custom metrics (PCM), they provide a detailed analysis and lessons to optimize the performance of HPA in the Kubernetes cluster.

The next article, "Balanced Leader Distribution Algorithm in Kubernetes Clusters" authored by Nguyen Dinh Nguyen and Taehong Kim [10], focuses on the stateful applications requiring a strong consistency of data among the replicas. The authors address that the leader-based consistency mechanisms may lead to a workload imbalance problem that a specific node with multiple concentrated leaders suffers from the heavy load due to its inherent design. They propose a balanced leader distribution algorithm to overcome the problem, and the experimental evaluations prove that distributing the leaders throughout nodes in the cluster improves the overall throughput of the cluster as well.

Funding: This work was supported by the National Research Foundation of Korea (NRF) grant funded by the Korea government (MSIT) (No. 2020R1F1A1048179).

Acknowledgments: The guest editors would like to thank all the authors for their submission to this Special Issue. We are also grateful to the reviewers for their valuable and constructive comments on the manuscripts.

Conflicts of Interest: The authors declare no conflict of interest.

References

1. Mijuskovic, A.; Chiumento, A.; Bemthuis, R.; Aldea, A.; Havinga, P. Resource Management Techniques for Cloud/Fog and Edge Computing: An Evaluation Framework and Classification. *Sensors* **2021**, *21*, 1832. [CrossRef]
2. Choi, J.; Ahn, S. Optimal Service Provisioning for the Scalable Fog/Edge Computing Environment. *Sensors* **2021**, *21*, 1506. [CrossRef] [PubMed]
3. Hossain, M.D.; Sultana, T.; Hossain, M.A.; Hossain, M.I.; Huynh, L.N.T.; Park, J.; Huh, E.-N. Fuzzy Decision-Based Efficient Task Offloading Management Scheme in Multi-Tier MEC-Enabled Networks. *Sensors* **2021**, *21*, 1484. [CrossRef] [PubMed]
4. Alenizi, F.; Rana, O. Dynamically Controlling Offloading Thresholds in Fog Systems. *Sensors* **2021**, *21*, 2512. [CrossRef]
5. Sheng, S.; Chen, P.; Chen, Z.; Wu, L.; Yao, Y. Deep Reinforcement Learning-Based Task Scheduling in IoT Edge Computing. *Sensors* **2021**, *21*, 1666. [CrossRef] [PubMed]
6. Park, B.; Seo, S.C. Efficient Implementation of NIST LWC ESTATE Algorithm Using OpenCL and Web Assembly for Secure Communication in Edge Computing Environment. *Sensors* **2021**, *21*, 1987. [CrossRef] [PubMed]
7. Hadzovic, S.; Mrdovic, S.; Radonjic, M. Identification of IoT Actors. *Sensors* **2021**, *21*, 2093. [CrossRef] [PubMed]
8. Naser Abdali, T.-A.; Hassan, R.; Mohd Aman, A.H.; Nguyen, Q.N.; Al-Khaleefa, A.S. Hyper-Angle Exploitative Searching for Enabling Multi-Objective Optimization of Fog Computing. *Sensors* **2021**, *21*, 558. [CrossRef] [PubMed]
9. Nguyen, T.-T.; Yeom, Y.-J.; Kim, T.; Park, D.-H.; Kim, S. Horizontal Pod Autoscaling in Kubernetes for Elastic Container Orchestration. *Sensors* **2020**, *20*, 4621. [CrossRef] [PubMed]
10. Nguyen, N.D.; Kim, T. Balanced Leader Distribution Algorithm in Kubernetes Clusters. *Sensors* **2021**, *21*, 869. [CrossRef] [PubMed]

sensors

MDPI

Review

Resource Management Techniques for Cloud/Fog and Edge Computing: An Evaluation Framework and Classification

Adriana Mijuskovic [1,*], Alessandro Chiumento [1], Rob Bemthuis [1], Adina Aldea [2] and Paul Havinga [1]

1 Department of Pervasive Systems, University of Twente, 7522 NB Enschede, The Netherlands; a.chiumento@utwente.nl (A.C.); r.h.bemthuis@utwente.nl (R.B.); p.j.m.havinga@utwente.nl (P.H.)
2 Department of Industrial Engineering and Business Information Systems, University of Twente, 7522 NB Enschede, The Netherlands; a.i.aldea@utwente.nl
* Correspondence: a.mijushkovikj@utwente.nl; Tel.: +315-3489-8227

Abstract: Processing IoT applications directly in the cloud may not be the most efficient solution for each IoT scenario, especially for time-sensitive applications. A promising alternative is to use fog and edge computing, which address the issue of managing the large data bandwidth needed by end devices. These paradigms impose to process the large amounts of generated data close to the data sources rather than in the cloud. One of the considerations of cloud-based IoT environments is resource management, which typically revolves around resource allocation, workload balance, resource provisioning, task scheduling, and QoS to achieve performance improvements. In this paper, we review resource management techniques that can be applied for cloud, fog, and edge computing. The goal of this review is to provide an evaluation framework of metrics for resource management algorithms aiming at the cloud/fog and edge environments. To this end, we first address research challenges on resource management techniques in that domain. Consequently, we classify current research contributions to support in conducting an evaluation framework. One of the main contributions is an overview and analysis of research papers addressing resource management techniques. Concluding, this review highlights opportunities of using resource management techniques within the cloud/fog/edge paradigm. This practice is still at early development and barriers need to be overcome.

Keywords: resource management; cloud computing; fog computing; edge computing; algorithm classification; evaluation framework

Citation: Mijuskovic, A.; Chiumento, A.; Bemthuis, R.; Aldea, A.; Havinga, P. Resource Management Techniques for Cloud/Fog and Edge Computing: An Evaluation Framework and Classification. *Sensors* **2021**, *21*, 1832. https://doi.org/10.3390/s21051832

Academic Editor: Taehong Kim

Received: 30 January 2021
Accepted: 24 February 2021
Published: 5 March 2021

1. Introduction

The Internet of Things (IoT) connects everyday devices with each other and with the larger Internet to bring more meaningful interactions between objects and people. The connection process typically brings together sensing, actuating, and control devices. Additionally, these devices conform to the necessary standard compliant communication protocols. IoT can realize the purpose of smart identifying, discovering, following, and controlling things in many efficient and diverse ways [1].

Thus, IoT is becoming popular in domains such as smart healthcare, transport, logistics, retail, industrial automation, and many others. For example, airports can operate in a significantly smarter manner. IoT can monitor the volume and flow of people at the airport. It can be applied in smart airfield lighting systems, to provide preventive maintenance and reduction of fuel consumption [2]. Additionally, improvement of the airport luggage delivery system can be completed by placing RFID tags and making use of smart sensors. That can be done to detect whether the luggage is transported to the proper person at the correct time and place [3]. These are a few instances that represent how the IoT technology can make the operational structure at the airport more efficient.

There are many other domains such as road and bridge monitoring, supply chain, healthcare, and water pipe monitoring, where IoT can be applied to improve the reliability of the specific information management systems.

The number of ubiquitous devices deployed in a geo-distributed manner is increasing at a rapid rate, and it is reaching up to billions. Smart devices produce an extensive amount of data, which needs to go through network infrastructures. Frequently, this can emerge as a problem. The generated data can be used to reinforce the working and evolution of smart environments. In existing cloud infrastructures, the data are sent to cloud servers for further processing and then returned to the devices. To this end, cloud computing has emerged, yet this paradigm is still commonly perceived as being at an exploratory phase. The National Institute of Standards and Technology (NIST) defines cloud computing as a design that allows sharing of many computing assets in format of services to clients. With this concept, users can efficiently modify their requirements at a low cost [4]. Another definition in a wider perspective [4] declares that services are provided by applications and systems' software in a data center.

However, cloud computing has certain limitations: the need to transport data from each single sensor to a data center over a network, process these data, and then send instructions to actuators. This represents a large limitation because: (a) the communication increases latency considerably; and (b) since sensors and actuators are often on the same physical device, control information might be outdated as well.

Fog and edge computing may aid cloud computing in overcoming these limitations. Fog computing and edge computing are no substitutes for cloud computing as they do not completely replace it. Oppositely, the three technologies can work together to grant improved latency, reliability, and faster response times. The geo-distributed nature of the fog layer and the edge devices also enable location awareness (see the next paragraphs). One of the key differences between fog and edge computing refers to where the intelligence and the processing power reside.

Fog computing employs many nodes between the cloud and the end devices in which intelligence can be located. These allocated smart nodes represent base stations or access points [5]. By bringing intelligence away from the cloud, fog computing can process the IoT data in close proximity to the data sources. Afterward, it can use resources from the cloud (only if needed) in a more effective mode than through individual devices. For instance, fog computing can move the intelligence to a Local Area Network (LAN) position in the network architecture and thereby provide support of data processing in a fog node or an IoT gateway [6].

Edge computing is about moving the intelligence, computing power, and intercommunication capabilities of an edge gateway straight to the devices. It typically does not associate with any types of cloud-based services but concentrates more on the IoT device-side [6]. An example includes mobile services, which need ultra-low latency and real-time access to a radio network. Edge computing can be seen as an approach to forward the computation and communication resources from the cloud towards the edge. That is done to enable services by avoiding latency and thereby provide swift message delivery to users [7].

In this paper, we focus on resource management techniques for cloud, fog, and edge environments. A considerable amount of research has been done on different techniques for resource management in cloud, fog, and edge computing. A proper resource management is important because task offloading can cause more expenses in terms of downtime and energy costs (e.g., due to required communication between sensing devices and servers). Furthermore, processing excessive resources at the servers can impact the task performance delay in a system that contains a vast number of users. Hence, efficient computational offloading is relevant when dealing with IoT resource management.

Studies already provided a classification of resource management algorithms and exploratory comparative analyses of applied algorithms in the cloud, fog, or edge scenarios. However, to our knowledge, limited literature exists on analyzing resource management techniques for cloud, fog, and edge computing while taking into account resource management metrics. Some of these metrics are: resource allocation, workload balancing, resource provisioning, and task scheduling.

In this paper, we aim to present an evaluation framework for applied algorithms for resource management focusing on cloud, fog, and edge computing. It can be useful to provide researchers and practitioners insights into how resource management techniques are used within the realm of cloud, fog, and edge computing. First, analyzing existing approaches can shed some light on the current state-of-the-art and act as a source of reference for future work. Second, presenting an overview of the studied algorithms attributes and characteristics can make it possible to: (1) identify specialized solutions tailored to specific user needs; and (2) generalize about the dispersed view on the cloud, fog, and edge computing paradigm.

To provide an answer to this challenge, we first address current challenges in cloud, fog, and edge computing with a focus on resource management. Consequently, we provide an analysis of solutions to these challenges from the existing literature. To this end, we identify and analyze 16 different resource management solutions and derive a taxonomy to evaluate them effectively. One of our key contributions refers to the classification and evaluation framework of resource management techniques. Another contribution is the analysis and discussion about the suitability of algorithms concerning a particular solution paradigm (i.e., cloud–fog, fog–edge, fog-only, and cloud-only solution). To make the functionality of the reviewed resource management techniques more explicit and present it in more detail, we provide a classification of the features given by the algorithms.

The remainder of this paper is organized as described below. Section 2 outlines the methodology that we follow in this research. In Section 3, we present architectural overviews on cloud, fog, and edge computing and address background information on resource management techniques. Section 4 discusses challenges and limitations in the cloud, fog, and edge computing related to resource management. Section 5 provides an evaluation framework for applied algorithms in cloud, fog, and edge scenarios. Section 6 presents a classification overview of the suitability of algorithms concerning a solution paradigm. Section 7 gives a discussion and an outlook for the limitations of this study. Section 8 concludes the study.

2. Methodology

We use the Design Science Research (DSR) methodology as discussed by Hevner [8] to structure the research in several steps (see Figure 1). The first phase is exploratory and discusses literature and challenges in the cloud, fog, and edge architectures with a focus on resource management. This represents the foundation for the development of the evaluation framework for applied algorithms in cloud/fog and edge scenarios, which is also completed in the first phase. The second phase includes the classification of resource management techniques, and a discussion of the findings.

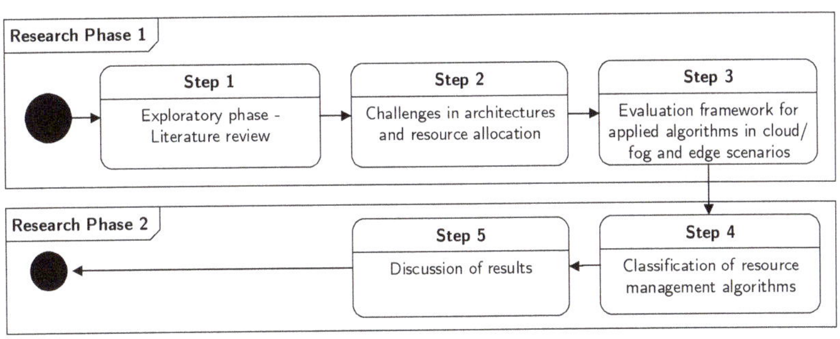

Figure 1. Research methodology followed throughout this paper.

The DSR is used as follows:

1. Research phase 1—Exploratory phase. This phase includes the DSR activities, which are necessary for the development of this research's artifact (the evaluation overview), including the literature, knowledge base, and research theory.
   (a) Collect research studies regarding the architectural overview for cloud, fog, and edge computing and research on existing resource management techniques.
   (b) Gather knowledge about challenges in architectures for cloud, fog, and edge computing.
   (c) Collect literature on algorithms applied for cloud, fog, and edge scenarios.
2. Research phase 2—Classification and discussion. This phase includes the design and development of the second artifact (classification of the resource management techniques).
   (a) Overview existing literature for attributes related to resource management.
   (b) Classify and compare the literature.
   (c) Examine which research challenges are addressed by the articles.

3. Background and Related Work

This section first introduces a high-level architectural study of cloud, fog, and edge computing. The section proceeds with discussing the role of resource management techniques for such architectures. Finally, we discuss some architectural overviews of cloud, fog, and edge computing applied to particular application domains.

3.1. High-Level Architectural Overview

Figure 2 presents a high-level architectural design of a typical IoT infrastructure including cloud, fog, and edge infrastructures, which can be applied in a smart pallet logistics case study. The architecture consists of a cloud network as the top layer, a fog network as the middle layer, and an edge layer as the bottom layer.

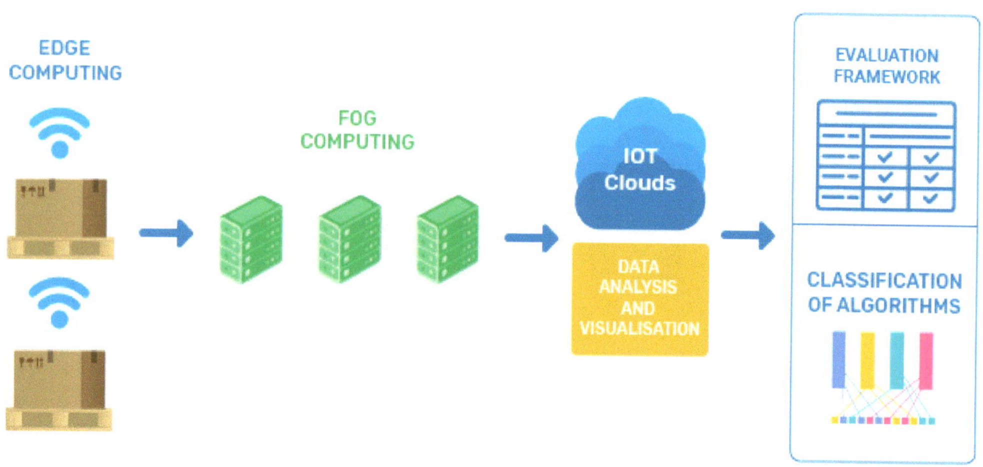

Figure 2. An example of cloud/fog/edge architecture for a smart pallet case study.

The concept of cloud computing is about enabling anything as a service such that services can be merged, shared, and monitored with minimum involvement [9]. Users can access services in a ubiquitous manner, through the network, and on request. There is a certain amount of time that is needed to accomplish the communication between the cloud and the existing IoT devices, which will be automatically added to the processing time. The accumulated time is captured by the cloud servers and it contributes towards the increase of a system's latency. Furthermore, this motivates the appearance of drastic effects on power and energy consumption [10]. As a result of the caused high latency, there can be indications of degradation in the Quality-of-Service (QoS) and Quality-of-Experience (QoE). Additionally, this will influence the reliability level of the system and generate delays in communication, capacity reduction, and excessive energy consumption. Some of the desired features for IoT infrastructures include modest latency, low response time, location awareness, low energy consumption, and portability support.

To accommodate some of these features, the computational paradigm fog computing was proposed [11]. In fog computing, data processing tasks are offloaded onto numerous middle-ware devices present in the network as a middle layer between the cloud and the end IoT devices. Each fog device is capable of processing the data that are being captured. This way, the overall latency is reduced, as the processing, happening locally, can lead to faster utilization, also locally, of the knowledge gathered. Fog computing represents the idea of broadening the cloud where the "things" are enhancing the application performance by removing the information processing within the cloud, and also by diminishing the bandwidth utilization in the network [12]. It has appeared as a promising technology that transports cloud applications in closer proximity to physical IoT devices. A fog node can also be seen as a mini-cloud, which is located near the edge layer of the network, and thus near the IoT devices connected to it [13]. A fog server represents a virtualized equipment, which contains on-board storage, computing, and communication capabilities. These features are meaningful when supporting the IoT application execution. Fog computing has been designed to deliver the following three core contributions: (1) diminish latency as the data are analyzed close to its sources from where it is initially gathered; (2) stabilizing network traffic, which is enabled by offloading gigabytes of network traffic from the core network connecting to the cloud; and (3) privacy and security support, which is enabled through proximity-by storing sensitive data in the nearby computer and network systems [12].

In edge computing, data processing is offloaded onto the edge devices [14]. Edge computing pushes the position of applications, services, and data to be close to the sources where such services are requested. In particular, the edge devices can be 'exploited' by fog computing nodes to handle some of the calculations, storage, and transmissions locally. Edge computing technologies are commonly deployed on single devices.

A limitation of using solely cloud computing is that a centralized cloud computing concept may not be sufficient for data processing and analyzing the vast amount of data gathered from IoT devices. One cause can be the (massive) data transfer which results in limited network performance. Edge computing is typically about transferring computing tasks from a centralized cloud to the edge layer (near the IoT devices). As a result, the transferred data are typically already pre-processed and much more compact than raw information [14].

The design of efficient allocation mechanisms for processing data among resources spread within various layers can be challenging. Especially in (near) real-time scenarios, a decision needs to be made quickly. Consider the example of having two data processing types: batch and stream data processing. Processing (big) batch data may happen (mostly) in the cloud, while most of the stream data processing may be more suitable for being distributed to fog or edge nodes. Depending on the design, a small set of stream data may also need further processing on the cloud. Likewise, some pre-processing might also be necessary at an edge node before transferring data to higher layers. System designs are vital for maximizing the potential of both computing paradigms effectively in real-time environments.

3.2. An Example of Resource Allocation in a Cloud/Fog System

Resource allocation strategies in cloud/fog/edge systems are responsible for assigning accessible resources to the system users [15]. It can be a challenge to assign resources efficiently to applications and their end users/consumers [16].

To give an example, consider the design model of [16]. This model administers the resource allocation in a fog environment (see the representation in Figure 3). The cloud–fog environment model is composed of three layers: a client layer, a fog layer, and a cloud layer. First, a solution for resource management is realized in the client and fog layers to accomplish the requirement of resources for clients. If there is no/limited availability of resources in the fog layer, then the request is passed to the cloud layer. The main functional components of this model are as follows:

- The fog server manager employs all the available processors to the client.
- Virtual machines (VMs) operate inquiries for the fog data server, process them, and then deliver the results to the fog server manager.
- Fog servers contain one fog server manager and virtual machines to conduct requests by using a 'server virtualization technique'.

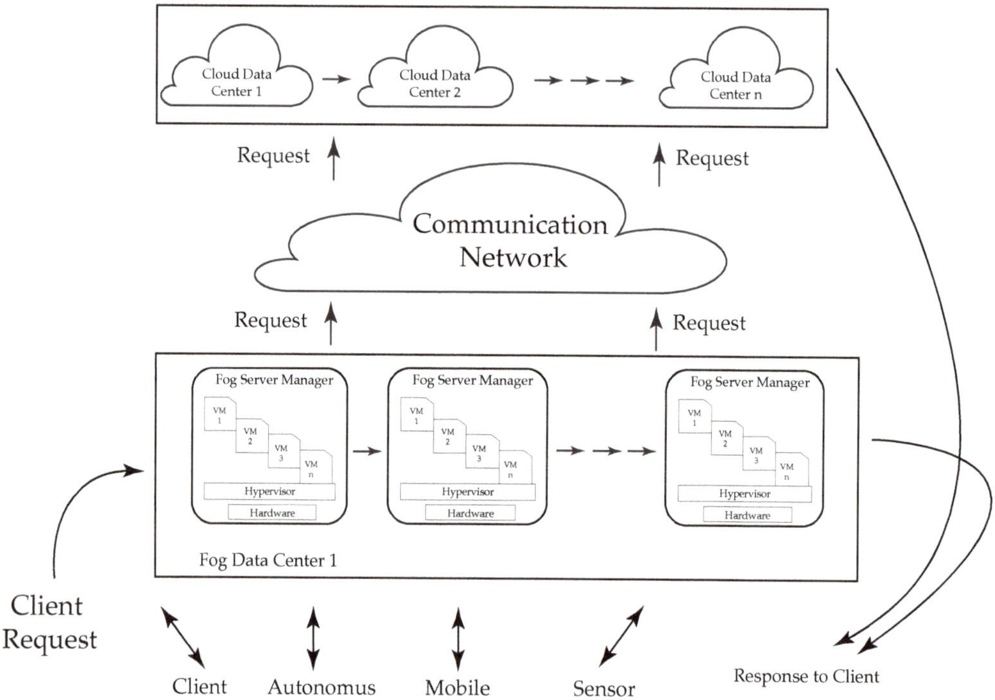

Figure 3. Three-layer architecture [16].

3.3. Some Application Domains

3.3.1. An Architecture Based on Cloud, Fog, and Edge Computing Paradigms in Real-Time Internets-of-EveryThings

According to Seal and Mukherjee [17], there are definite tiers of a universal fog computing architecture. Tier 1 depicts the 'Edge Tier', which consists of multiple Terminal Nodes (TNs). TNs represent mobile and smart nodes that are capable of detecting various location parameters and then transmitting them to the upper layer. Tier 2 is known as the middle layer or the 'Fog Layer'. This layer is composed of smart devices: routers, gateways, switches, and access points that can contribute to data computation, data storage, routing and packet delivery. Tier 3 is known as the cloud computing layer, which contains personal computers and servers.

Virtual Clusters (VCs) are defined as location-based parameters, which are composed of IoE devices often known as TNs. In this specific case, the role of TNs is to examine their environment and then transfer the data to the fog layer. Each Fog Instance (FI) monitors its own VC. The fog computing architecture can be further categorized into two sub-components: (a) the fog abstraction layer; and (b) the fog orchestration layer [17]. The first one deals with the management of fog resources, virtualization support, and configures tenant privacy, while the second layer contains the fog properties. Some of the fog properties are: heterogeneity, edge location, geographical distribution, support for mobility, real-time interactions, and interoperability. The fog orchestration layer consists of a software agent known as foglet, dedicated to monitoring the condition of the terminal devices. A decentralized database is used for scalability and fault tolerance, and the service orchestration module's role is to be responsible for the policy-based routing of application requests. The orchestration module also needs to decide whether it will transmit to cloud centers.

The fog devices' utilization is for limited semi-permanent storage, which facilitates provisional data storage and handles applications that are sensitive to latency. The cloud is accountable for the storage of large data chunks within its data centers. Those data centers typically contain massive computational abilities. The fog layer enables the cloud to be accessed and applied in an efficient and controlled manner.

3.3.2. An Architecture for Smart Manufacturing Based on Cloud, Fog, and Edge Paradigm

Cloud computing reinforces ubiquitous and on-demand network access to a distributing pool of computing resources (e.g., processing and storage facilities, appliances, services, etc.). By using virtualization technology, cloud computing shelters the diversification of basic devices and provides different services in a transparent way to the users, including IaaS (Infrastructure-as-a-Service), PaaS (Platform-as-a-Service), and SaaS (Software-as-a-Service). Because of the expansion of various access devices, cloud computing can encounter obstacles in bandwidth, latency, network unavailability, privacy, and security. Fog computing is viewed as an expansion of cloud computing to the edge network, conducting services (e.g., computation, storage, and network) in close proximity to the end-user devices (e.g., network routers), instead of transferring data to the cloud [18]. In a fog computing concept, data storage and processing largely depend on local devices, rather than on a cloud system.

Complementary to fog computing, edge computing grants computation to be completed at the edge of the network and an approaching environment to the data sources [18]. The crucial divergence between fog and edge computing is that fog depends on the interconnection amid nodes, while edge computing operates in the segregated edge nodes.

Edge computing administers services nearby the data sources to meet the critical requirements on privacy, security, agile association, and real-time optimization [18]. In pursuance of enlarging the application of smart manufacturing, by utilizing the cloud and administering future aspects of smart solution applications, a reference architecture based on cloud/fog/edge computing for smart manufacturing has been recommended.

3.4. *Techniques for Handling Resource Management*

There are many articles discussing mechanisms for handling resource management (e.g., allocation, provisioning, workload balance, and task scheduling). Specific goals relate to reduction of the overall energy consumption, the latency, or the overall communication costs. This section presents a concise literature overview on different techniques for handling resource management in cloud/fog and edge computing. The metrics under which we evaluate the reviewed research are algorithm classification type, deployment scenario, resource management criteria (resource allocation, resource provisioning, workload balance, and task scheduling), QoS, energy management, and environment. In the appointed evaluation, we focus on analyzing 16 state-of-the-art methods.

In [19], the authors focused on resource management which is done in the fog layer, aiming to minimize latency and enhance reliability. There is a consumer layer where its users can accomplish their specific current demands via fog and cloud. Requests per hour, response time, and processing time parameters are considered by using the round-robin algorithm, equally spread execution algorithm, and a proposed algorithm. Their focus is on considering a fog and a cloud environment together for resource optimization. The authors implemented the Fog-2-Cloud framework for the management of customers' requirements by utilizing six fog nodes and twelve MicroGrids in residential buildings. Fog servers helped in storing consumers' private data. Their used performance parameters were response time, requests per hour, and computing time that can be improved by using the Shortest Job First (SJF) algorithm. This algorithm is compared with other techniques Round Robin (RR) and Equally Spread Current Execution (ESCE), which outperformed the other two algorithms.

According to da Silva and d. Fonseca [20], fog and cloud can cooperate to advance their service distribution to the clients. This study is about a Gaussian Process Regression for Fog–Cloud Allocation (GPRFCA). It describes a mechanism that chooses where to allocate tasks based on the specific application requirements. The infrastructure is composed of a fog and cloud layer. The GPRFCA technique [20] decides where to appoint an assignment that needs to be computed while considering the availability of resources and latency costs. To advance the utilization of fog resources, GPRFCA is employed to predict the arrival of future requests based on the historical information. Such a prediction can support resource provisioning to future requests. That stands especially for real-time application requests which can only be processed within the fog. A simulation was performed, and its results represent that the given solution stabilizes the assignments between fog and cloud and the trade-off among latency, blocking, and energy consumption.

Fog-based computing and storage offloading for data synchronization in IoT create a large amount of data, which is partly due to the increase in IoT devices that are connected [21]. If, at a later stage, IoT devices transmit data to the cloud, then the data privacy can become a challenge. To address this, Wang et al. [21] proposed an architecture for data synchronization based on fog computing. It is achieved via offloading computing parts and storage work towards the fog servers and then data privacy can be better guaranteed. Additionally, a differential coordination founded on fog computing is recommended. The benefits of their composed architecture are: (a) data chunks can be stored in the fog server for enabling security; (b) the fog server facilitates the computation offloading and storage, which formerly belonged to the cloud and user's devices; and (c) the transmission overload is minimized.

In [22], a method named Dynamic Resource Allocation Method (DRAM) is presented. This method relies upon static resource allocation and dynamic resource scheduling to achieve dynamic load balancing. Agarwal et al. [16] presented the Efficient Resource Allocation (ERA) method, which minimizes the response time and maximizes the throughput of resources. In [23], the authors discussed a method in which the resources are allocated according to a different priority. Taneja and Davy [24] proposed an iterative algorithm to reduce latency and energy consumption. Section 5 provides more details about the resource management techniques.

3.5. Related Work

Building further on the work discussed in [6], we provide an evaluation framework for resource management techniques applied in cloud/fog and edge computing scenarios. Naha et al. [6] provided a summary of research on resource allocation and scheduling only in the fog. It can be concluded that many articles addressed mainly the role of resource allocation in the fog environment. However, further investigation on QoS, load balancing and energy efficiency needs to be considered [6]. There are several identified limitations regarding the use of fog computing. One challenge refers to the synthetic work done regarding the validation of methods [6]. Another challenge refers to the presentation of only cloud-based simulations, which are not completely suitable for the fog computing concept (which are typically dynamic environments).

Bendechache et al. [25] presented some research articles focused on resource allocation. Some of the explored resource management metrics were divided into two sections: resource provisioning and resource scheduling. Their provisioning metrics are detection, selection, and mapping, whereas the scheduling metrics are allocation, monitoring, and load-balancing. Additionally, several variables or Key Performance Indicators (KPIs) were investigated such as scalability, latency, VM placement, failure rates, accuracy, resource utilization, energy consumption, cost, efficiency, Service Level Agreement (SLA), and QoS. The contribution of this research survey is quite detailed, but it presents limited research articles that are focused on cloud, fog, and edge computing. Therefore, this represents a motivation to study resource management techniques for cloud, fog, and edge computing.

In [26], three types of taxonomies are demonstrated: (i) a classification of performance metrics for evaluating cloud, fog, and edge computing; (ii) metrics based on cloud models; and (iii) classification of identified metrics based on a concept known as MAPE-K. Based on the collected literature, the authors identified that the common performance metrics for cloud, fog, and edge computing include throughput, network congestion, fault-tolerance, statistical analysis measurements, scalability, cost/profit, and SLA violation. The taxonomy of metrics based on cloud models suggests the use of the following groups: private, public, hybrid, single-provider, multi-provider, and federated. According to a MAPE-K loop, there are four categories of parameters, including monitoring, analyzing, planning, and executing. Their results represent a mapping between the proposed taxonomy and existing literature on the cloud, fog, and edge computing paradigm. However, the study could be further extended by providing a proper detailed list of proposed solutions from the reviewed literature, and respectively their classification.

Ghobaei-Arani et al. [27] provided a taxonomy of resource management approaches in fog computing. The taxonomy considers the following categories: resource provisioning, application placement, resource scheduling, task offloading, load balancing, and resource allocation. They focused on structuring the literature according to resource management approaches. For each resource management approach, they provided details about the case study, utilized technique, used performance metric, evaluation tool, advantages, and weaknesses. Overall, this study provided knowledge about existing articles for each resource management approach, but only considering fog computing, while it would be also interesting to include edge computing. Additionally, the article only addresses the solution approaches in an exploratory manner. In other words, the research work represents an analytical examination and discussion on existing studies about resource management.

Lastly, Salaht et al. [28] delivered a list of optimization metrics to address resource management and service placement problems. The considered metrics are latency, resource utilization, cost, energy consumption, quality of experience, congestion ratio, and blocking probability. Based on the findings of Salaht et al. [28], further research work should be done on challenges regarding service placement problems, optimization strategies, and evaluation environments.

4. Challenges in Resource Allocation for Cloud, Fog, and Edge Computing

There exist several challenges regarding cloud, fog, and edge architectures, such as the deployment of 5G, serverless computing, resource allocation, optimization, energy consumption, data management, applying federation concepts to fog computing, trust models, business and service models, mobility, and industrial IoT [29]. A challenge in 5G includes realizing the concept of network shredding to backup a service collection with certain performance requirements requests. Some of them are: resource management throughout, fog nodes, wireless, optical packets, and cloud domains [29]. Recent developments in network virtualization grant guidelines for network shredding, but they do not provide a unified and general collection of resources over various domains. Based on the reviewed literature, we present the challenges in Table 1.

Table 1. Challenges in architecture for cloud, fog, and edge computing.

Challenges	References
Serverless computing	[29]
Energy consumption	[29]
Data management and locality	[29]
Orchestration in fog for IoT	[29]
Business and service models	[29]
Load balancing	[30]
Security and efficiency issues	[21]
Data integrity and availability	[21]
Cloud-based synchronization	[21]
Dynamic scalability	[24]
Efficient network processing	[24]
Latency sensitivity	[24]

In terms of *serverless computing* [29], to achieve micro-services management through the cloud/fog/edge hierarchy, there are challenges regarding the flow of services among cloud, fog, and edge computing devices. The automatic administration of the micro-services must audit the deployment location and context; in addition, the resource constraints that may exist in the fog need to be taken into account. Additionally, the diversity of the system across an IoT cloud–fog ecosystem can be challenging for the deployment of micro-services and reconfiguration.

Furthermore, the network topology can be expected to change regularly due to devices mobility and changing application requirements. The high levels of heterogeneity in IoT devices and the variability of the environment call for active and dynamic system management based upon multi-criteria *resource allocation* [29]. Resource management systems and multi-criteria schedulers may instantly enhance resource allocation in terms of handling dynamic behaviours. That can be challenging since the number of variables can largely expand the search area and that can consequently lead to long scheduler execution times.

A substantial demanding route for prospect research is in diminishing *energy consumption* [29] where the target should be researching the aspect and significance of data in the cloud/fog/edge ecosystem, onward with the definition of 'economical data management'. The objective behind this is monitoring in detail the implication of various data types and whether the data are required most of the time.

In recent years, there is an expansion in production and use of data, and that accomplished a few remarkable rates. Concerning *data management and locality* [29] in IoT cloud–fog computing systems, accessibility problems have to be considered. Computing systems consist of several networking technologies, such as mobile, wireless, or wired. When the resources are centralized within the cloud, certain networking challenges, such as availability, scalability, and interoperability, might be partially addressed. However, some of the innovative problems (e.g., network bottlenecks and latency) can be addressed by using fog and edge computing. A particular challenge is how to quantify the trade-off among data distribution and services at the fog or cloud layers. One way towards approaching this issue is via smart service placement. More specifically, this can be done by data locality, which is achieved by placing the needed services closer to the data that they administer. Suitable candidates, according to Bittencourt et al. [29], are applications that do not need high computation power and are capable of analyzing large data volume.

In terms of *orchestration in fog for IoT* [29], privacy requires to be tackled according to the European Union General Data Protection Regulation (GDPR) and other similar regulations imposed all around the world. Privacy regulations are important because, when fog nodes are placed close to the end-users, one may attempt to gather, process, and store data, and that can violate users' privacy. The performance of fog orchestration for the IoT deals with several challenges, related to 5G networks such as an increase in density of devices combined with latency and reliability requirements of demanding applications along with the mobility of nodes, which boost important problems concerning the system monitoring, and that is significant for proper resource management. Other fundamental aspects that directly impact the performance of (dynamic) fog orchestration are component selection and placement, which need to be additionally investigated in the future, as well as research on efficient techniques to prevent (minimize or stop) the overloading and avoid delays.

Another challenge is *business and service models*. The fog can be deployed as a hybrid cloud, where specific local resources can be extended with resources from the cloud. Additionally, when different stakeholders are incorporated in a specific hierarchy from IoT to the cloud, this can create a scenario in which different elements of the overall systems are owned or managed by completely unrelated entities or stakeholders, e.g., IoT devices can be owned by the state, while fog nodes are owned by a cloud company. It is challenging to determine how IoT services can be combined with services from fog and cloud computing, and then how they can be monitored and administered when many players at different levels are participating.

In [30], the challenges with IoT appliances in cloud, fog, and edge computing are related to replying to resource requirements and *load balancing*. In this article, load balancing is considered as one of the meaningful strategies to accomplish efficient usage of resources and reduce or avoid congestion. Therefore, it is a distinguishing challenge to obtain load balance for the processing nodes in a fog environment all along with an IoT application execution. According to [20], the determined challenge was regarding minimizing latency as well as balancing the workload to reduce energy consumption.

The limitations considered in [21] are related to cloud computing and *cloud-based synchronization*, which is a particular core service in the cloud computing area. The IoT devices synchronize most of the data to the cloud. There are two challenges in this specific scheme referring to *security and efficiency issues*. The security issues regarding cloud storage revolve around the following aspects: *integrity, privacy, and availability of data*. The selected established security threats are data exposure, data deficiency, malicious user handling, wrong use of cloud computing and its services, and possibly session stealing during data accessing. Problems such as connection cost and latency between the cloud system and edge layer devices are not tolerable in detention-sensitive applications. While there are, for example, some synchronization tools such as MicrosoftActiveSync and Botkinds AllwaySync, the drawback is that they regularly transmit an entire system file even when there is a small change occurrence. This coordination type may cause redundant communication and latency issues, where users frequently modify the data. It can be concluded that traditional coordination among cloud and IoT devices has certain disadvantages such as when the IoT devices fail to secure confidential data, and/or when common data changes cause high data and communication redundancy.

In [22], limitations of resource requirements and load balancing for IoT appliances in cloud, fog, and edge computing are presented. Load balancing is an important factor that is valuable to increase resource efficiency by avoiding bottlenecks, overload, and low load situations. Accordingly, it is an obstacle to accomplish load balance for the computing nodes in a fog environment at the same time as the IoT application execution occurs. According to Taneja and Davy [24], cloud computing offers many assets, but with expansion in more ubiquitous mobile sensing devices coupled with technological upgrades, the imminent IoT ecosystem demands the computing network architecture of the cloud. A few of the requirements that need to be met are *dynamic scalability, efficient-in-network processing, and latency-sensitive communication*; these are the requirements for IoT application which drove the evolution of fog computing.

5. Evaluation Framework for Resource Management Algorithms in Cloud/Fog and Edge Scenarios

Table 2 shows an overview of selected techniques used in the reviewed literature about resource management in cloud/fog and edge-based scenarios. The resource management algorithms are summarised in the table and evaluated according to several metrics that are discussed below. Resource management is about achieving coordination of resources that is highlighted by supervision (management) actions and performed by service providers and users [31]. It considers the resource allocation process from resource providers to the users. The algorithms discussed in the following subsections employ different resource management metrics which are examined as well and can be used for further evaluation.

5.1. Resource Allocation

Resource allocation represents a technique that is used to optimize the utilization of resources and reduce the required costs for processing [32]. Fulfillment time of a task is an important aspect that should be considered since it can impact the completion of resource allocation [33]. As indicated in Table 2, RR, ESCE, SJF, GPRFCA, ERA, Priority-based Resource Allocation algorithm (PBSA), and Feedback-Based Optimized Fuzzy Scheduling algorithm (FOFSA) use resource allocation techniques.

5.2. Workload Balance

Workload balancing is an important factor used to manage energy effectiveness and also avoid congestion, low-load resource management, and overload. Currently, this represents a challenge for the processing nodes, which are placed in the fog environment. For instance, in [34], a workload balancing algorithm is proposed for fog computing, aiming to reduce the data flow latency in the transmission procedures by connecting IoT devices to the appropriate base stations (BSs). The article discusses several workload balancing algorithms from the literature: RR, SJF, ESCE, GPRFCA, DRAM, ERA, PBSA, FOFSA, Hill Climbing algorithm (HCLB), Efficient Load Balancing algorithm (ELBA min-min), and Tabu Search algorithm.

5.3. Resource Provisioning

Resource provisioning represents an approach (solution) that shows how to administer requests for tasks and data among fog nodes [35]. Resource provisioning is a further step in resource allocation. As discussed above, resource allocation deals with just assigning a set of resources to a task, while resource provisioning deals with the activation of the allocated resources. Remote Sync Differential Algorithm (RSYNC), Fog Sync Differential Algorithm (FSYNC), Reed–Solomon Fog Sync (RS-FSYNC), ERA, and Energy-aware Cloud Offloading (ECFO) are the algorithms that deal with resource provisioning.

5.4. Task Scheduling

To manage a large set of tasks that are working together and are dependent on a certain set of resources, task scheduling algorithms have been proposed to define a schedule to service tasks to avoid conditions such as deadlocks [36]. Table 2 shows a few algorithms that manage resources based on task scheduling: RR, SJF, ESCE, GPRFCA, DRAM, PBSA, FOFSA, ELBA, Tabu, and ECFO.

Table 2. Evaluation framework for applied algorithms in fog–cloud and edge scenarios.

Author & Year	Algorithm	Deployment	Classification	Resource Management				Additional Classification		Environment
Resource Management Techniques in Fog/Cloud Edge Scenarios				Resource Allocation	Workload Balance	Resource Provisioning	Task Scheduling	QoS	Energy Management	
Javaid, S. et al. (2018) [19]	RR	Simulation (Cloud Analyst)	Discovery	✓	✓		✓		✓	Cloud–Fog
Javaid, S. et al. (2018) [19]	ESCE	Simulation (Cloud Analyst)	Discovery	✓	✓		✓		✓	Cloud–Fog
Javaid, S. et al. (2018) [19]	SJF	Simulation (Cloud Analyst)	Discovery	✓	✓		✓		✓	Cloud–Fog
Da Silva, R.A.C. et al. (2018) [20]	GPRFCA	Simulation iFogSim [12]	Discovery & Load-balancing	✓	✓		✓		✓	Cloud–Fog
Wang, T. et al. (2019) [21]	RSYNC	Experiments in different conditions, two situations of synchronization	Discovery & Off-loading			✓				Fog
Wang, T. et al. (2019) [21]	FSYNC	Experiments in different conditions, two situations of synchronization	Off-loading			✓				Fog
Wang, T. et al. (2019) [21]	RS - FSYNC	Experiments in different conditions, two situations of synchronization	Off-loading			✓				Fog
Xu et al. (2018) [22]	DRAM	Evaluation done with three different types of computing nodes	Load-balancing		✓		✓			Fog
Agarwal et al. (2016) [16]	ERA	Simulation (Cloud Analyst)	Load-balancing	✓	✓	✓				Cloud–Fog
Savani et al. (2014) [23]	PBSA	Simulation (CloudSim 3.0.3)	Load-balancing	✓		✓				Cloud
Taneja et al. (2017) [24]	Iterative Algorithm	Evaluation done in three different topologies with different workloads	Placement					✓	✓	Cloud–Fog
Arunkumar et al. (2020) [37]	FOFSA	Simulation iFogSim	Load-balancing	✓	✓		✓	✓	✓	Fog
Chandak et al. (2018) [38]	HCLB	Simulation CloudAnalyst tool	Load-balancing		✓					Cloud–Fog
Manju et al. (2019) [39]	ELBA (min-min)	Simulation CloudAnalyst tool	Load-balancing		✓		✓		✓	Cloud–Fog
Téllez et al. (2018) [40]	Tabu Search	Simulation Cloudlet Tool	Load-balancing		✓		✓			Cloud–Fog
Jiang et al. (2019) [41]	ECFO	Cloud server and three Raspberry Pi3 devices	Off-loading			✓	✓	✓	✓	Fog–Edge

6. Classification of Resource Management Algorithms Applied in Cloud/Fog and Edge Scenarios

To compare the various state-of-the-art algorithms presented in several papers, inspired by Hong and Varghese [42], we classify the selected algorithms into six categories. Classification helps in terms of the identification of existing solutions and understanding their diversity as well. It can support researchers and practitioners in the process of learning about different algorithmic solutions to understand their features, differences, and similarities. The reviewed solutions consider how resources are handled among cloud, fog, and edge devices. In this paper, we briefly overview these 16 algorithms. They represent the basis for building the evaluation framework (Table 2), which is the foundation of this paper and the emerging classification of the algorithms are presented in Figure 4. We created this classification to address the key contributions in the area of resource management.

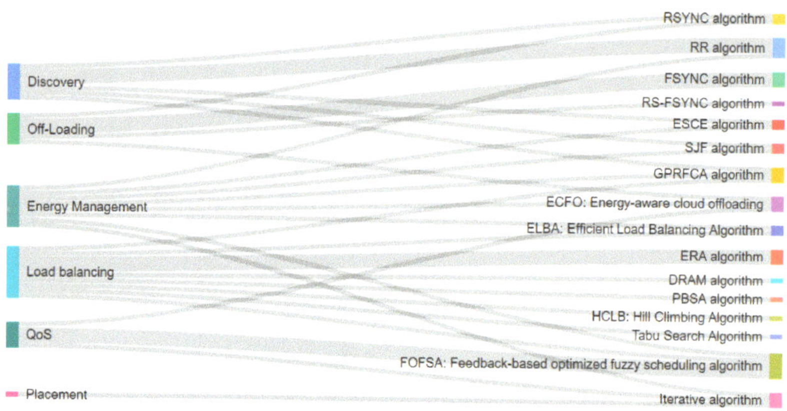

Figure 4. Classification of resource management techniques.

6.1. Discovery

Discovery is used to find available resources from the cloud, fog, or edge layers, based on workload requirements, to identify where they can be deployed efficiently. Fog servers have to use as many resources as desirable through accepting a high volume of tasks as possible. A manner of doing this is by using a manager or master entity that has an overall view of the resources. Afterward, based on the workload's requirements, it can allocate resources properly among fog and cloud layers. According to Hong and Varghese [42], in the edge/fog computing concept, the discovery algorithms stand for determining resources in the edge network that can be employed for further distributed processing.

For example, according to Javaid et al. [19], the algorithms RR, ESCE, and SJF belong in this category. GPRFCA and RSYNC belong in this group as well.

- RR—Round Robin Algorithm: According to the authors of [19], the RR algorithm for cloud computing has been adopted on the basis of defining time schedules. The scheduler creates certain specifics of VMs in an assignment table. Then, it assigns jobs that are received for data centers (DCs) to a set of VMs. Initially, a VM is initialized with an ID of a current VM variable and then the demanded job is mapped with the current VM variable.
- ESCE—Equally Spread Current Execution: The ESCE algorithm enforces the spread spectrum approach and collaborates with a large number of active duties on VMs at any specific time segment [19]. By using ESCE, the scheduler can register the VMs assignment table, and then keep up a list of VMs' IDs and their operating tasks on any VM. Once the task is performed, at any specific time interval, the VM table can be changed. In the beginning, the active task count is 0; on the occurrence of a new job the scheduler determines the VM having the minimum task count. If many tasks are

assigned to many VMs that are with the minimum count, then the first VM will be selected for the task processing.

- SJF—Shortest Job First: The SJF algorithm executes tasks by labeling the task size as a priority, and the priority is further controlled by the size of consumers' requests [19]. SJF can allocate tasks to VMs based on their fogs, the priority of distances, and size. The scheduler can be used to distribute the job on VMs based on the spread spectrum approach. SJF schedules the jobs by enabling minimum completion time, higher efficiency, and minimum turn-around time.
- GPRFCA—Gaussian Process Regression for Fog–Cloud Allocation: The GPRFCA mechanism is used to discover predictions to govern work activities on fog nodes while reducing latency [20]; as such, it belongs to the discovery group. Generally, it investigates the history of formerly sent requests for future arrivals' predictions of VMs, which are by rigorous latency demands [20]. By adopting these predictions, this technique can store the required resources within the fog nodes for future requests. Consequently, they should be completed within the fog layer itself, and then tasks that are not vulnerable to delays are assigned in the cloud. This leads towards an increase in fog nodes' utilization [20].
  CPU and RAM are important assignable resources for this mechanism [20]. The algorithm starts with the calculation of the number of VMs which can be still executed by the fog node (this is done by taking into consideration CPU and RAM). Furthermore, the Gaussian Process regression is then called (Line 4) to predict the VMs number, ***future VMs***, which should be incorporated also in the fog, but at the next interval.
- RSYNC—Remote Sync Differential Algorithm: RSYNC is one of the first algorithms to face the problem of complete synchronization whenever an update (change in file) is performed [21]. As the name implies, this differential algorithm is used to transmit only that particular part of the data that experiences an update. Since every instance of synchronization sends a small piece of information, the communication cost and latency decreases when compared with previous algorithms. Nevertheless, RSYNC is more suitable for establishing a communication path between IoT devices and the cloud layer. Although it sends only the updated data, it still needs to send a synchronization request every time that IoT device does an update.

6.2. Off-Loading

Off-loading is accountable for the resource provisioning tasks. It concentrates on storage provisioning instead of computation. It determines where data should be stored to lower transmission expense and the delay between the cloud computing layer and IoT (edge) devices [41]. Following Wang et al. [21], we identify two main differential synchronization algorithms, RSYNC and FSYNC, since they are focused on where the storage is provisioned properly. FSYNC decreases the latency and the communication costs considerably due to the use of the fog layer and a specific defined threshold. The threshold refers to the number of trivial changes that can be saved in the fog.

- RSYNC: This algorithm is explained in the previous subsection.
- FSYNC—Fog Sync Differential Algorithm: The FSYNC algorithm deals with the RSYNC issue [21]. The issue refers to the case that there are many requests when the edge device is modified. During each request, new data are generated, which lead to the creation of additional load on the cloud server. It differentiates by adding two elements, a fog computing layer, and a threshold. It establishes a threshold, and then, when the IoT device updates, the algorithm will send only the part of the data that has changed to the fog layer. The difference is that there are no requests and data being sent to the cloud. Additionally, only when the threshold is reached the fog servers will send a complete synchronization of the data. Otherwise, the following updates will be done between fog servers and IoT devices.
- RS-FSYNC Differential Algorithm: RS-FSYNC is a (Reed–Solomon Fog Sync) differential algorithm [21]. By applying the Reed–Solomon code, the security of the user's

data can be enhanced. The Reed–Solomon code is included in the FSYNC algorithm. Additionally, it uses an advantage from the storage capacity of the fog server to handle an encryption problem. Furthermore, it represents a variant of erasure code that was used within the distributed storage field. The objective is to revise errors created by the redundant data, which is generated by the original data.

- ECFO—Energy-aware cloud offloading: The energy expenditure of a local device can be accordingly diminished by offloading computational tasks to a remote device. Although supplementary transmission energy and communication latency may happen due to the appearance of data transmission between the remote device and local system [41], the specific challenge addressed by ECFO is how to distribute multiple tasks to and from multiple fog devices taking into account each device computational ability and the overall communication constraints [41]. To solve this problem, the ECFO algorithm tracks the bandwidth and schedules queues between devices to detect the energy consumption and provide an offloading decision. The process is dedicated to scheduling offloading activities into a two-phase deadline in order to dynamically adapt to changes in run-time network bandwidth. In the end, it also plans setbacks, which are caused by devices with multiple tasks.

6.3. Load-Balancing

Load-balancing distributes the workload to resources to make the operations more efficient by avoiding congestion, low load, and overload [22]. The considered algorithms based on load-balancing are DRAM [22], ERA [16], PBSA [23], GPRFCA, FOFSA, HCLB, ELBA, and Tabu Search algorithm.

- DRAM—Dynamic Resource Allocation Method: DRAM [22] is a dynamic resource allocation method that consists of the following steps:
  - Fog service partition: This is pre-processing in which the fog services can be categorized according to the resource requirement of each node type [22].
  - Spare space detection: To decide whether a node is portable to accommodate a fog service, identifying the extra space of all processing nodes is needed [22].
  - Static resource allocation for the fog service subset: For services within the fog that belong to the same subset of services, the appropriate processing nodes are selected to accommodate them [22]. When allocation starts, the node with the lowest extra space is selected.
  - Load-balance global resource allocation: The dynamic resource allocation strategy is executed to achieve load balance [22].
- ERA—Efficient Resource Allocation Algorithm: The ERA algorithm in [16] was designed to achieve effective resource allocation in the fog layer. The client makes a request and this request can be accepted only by the fog layer. If the fog does not process the request within a given time frame, then the process is transmitted towards the cloud [16]. With this method, the response period is diminished and the throughput is increased.
- PBSA—Priority based Resource Allocation Algorithm: In PBSA [23], batches of user's requirements are created according to the type of the task, the processing amount, and the time that the clients need [23]. If the specific resources that the user needs are not there, then the client needs to wait until they become available. If two identical requirements have a particular request with the same priority, then the method of 'first comes, first served' is used.
- GPRFCA: The GPRFCA algorithm belongs to this category as well.
- FOFSA—Feedback-Based Optimized Fuzzy Scheduling Algorithm: The Feedback-Based Optimized Fuzzy Scheduling Algorithm (FOFSA) is proposed in [37]. FOFSA works with two procedures: multilevel queue scheduling and multilevel feedback queues. The job activities are enrolled in different levels of queues. The queues are managed based on the concept of 'first come, first served'. The job activities can be appointed to resources per specific priority. If the job activity is not assigned to a

particular resource, then the job is simply removed from the waiting sequence. A task's priority can be decided by the fuzzy inference system procedure presented in [37]. Additionally, an architecture of the fuzzy-based scheduling is introduced in [37]. The proposed methodology was tested with iFogSim and analyzed with different existing dynamic algorithms. It was justified by the fact that it contains an effective scheduling strategy and upgrades the QoS parameters. The suggested methodology achieved a reduction in power utilization and enforcement time.

- HCLB—Hill Climbing Algorithm: HCLB algorithm is defined as a mathematical optimization technique that is used for searching and monitoring the loads among VMs [38]. This technique is established on a random solution to discover accessible VMs. The goal of the algorithm is to find a solution to the problem of discovering accessible VMs, and the searching loop executes only when the appropriate solution is found [38]. When the nearest VM is detected, the loop is increased in HCLB [38]. Then, the best VM is selected, and a request is assigned to it for further processing.
- ELBA—Efficient Load Balancing Algorithm: The min-min algorithm is implemented in the fog where fog nodes are divided in clusters and the algorithm determines the task which has minimum enforcement time and appoints it a particular node. That node is able to process it in a faster manner [39]. When a cluster is busy, the controller node inspects neighborhood clusters that contain 'inactive' fog nodes and sends activity to the node which presents lowest latency. Afterwards, the cluster shall send the activity with the favorable latency. If the cluster with 'inactive' fog node is located far away, then the particular task should be instantly sent to a cloud system for further processing. It could be effective to process the activity in the cloud or, instead, leave it to have a delay due to pre-processing at the fog nodes. In another situation, where two or more neighboring 'inactive' nodes are accessible, the node with the smallest latency can transmit the job activity [39]. Two factors need to be deliberated to calculate latency: one refers to the number of stand-by requests that need to be supplied in the clusters and the other refers to the inactive node's distance from the task originator. Calculation of the lowest distance between the source node and a fog node or a cloud data center can be determined by using Equation (1) [39]. N represents minimum latency, S is the source from where a particular activity is re-transmitted, C is the nearest cloud data center, and n depicts the number of fog nodes.

$$N = min\left[[d(s,c)], min\sum_{i=1}^{n}[d(s,ni)]\right] \quad (1)$$

- Tabu Search Algorithm: Tabu search is used to determine an optimal solution regarding the distribution of tasks between nodes that belong in the cloud and fog layers. It is done by utilizing search which frequently moves towards an improved solution every time [40]. The searching process will be terminated the moment a stopping condition is detected. Optimal load balancing is one of the biggest concerns in fog computing. To accomplish optimal load balancing, [40] used Tabu search in fog computing for load balancing. In this study, a bi-objective cost function was considered to achieve online computations, where the initial one implies the computation cost of computing tasks in the fog nodes, and the second one supports it in the cloud.

6.4. Placement

Placement is used to determine the suitable resources to satisfy the required workload. The main purpose is to distribute the incoming computation tasks to the appropriate fog/edge resources.

- Iterative Algorithm based on resource placement: [24] proposed an iterative method that is based on resource deployment of IoT applications in a cloud–fog computing setting. This method is composed of three algorithms. The first algorithm sorts the network nodes and application modules according to their requirements and capacity (CPU, RAM, and network bandwidth). The second algorithm looks for an eligible network mode that meets the module's requirement. The last algorithm is responsible for ensuring the requirement check, which is done by using the COMPARE function [24].

6.4.1. QoS

We distinguish QoS as one of the classification categories of resource management techniques. Additionally, it can be used as a feature that may be used for further evaluation of the reviewed algorithms. When taking into consideration the use of cloud computing as a solution concept, we should be aware that the data transfer between cloud and clients will contribute towards the increase in feedback latency [43]. This will lead to restrain the cloud service to provide quality of service to clients [43]. The QoS concept is defined in the ITU-T Recommendation E.800 and refers to the following [44]:

> *"The collective effect of service performance, which determines the degree of user's satisfaction of the service."*

The QoS consists of a set of parameters that pertain to the traffic performance of the network, but, in addition to this, the QoS also includes additional concepts. Therefore, they can be summarised as:

- Service support performance
- Service operability performance
- Serviceability performance
- Service security performance

The following group of reviewed algorithms belong in this category: Iterative Algorithm based on resource placement, FOFSA, and ECFO.

6.4.2. Energy Management

Enormous amount of energy savings can be obtained by taking into consideration energy consumption and energy management, which are associated with IoT and the cloud, fog, and edge paradigms [29]. Various methods can be used to address these concerns such as: (1) algorithms for energy-aware data transfer; (2) algorithms that limit the amount of data which is transferred within the network by utilizing certain criteria (thresholds); and (3) algorithms which exchange processing with communication, by using concrete objectives to achieve a balanced trade-off [29]. Based on the reviewed literature, energy management is selected as one of the classification categories for resource management techniques, to which we consider that the following algorithms belong: RR, SJF, ESCE, GPRFCA, Iterative algorithm based on resource placement, FOFSA, ELBA(min-min), and ECFO.

7. Discussion and Limitations

One of the key contributions of this paper is to provide an evaluation and classification overview of applied algorithms for resource management that address cloud/fog and edge environments. To support researchers in the further evaluation analysis process, they may initially need to understand the cloud/fog/edge architecture concept, and then learn about the potential challenges. In the end, researchers can finally explore in detail the existing resource management techniques that can address some of the potential challenges. Conforming to the conducted literature review, we identify a few solutions out of the 16 algorithms that can respond to some of the challenges, as shown in Table 3.

There exist certain challenges regarding resource allocation on a cloud/fog/edge network. When data are processed and then saved in a cloud system and if data centers are positioned far away from the devices, the complete process of data storage and processing may take a long time. Then, tasks need to be distributed in a manner that the entire network of devices inside a fog computing infrastructure can be completely utilized. If they are concentrated only in one particular area of the network, it will replicate a traditional cloud computing model which is not a desired factor. The distributed task allocation is focused on diminishing the average latency of service while lowering the overall quality loss.

Table 3. Addressed challenges.

	Algorithm	References
Load Balancing [30]	GPRFCA	[20]
	ERA	[16]
	DRAM	[22]
	PBRA	[23]
	HCLB	[38]
	ELBA(min-min)	[39]
	Tabu Search	[40]
	FOFSA	[37]
Energy Consumption (Management) [29]	RR	[19]
	SJF	[19]
	ESCE	[19]
	Iterative algorithm	[24]
	FOFSA	[37]
	ELBA(min-min)	[39]
	ECFO	[41]
	GPRFCA	[20]

The analyzed algorithms are grouped per type of solution paradigm, as represented in Figure 5. It clearly illustrates which algorithms belong to a specific type of solution paradigm: cloud–fog, fog–edge, fog-based, or cloud-based. The majority of analyzed techniques (nine of them) belong to the cloud/fog paradigm, while five are only fog-based solutions, one is fog/edge type, and one technique is only a cloud-based solution. Additionally, this indicates that researchers and experts in IoT could focus on developing an algorithm that will address resource management challenges in the complete cloud/fog/edge paradigm.

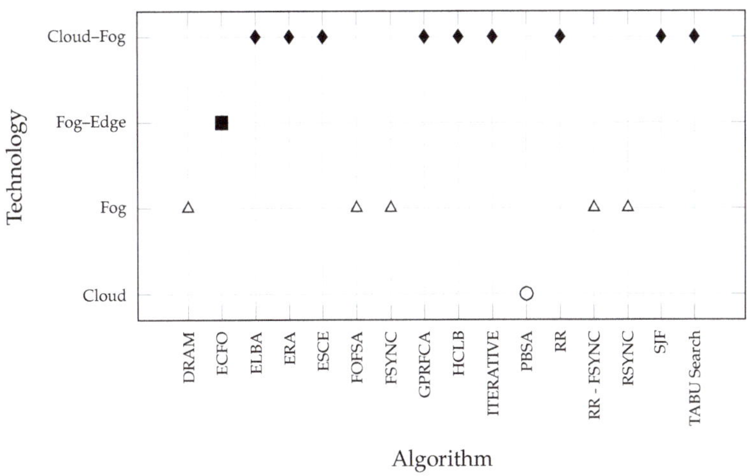

Figure 5. Reviewed algorithms per type of solution paradigm.

Furthermore, this review proposes that algorithms can be classified according to their characteristics in six classes: discovery, load-balancing, off-loading, placement, QoS, and energy management. The discovery group finds available resources from either the cloud or fog layers based on specific workloads requirements, to identify where they can be deployed efficiently. The following algorithms belong to the discovery group: RR, ESCE, SJF, GPRFCA, and RSYNC algorithm. The offloading group is responsible for resource provisioning tasks, with the corresponding algorithms: RSYNC, FSYNC, RS-FSYNC, and ECFO algorithm. The load-balancing group handles the distribution of workload to resources. The algorithms which are considered in this group are DRAM, ERA, PBSA, GPRFCA, FOFSA, HCLB, ELBA (min-min), and Tabu Search Algorithm.

The placement group refers to finding the suitable resources to deploy the workload. Placing the incoming computation tasks to appropriate fog/edge resources is important. The only found algorithm that belongs to this group is the Iterative Algorithm based on resource placement.

Most of the algorithms were evaluated by using CloudAnalyst, SimCloud, Cloudlet tool, OMNET++, or iFogsim tools. Some studies in resource scheduling have experienced low scalability and proposed centralized topology in several case studies. One of the most important factors is scalability in resource management of fog computing, which should be improved in the scheduling scenarios. In addition, self-adaptive resource scheduling is one of the key issues in resource management of fog computing that few research studies have considered.

The provided evaluation of resource management techniques is limited to the features provided in Table 2. From a particular group/category of resource management techniques, we assume that 'the most appropriate' responsive algorithm provides all the features. For instance, all discovery algorithms can be 'appropriate' (except for RSYNC), since all these algorithms support the same metrics: resource allocation, workload balance, and task scheduling (refer to Table 2). From the reviewed load-balancing algorithms, the 'most' responding algorithms to our criteria are FOFSA and ERA. FOFSA pillars resource allocation, workload balance, and task scheduling, while ERA supports resource allocation, workload balance, and resource provisioning. An offloading algorithm that meets most of the specified criteria is ECFO, which performs resource provisioning and task scheduling.

This evaluation framework can be extended by applying a multi-criteria decision-making method, which could help the readers in the decision process to select a resource management algorithm. One of the limitations of this paper is that it does not provide any experiment to test the application of the researched algorithms to verify their usability and competitiveness. Another limitation in this research is that we propose an evaluation framework for resource management techniques that can be applied in cloud/fog and edge environments, but there is not a proper comparison analysis of the indicated approaches.

8. Conclusions

The goal of this paper is to provide an evaluation framework and classification of different resource management techniques that can be applied in cloud/fog and edge scenarios. It is useful for cloud/fog/edge architects to have a concise representation of the various challenges in resource management.

Cloud, fog, and edge computing govern a paradigm that can offer a solution for IoT applications that are sensitive to delay. Besides, fog nodes usually have higher repository capacity and data processing, which can be used for improving performance and reducing cost communication and latency. To be able to evaluate the state-of-the-art algorithms used in multiple research articles, in this paper, we analyze algorithms that can be classified into six categories. Thereafter, we consider how resources can be handled among cloud, fog, and edge devices.

In future work, the focus can be on making an analysis and comparison (e.g., through simulations) between them rather than an evaluation overview. Furthermore, some of the reviewed algorithms can be used in the simulation of a cloud/fog/edge architecture suitable for a particular application domain (e.g., smart logistics). We suggest research on case studies, preferably from a variety of domains.

Additional research work can also be done in terms of investigating (new) algorithms that not only deal with resource management but also address other challenges in cloud/fog/edge computing environments. We also recommend further research on validating and extending the evaluation framework and classification method, for example by conducting a systematic literature search.

Author Contributions: Conceptualization, A.M., R.B., A.C., A.A. and P.H.; Formal analysis A.M.; Methodology A.M. and A.A.; Supervision A.C., A.A. and P.H.; Visualization A.M. and R.B.; Writing—review and editing, A.M. and R.B. All authors have read and agreed to the published version of the manuscript.

Funding: This work is supported by OP Oost, project CountDown. The authors are thankful regarding the collaboration to all project partners.

Acknowledgments: We would like to thank Yanqui Huang and Henk Soppe for their discussions and effective comments on different sections of the review. Additionally, we would like to thank Maja Piponska for her suggestions regarding the visualization of Figure 2.

Conflicts of Interest: The authors declare no conflict of interest.

References

1. Chen, S.; Xu, H.; Liu, D.; Hu, B.; Wang, H. A Vision of IoT: Applications, Challenges, and Opportunities With China Perspective. *IEEE Internet Things J.* **2014**, *1*, 349–359. [CrossRef]
2. Mijuskovic, A.; Bemthuis, R.; Aldea, A.; Havinga, P. An Enterprise Architecture based on Cloud, Fog and Edge Computing for an Airfield Lighting Management System. In Proceedings of the 2020 IEEE 24th International Enterprise Distributed Object Computing Workshop (EDOCW), Eindhoven, The Netherlands, 5 October 2020; pp. 63–73. [CrossRef]
3. Muhammad, S.; Muhammad, S. The Internet of Things Architecture, Feasible Applications and Fundamental challenges. *Int. J. Comput. Appl.* **2018**, *179*, 975–8887. [CrossRef]
4. Gai, K.; Li, S. Towards Cloud Computing: A Literature Review on Cloud Computing and Its Development Trends. In Proceedings of the 2012 Fourth International Conference on Multimedia Information Networking and Security, Uttar Pradesh, India, 3–5 November 2012; pp. 142–146. [CrossRef]
5. Karagiannis, V.; Schulte, S. Comparison of Alternative Architectures in Fog Computing. In Proceedings of the 2020 IEEE 4th International Conference on Fog and Edge Computing (ICFEC), Melbourne, Australia, 11–14 May 2020; pp. 19–28. [CrossRef]
6. Naha, R.K.; Garg, S.; Georgakopoulos, D.; Jayaraman, P.P.; Gao, L.; Xiang, Y.; Ranjan, R. Fog Computing: Survey of Trends, Architectures, Requirements, and Research Directions. *IEEE Access* **2018**, *6*, 47980–48009. [CrossRef]
7. Deng, S.; Zhao, H.; Fang, W.; Yin, J.; Dustdar, S.; Zomaya, A.Y. Edge Intelligence: The Confluence of Edge Computing and Artificial Intelligence. *IEEE Internet Things J.* **2020**, *7*, 7457–7469. [CrossRef]
8. Hevner, A.R.; March, S.; Park, J.; Ram, S. Design Science in Information Systems Research. *Manag. Inf. Syst. Q.* **2004**, *28*, 75–105. [CrossRef]
9. Elazhary, H. Internet of Things (IoT), mobile cloud, cloudlet, mobile IoT, IoT cloud, fog, mobile edge, and edge emerging computing paradigms: Disambiguation and research directions. *J. Netw. Comput. Appl.* **2019**, *128*, 105–140. [CrossRef]
10. Tedeschi, P.; Sciancalepore, S. Edge and Fog Computing in Critical Infrastructures: Analysis, Security Threats, and Research Challenges. In Proceedings of the 2019 IEEE European Symposium on Security and Privacy Workshops (EuroS & PW), Stockholm, Sweden, 17–19 June 2019. [CrossRef]
11. Bar-Magen, J. Fog computing: introduction to a new cloud evolution. In *Escrituras Silenciadas: Paisaje Como Historiografía*; Forniés Casals, J.F., Numhauser, P., Eds.; Universidad de Alcalá: Madrid, Spain, 2013; pp. 111–126.
12. Masip-Bruin, X.; Marin-Tordera, E.; Jukan, A.; Ren, G.J. Managing resources continuity from the edge to the cloud: Architecture and performance. *Future Gener. Comput. Syst.* **2018**, *79*, 777–785. [CrossRef]
13. Tordera, E.M.; Xavi, M.B.; Alminana, J.; Jukan, A.; Ren, G.J.; Zhu, J.; Farré, J. What is a fog node a tutorial on current concepts towards a common definition. *arXiv* **2016**, arXiv:1611.09193.
14. Li, H.; Ota, K.; Dong, M. Learning IoT in Edge: Deep Learning for the Internet of Things with Edge Computing. *IEEE Netw.* **2018**, *32*, 96–101. [CrossRef]
15. Premsankar, G.; Francesco, M.; Taleb, T. Edge computing for the Internet of Things: A case study. *IEEE Internet Things J.* **2018**, *5*, 1275–1284. [CrossRef]
16. Agarwal, S.; Yadav, S.; Yadav, A.K. An efficient architecture and algorithm for resource provisioning in fog computing. *Int. J. Inf. Eng. Electronic Bus. (IJIEEB)* **2016**, *8*, 48–61. [CrossRef]
17. Seal, A.; Mukherjee, A. On the Emerging Coexistence of Edge, Fog and Cloud Computing paradigms in Real-Time Internets-of-EveryThings which operate in the Big-Squared Data space. In Proceedings of the SoutheastCon 2018, Huntsville, AL, USA, 19 April 2018; pp. 1–9. [CrossRef]
18. Qi, Q.; Tao, F. A Smart Manufacturing Service System Based on Edge Computing, Fog Computing, and Cloud Computing. *IEEE Access* **2019**, *7*, 86769–86777. [CrossRef]

19. Javaid, S.; Javaid, N.; Saba, T.; Wadud, Z.; Rehman, A.; Haseeb, A. Intelligent resource allocation in residential buildings using consumer to fog to cloud based framework. *Energies* **2019**, *12*, 815. [CrossRef]
20. da Silva, R.A.C.; da Fonseca, N.L.S. Resource Allocation Mechanism for a Fog-Cloud Infrastructure. In Proceedings of the 2018 IEEE International Conference on Communications (ICC), Kansas City, MO, USA, 20–24 May 2018; pp. 1–6. [CrossRef]
21. Wang, T.; Zhou, J.; Liu, A.; Bhuiyan, M.Z.A.; Wang, G.; Jia, W. Fog-Based Computing and Storage Offloading for Data Synchronization in IoT. *IEEE Internet Things J.* **2019**, *6*, 4272–4282. [CrossRef]
22. Xu, X.; Fu, S.; Cai, Q.; Tian, W.; Liu, W.; Dou, W.; Sun, X.; Liu, A.X. Dynamic resource allocation for load balancing in fog environment. *Wirel. Commun. Mob. Comput.* **2018**, *2018*. [CrossRef]
23. Buchade, A.; Nirav, M.S. Priority Based Allocation in Cloud Computing. *Int. J. Eng. Res. Technol.* **2014**, *3*, 855–857.
24. Taneja, M.; Davy, A. Resource aware placement of IoT application modules in Fog-Cloud Computing Paradigm. In Proceedings of the 2017 IFIP/IEEE Symposium on Integrated Network and Service Management (IM), Lisbon, Portugal, 8–12 May 2017; pp. 1222–1228. [CrossRef]
25. Bendechache, M.; Svorobej, S.; Takako Endo, P.; Lynn, T. Simulating Resource Management across the Cloud-to-Thing Continuum: A Survey and Future Directions. *Future Internet* **2020**, *12*, 95. [CrossRef]
26. Aslanpour, M.S.; Gill, S.S.; Toosi, A.N. Performance evaluation metrics for cloud, fog, and edge computing: A review, taxonomy, benchmarks and standards for future research. *Internet Things* **2020**, *12*, 100273. [CrossRef]
27. Ghobaei-Arani, M.; Souri, A.; Rahmanian, A.A. Resource management approaches in fog computing: a comprehensive review. *J. Grid Comput.* **2019**, *18*, 1–42. [CrossRef]
28. Salaht, F.A.; Desprez, F.; Lebre, A. An Overview of Service Placement Problem in Fog and Edge Computing. *ACM Comput. Surv.* **2020**, *53*, 1–35. [CrossRef]
29. Bittencourt, L.; Immich, R.; Sakellariou, R.; Fonseca, N.; Madeira, E.; Curado, M.; Villas, L.; DaSilva, L.; Lee, C.; Rana, O. The Internet of Things, Fog and Cloud continuum: Integration and challenges. *Internet Things* **2018**, *3-4*, 134–155. [CrossRef]
30. Agarwal, S.; Yadav, S.; Yadav, A.K. An architecture for elastic resource allocation in Fog Computing. *Int. J. Comput. Sci. Commun.* **2015**, *6*, 201–207. [CrossRef]
31. Madni, S.H.H.; Latiff, M.S.A.; Coulibaly, Y.; Abdulhamid, S.M. Recent advancements in resource allocation techniques for cloud computing environment: a systematic review. *Clust. Comput.* **2017**, *20*, 2489–2533. [CrossRef]
32. Rafique, H.; Shah, M.A.; Islam, S.U.; Maqsood, T.; Khan, S.; Maple, C. A Novel Bio-Inspired Hybrid Algorithm (NBIHA) for Efficient Resource Management in Fog Computing. *IEEE Access* **2019**, *7*, 115760–115773. [CrossRef]
33. Ni, L.; Zhang, J.; Jiang, C.; Yan, C.; Yu, K. Resource Allocation Strategy in Fog Computing Based on Priced Timed Petri Nets. *IEEE Internet Things J.* **2017**, *4*, 1216–1228. [CrossRef]
34. Fan, Q.; Ansari, N. Towards Workload Balancing in Fog Computing Empowered IoT. *IEEE Trans. Netw. Sci. Eng.* **2020**, *7*, 253–262. [CrossRef]
35. Skarlat, O.; Schulte, S.; Borkowski, M.; Leitner, P. Resource Provisioning for IoT Services in the Fog. In Proceedings of the 2016 IEEE 9th Conference on Service-Oriented Computing and Applications (SOCA), Macau, China, 4–6 November 2016; IEEE Computer Society: Los Alamitos, CA, USA, 2016; pp. 32–39. [CrossRef]
36. Yin, L.; Luo, J.; Luo, H. Tasks Scheduling and Resource Allocation in Fog Computing Based on Containers for Smart Manufacturing. *IEEE Trans. Ind. Inform.* **2018**, *14*, 4712–4721. [CrossRef]
37. Doddi, A.; Krishna, P.; Mallikarjuna, B. Feedback-based fuzzy resource management in IoT using fog computing. *Evol. Intell.* **2020**, 3. [CrossRef]
38. Zahid, M.; Javaid, N.; Ansar, K.; Hassan, K.; Khan, M.K.; Waqas, M. Hill Climbing Load Balancing Algorithm on Fog Computing. In Proceedings of the International Conference on P2P, Parallel, Grid, Cloud and Internet Computing, Taichung, Taiwan, 27–29 October 2018. [CrossRef]
39. Manju, A.B.; Sumathy, S. Efficient Load Balancing Algorithm for Task Preprocessing in Fog Computing Environment. In *Smart Intelligent Computing and Applications*; Satapathy, S.C., Bhateja, V., Das, S., Eds.; Springer: Singapore, 2019; pp. 291–298. [CrossRef]
40. Téllez, N.; Miguel, J.; Salazar, A.; Nino-Ruiz, E. A Tabu Search Method for Load Balancing in Fog Computing. *Int. J. Artif. Intell.* **2018**, *16*, 78–105.
41. Jiang, Y.; Chen, Y.; Yang, S.; Wu, C. Energy-Efficient Task Offloading for Time-Sensitive Applications in Fog Computing. *IEEE Syst. J.* **2019**, *13*, 2930–2941. [CrossRef]
42. Hong, C.H.; Varghese, B. Resource Management in Fog/Edge Computing: A Survey. *arXiv* **2018**, arXiv:1810.00305.
43. Lai, C.; Song, D.; Hwang, R.; Lai, Y. A QoS-aware streaming service over fog computing infrastructures. In Proceedings of the 2016 Digital Media Industry Academic Forum (DMIAF), Santorini, Greece, 4–6 July 2016; pp. 94–98. [CrossRef]
44. E.800: Terms and Definitions Related to Quality of Service and Network Performance Including Dependability. ITU-T Recommendation. Available online: https://www.itu.int/rec/T-REC-E.800-199408-S/en (accessed on 3 March 2021).

sensors

MDPI

Article

Optimal Service Provisioning for the Scalable Fog/Edge Computing Environment

Jonghwa Choi and Sanghyun Ahn *

Department of Computer Science and Engineering, University of Seoul, Seoul 02504, Korea; David13@uos.ac.kr
* Correspondence: ahn@uos.ac.kr; Tel.: +82-(0)2-6490-2446

Abstract: In recent years, we observed the proliferation of cloud data centers (CDCs) and the Internet of Things (IoT). Cloud computing based on CDCs has the drawback of unpredictable response times due to variant delays between service requestors (IoT devices and end devices) and CDCs. This deficiency of cloud computing is especially problematic in providing IoT services with strict timing requirements and as a result, gives birth to fog/edge computing (FEC) whose responsiveness is achieved by placing service images near service requestors. In FEC, the computing nodes located close to service requestors are called fog/edge nodes (FENs). In addition, for an FEN to execute a specific service, it has to be provisioned with the corresponding service image. Most of the previous work on the service provisioning in the FEC environment deals with determining an appropriate FEN satisfying the requirements like delay, CPU and storage from the perspective of one or more service requests. In this paper, we determined how to optimally place service images in consideration of the pre-obtained service demands which may be collected during the prior time interval. The proposed FEC environment is scalable in the sense that the resources of FENs are effectively utilized thanks to the optimal provisioning of services on FENs. We propose two approaches to provision service images on FENs. In order to validate the performance of the proposed mechanisms, intensive simulations were carried out for various service demand scenarios.

Keywords: fog/edge computing; service provisioning; service placement; service offloading; Internet of Things (IoT)

Citation: Choi, J.; Ahn, S. Optimal Service Provisioning for the Scalable Fog/Edge Computing Environment. *Sensors* **2021**, *21*, 1506. https://doi.org/10.3390/s21041506

Academic Editor: Taehong Kim

Received: 31 January 2021
Accepted: 19 February 2021
Published: 22 February 2021

1. Introduction

The Internet of Things (IoT) is a new wave significantly affecting our lives in various areas including smart energy grids, smart factories, smart cities, and smart farms [1]. IoT is accomplished by sensing (monitoring) the target environment, collecting and analyzing the sensory information and actuating based on the feedback information from the analysis. The data produced in the IoT environment may be enormous and require intensive computation such as for deep learning-based prediction. Hence, the IoT data tend to be transmitted to the remote cloud data center (CDC) for storage and computation [2–4]. Many IoT applications have stringent timing requirements which may not be met by cloud computing because of long-distance message transmissions through the network between CDC and end devices (aka IoT devices). For the reduced latency between CDC and end devices, the concept of fog computing has been introduced by Cisco [5]. In fog/edge computing (FEC), fog/edge nodes (FENs) located in the proximity of end devices provide services to them in lieu of CDC without causing a long delay by executing the lightweight virtualized service images pre-allocated from CDC [5–7]. However, due to the limited capacity (CPU, storage, etc.) of FENs, only a subset of service images can be placed on each FEN [8,9].

In the previous work on service provisioning (or placement) in FEC [10,11], a service is placed on the FEN satisfying the given service requirement for the execution of the service, assuming that the corresponding service image is already installed in the FEN.

Service provisioning, service placement and service offloading are interchangeably used for installing service images and executing them on FENs or CDC instead of end devices. However, due to the limited capability of a FEN, it is infeasible to place all the services on an FEN. Even in the case that FENs install service images on demand for offloaded services, the resource of FENs may not be efficiently utilized because of patchwork-like service image placement [12]. In addition, depending on the IoT applications, the type of services requested may vary. For example, in an area occupied by smart factories, the service analyzing the sensed data from a specific machine for the detection of malfunctions will be requested by the factories equipped with the machine. Placing service images on FENs in the per-service request-based manner may result in the improper resource usage of FENs due to redundantly placed service images. If we can optimally provision service images on FENs based on the pre-obtained service demands (service demands and service requests are interchangeably used in this paper), the resource utilization of FENs can be optimized. To our knowledge, this issue has not been addressed to date and thus, we formulate the problem of optimal service provisioning (SP) in the FEC environment. In our work, we adopted "service provisioning" instead of service placement in the sense that service provisioning means service placement with planning.

For the scalability of the FEC environment, FENs are located hierarchically near to end devices and in between end devices and CDC which is the last resort of providing services to end devices. The following two cases (c1) and (c2) are not desirable from the perspective of resource utilization and latency, respectively:

(c1) For a sporadically requested service, the corresponding service images are redundantly placed on multiple FENs.
(c2) For the coverage of multiple requests of a service, the corresponding service image is provisioned on a FEN at a high level of the FEN hierarchy.

If we consider both the resource usages and locations of FENs and the locations of end devices requesting services, we could avoid the above-mentioned cases. The consideration of only the case (c1) (i.e., the resource utilization of FENs) may result in the case (c2) (i.e., long delay), and vice versa. Therefore, in provisioning service images in the FEC environment, we must consider both the resource usages and locations of FENs and the delay requirements and locations of service requesters. In this paper, we propose two SP approaches by extending our previous work [12]; the first one is on the basis of the number of service requests from end devices and the second one is based on the locations of end devices requesting services. We can assert that the proposed FEC environment is scalable because more service images can be accommodated by FENs thanks to the optimal provisioning of service images on them. For the performance evaluation of the proposed mechanisms, we perform simulations for various service requesting scenarios and analyze the performance in terms of the total number of service images provisioned on FENs and the total number of service requests not accommodated by FENs.

The rest of the paper is organized as follows. In Section 2, we describe the related work on the service provisioning, offloading or placement on FENs. In Section 3, the problem of service provisioning FENs is formulated as a 0–1 integer linear programming problem for the proof of the NP-hardness of the problem and then, our two approaches are described in detail. The performance of our mechanisms is evaluated in Section 4 from the simulation results. Finally, Section 5 concludes this paper.

2. Related Work

Services can be offered to end devices via computation offloading in which CDC or FENs perform computation in lieu of end devices. SP on FENs is the pre-allotment of the resources, like CPU, RAM and storage, of FENs to service images for computation offloading [6–8]. For the sake of effective computation offloading, SP on FENs must be properly carried out beforehand.

In the FEC environment, the service provisioning, placement or service offloading has been studied by many researchers with different objectives and considerations, like power consumption [13], quality of experience (QoE) [14–16], migration [17], network perspective [18], service atomization [19], etc. The authors of [19] proposed strategies for offloading service executions by considering the resources of CDC and FENs. For that, they introduced the concept of service atomization and parallel resource allocation. By service atomization, complex IoT services are divided into smaller atomic services which, then, can be executed on multiple FENs sequentially or parallelly in a distributed way. In their scheme, end devices offload service executions to CDC or FENs according to network and processing demands.

In [18], the application provisioning problem in the fog-cloud environment is studied from a network perspective to guarantee the quality of service (QoS) of application (service) data streams in terms of transmission delay and bandwidth for each application. Application provisioning is to find the host node (i.e., FEN) and data routing for a specific application. The authors formulate the issue as the single application provisioning (SAP) and the multi-application provisioning (MAP) problems. SAP determines the path to the FEN satisfying the bandwidth and delay requirements of data traffics from end devices for one application. MAP is the extension of SAP for multiple applications from the perspective of network link capacity (bandwidth) usage.

In [11], the deployment of multi-component application in the fog hierarchy is defined as the components deployment problem (CDP) which determines candidate FENs to be deployed with the components of an application according to application-specific QoS requirements of end devices on latency and bandwidth, software and hardware capabilities of FENs and business policies. They target to deploy large-scale applications composed of multiple components that can be independently deployable and work together for the infrastructure and prove that the CDP problem is NP-hard by the reduction from the subgraph isomorphism problem (SIP) [20]. The CDP problem is to find the optimal deployment of a single application (service) based on the service requests of end devices in a centralized manner.

Both [18] and [11] determine the FENs to be provisioned with services in a centralized way for a given set of service requests, similar to ours. However, theirs take account of each individual FEN, not of entire FENs, from the perspective of resource usage. The consideration on the capacity of each FEN is just for checking the relevant constraint, but the consideration on the total resource usage of all the FENs is for optimizing the overall resource usage of FENs. Thus, the latter is more appropriate for enhancing the scalability of the fog infrastructure in terms of resource usage.

Related to the resource provision of FENs, the authors of [14,15] proposed a FEC architecture composed of fog colonies each consisting of one fog orchestration control node and multiple fog cells. The fog orchestration control node is a fog cell with extended functionalities like managing fog cells in the same fog colony and other connected control nodes. The fog cell is the software components running on an FEN. The fog orchestration control node keeps all the service implementations (i.e., service images) in its service registry, located in the low-cost abundant storage unit, and deploys the corresponding service image to the fog cell covering the service requester on demand. In this way, the constrained resources of FENs are effectively utilized. The fog cell for a service request is determined based on the resources of FENs and the resource and delay requirements of the service request such that the resource requirement is satisfied with the minimum delay. In the scheme, all the service images are kept in the storage of the fog orchestration control node and deployed on FENs per service request by the control node. This will incur overwhelming communication overhead and delay in deploying service images on demand from fog orchestration control nodes to fog cells, and the requirement on the fog orchestration control node to be installed with all the service images is infeasible or, at least, inefficient if plenty of services are to be deployed in the fog.

In this paper, we aim to formulate and resolve the problem of minimizing the total number of service images provisioned on FENs in order to optimize the overall resource usage of FENs.

3. Service Provisioning for Fog/Edge Nodes

3.1. System Environment and Problem Description

For the simplicity of problem formalization, we assume that there is one CDC and all the service demands are accommodated by the FENs. Each service demand is assumed to require the same amount (e.g., one capacity unit) of resources even though service demands are heterogeneous in reality. The network link capacity is assumed to be sufficient for the delivery of the data generated by the service demands (i.e., we do not consider the network link capacity as a constraint in the problem).

We define the notations needed for the problem formalization in Table 1. S is the set of the services, $\{s_1, \ldots, s_\alpha\}$, and F is the set of the FENs, $\{f_1, \ldots, f_\beta\}$, and U is the set of the end devices, $\{u_1, \ldots, u_\gamma\}$, and D is the service demand matrix, $[d_{ik}]$, where d_{ik} is 1 if the service s_i is demanded by the end device u_k.

Table 1. Notations for the definition of the optimal service provisioning (SP) problem.

Notation	Description
S	The set of services; $S = \{s_1, \ldots, s_\alpha\}$
F	The set of FENs; $F = \{f_1, \ldots, f_\beta\}$
U	The set of end devices; $U = \{u_1, \ldots, u_\gamma\}$
X	The service placement matrix; , $X = [x_{ij}]$
x_{ij}	The binary variable indicating whether a service image of the service s_i is provisioned in the FEN f_j or not; $x_{ij} = \begin{cases} 1, \text{ if } s_i \text{ is provisioned on } f_j \\ 0, \text{ otherwise} \end{cases}$
D	The given service demand matrix consisting of the service demands, d_{ik}'s
d_{ik}	The element of D indicating whether the service s_i is demanded by the end device u_k or not; $d_{ik} = \begin{cases} 1, \text{ if } s_i \text{ is demanded by } u_k \\ 0, \text{ otherwise} \end{cases}$
c_j	The maximum capacity of the FEN f_j
T_i	The maximum delay requirement of the service s_i
τ_{jk}	The delay from the FEN f_j to the end device u_k
b_{ik}	The binary variable indicating whether the delay requirement of the service demand d_{ik} is satisfied or not; $b_{ik} = \begin{cases} 1, \text{ if the delay requirement of } d_{ik} \text{ is satisfied} \\ 0, \text{ otherwise} \end{cases}$

The problem of the optimal service provisioning (SP) in the FEC environment can be defined as follows:

$$\text{Minimize } \sum_{i\in\{1,\ldots,\alpha\}} \sum_{j\in\{1,\ldots,\beta\}} x_{ij} \quad (1)$$

$$\text{Subject to}: \sum_{i\in\{1,\ldots,\alpha\}} x_{ij} \le c_j, \forall\, j \in \{1,\ldots,\beta\} \quad (2)$$

$$x_{ij} = \begin{cases} 1, \text{ if a service image of } s_i \text{ is provisioned on } f_j \\ 0, \text{ otherwise} \end{cases} \quad (3)$$

$$x_{ij} \in \{0, 1\}, \forall\, i \in \{1,\ldots,\alpha\},\ j \in \{1,\ldots,\beta\} \quad (4)$$

$$\sum_{i\in\{1,\ldots,\alpha\}} \sum_{k\in\{1,\ldots,\gamma\}} d_{ik} = \sum_{i\in\{1,\ldots,\alpha\}} \sum_{k\in\{1,\ldots,\gamma\}} b_{ik} \quad (5)$$

$$b_{ik} = \begin{cases} 1, \text{ if } \exists\, j \in \{1,\ldots,\beta\} \text{ satisfying } 0 < \tau_{jk} \times x_{ij} \times d_{ik} \le T_i \\ 0, \text{ otherwise} \end{cases}, \quad \forall\, i \in \{1,\ldots,\alpha\} \text{ and } k \in \{1,\ldots,\gamma\} \quad (6)$$

$$b_{ik} \in \{0, 1\}, \forall\, i \in \{1,\ldots,\alpha\},\ k \in \{1,\ldots,\gamma\} \quad (7)$$

Equation (1) is the objective function of minimizing the total number of the service images provisioned on the FENs. Equation (2) is for the capacity constraint, termed Condition-C, and Equations (5) and (6) are for the delay constraint, termed Condition-D. The Equations (3), (4), (6) and (7) define the binary variables b_{ik} and x_{ij}. $x_{ij} = 1$ indicates that a service image of s_i is deployed in f_j. Equation (2) checks whether each FEN accommodates service demands within its maximum capacity c_j. Equation (5) checks for whether the delay requirements of all the service demands in D are satisfied. In Equation (6), the binary variable b_{ik} is set to 1, for the service demand d_{ik} with 1, if there exists at least one FEN with a service image of s_i that satisfies the maximum delay requirement of T_i, i.e., $0 < \tau_{jk} \times x_{ij} \times d_{ik} \leq T_i$.

Because the optimal SP problem was formulated as a 0–1 integer linear programming problem, the problem is NP-hard. Thus, we propose two heuristic mechanisms that determine how to deploy service images on FENs according to the collected service demands from end devices.

3.2. Logical Fog Network

CDC determines the FENs to be installed with specific service images based on the service demands of the end devices. The locations of the FENs can be anywhere in the given network, determined by the deployment strategy which is out of the scope of this paper. In general, the physical network topology of the FEC environment is a mesh which increases the computational complexity of determining the optimal FENs to be provisioned with service images. Therefore, in order to simplify the problem, we form a logical tree topology of the FENs rooted at CDC, called a logical fog network [12], N_{fog}, in a hierarchical manner, as shown in Figure 1b for the physical network in Figure 1a in which CDC is labeled as f_0 for the sake of convenience. Because N_{fog} is a subnetwork of the given physical network and the tree topology is a special case of the mesh topology, the SP problem for N_{fog} is also NP-hard.

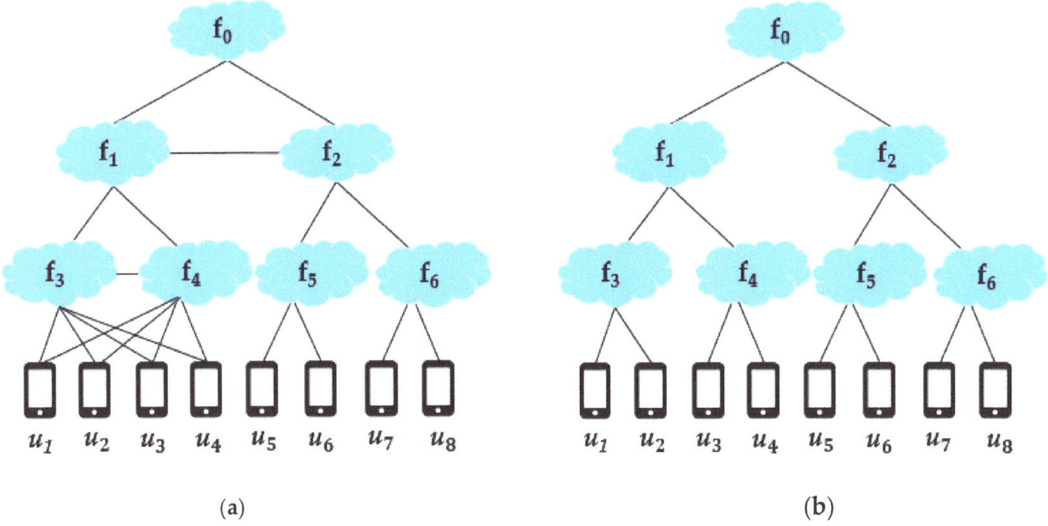

Figure 1. Logical fog network: (**a**) an example of a physical network; (**b**) the logical fog network of (**a**).

In N_{fog}, if an end device is in the coverage area of a FEN, we say that the end device is in the direct coverage of the FEN or the FEN directly covers the end device. If an FEN is an ancestor (including the parent) of the FEN directly covering an end device, the ancestor FEN is said to cover the end node indirectly. In N_{fog}, all the FENs on the path from CDC to the directly covering FEN of an end device cover the end device.

A service requester is an end device requesting the corresponding service. The data from a service requester can be handled by any FEN, with the service image and satisfying the requirements, both Condition-C and Condition-D, of the service requester on the path from the directly covering FEN up to CDC in N_{fog}. Each FEN, located in between the service requester and the service handling FEN, just delivers the data from the service requester to its parent FEN for the proper processing of the data.

We assume that CDC has the information on N_{fog} including the end devices directly covered by the FENs and the information of the locations and the capacities of the FENs. For the provisioning of service images in N_{fog}, we assume that CDC has unlimited capacity such that any requirements of the end devices, except for latency, can be accommodated—that is, if there exist any service requests not accommodated by any FENs, CDC can cover those service requests.

The logical fog network, N_{fog}, is constructed as follows:

[Step N1] The initial N_{fog} is null and the FENs are sorted in the decreasing order of the capacity.

[Step N2] CDC is first included in N_{fog} as the root and the level of CDC is set to 0.

[Step N3] From the sorted list of the FENs obtained at [Step N1], the first k FENs are selected as the child nodes of CDC. Γ is the set of the child FENs of CDC. The value of the parameter k has to be properly decided by the fog network designer.

[Step N4] The delay of the link between an FEN in Γ and CDC is adjusted to a value less than its original delay (e.g., a half of the original delay).

[Step N5] Each FEN in $F - \Gamma$ is added to N_{fog} via the minimum delay (shortest) path to CDC.

In [Step N3], the criterion for selecting the FENs in Γ is the capacity because an FEN with more capacities can have more service images installed and a higher possibility of covering more end devices. In [Step N5], the criterion for adding the FENs in $F - \Gamma$ to N_{fog} is the latency because the main purpose of FEC is reducing the latency from end devices. After [Step N5], the obtained N_{fog} is a logical tree rooted at CDC with the FENs in Γ directly connected to CDC and with each FEN in $F - \Gamma$ connected to CDC via the shortest path. The reason for the link cost (delay) adjustment in [Step N4] is to make the possibility of the FENs in $F - \Gamma$ to go through the FENs in Γ to get to CDC. From this, we can achieve the load balancing effect by making high-capacity FENs share the burden of computing load with CDC. Through the procedure of [Step N1]~[Step N5], we can obtain a logical fog network that is appropriate for less service images provisioned thanks to the tree topology and for lower delays from end devices to FENs thanks to the shortest path branches. In the following Section 3.3, we described two mechanisms finding the right FENs, based on N_{fog}, that can accommodate the service images demanded by the end devices with the aim of minimizing the number of service images provisioned.

3.3. Service Provisioning Based on Service Demands

In this subsection, we propose two heuristic SP mechanisms, based on N_{fog}, in which the amount of service demands for a FEN is the major factor to be considered in placing a service image on the FEN. The first SP mechanism provisions the highest-level FEN, satisfying the conditions, Condition-C and Condition-D, of the most-demanding service, with the service so that the coverage of the FEN for the service can be maximized. Thus, the first SP mechanism is named as the maximal coverage-SP (MC-SP) mechanism. The MC-SP mechanism performs well if the distribution of the end devices demanding a specific service is uniform in the logical fog network. However, if we consider an industrial area with many smart factories requesting a specific service, the MC-SP mechanism may not perform optimally. In this situation, it is desirable to place the corresponding service image near the area. Thus, in our second mechanism, we take into consideration the locations of the end devices demanding a specific service in finding the right FEN to be placed with service images. The second mechanism is named as the flexible coverage-SP (FC-SP) mechanism. In the Sections 3.3.1 and 3.3.2, the MC-SP and the FC-SP mechanisms are described in detail, respectively. For the simplicity of the description of the mechanisms, we assume that, in N_{fog}, there exists at least one FEN satisfying Condition-C and Condition-D of each service in S.

3.3.1. Maximal Coverage-Service Provisioning Mechanism

Before we describe the procedure of the MC-SP mechanism, the FENs in F are rearranged to $\check{F} = \langle f'_1, \ldots, f'_\beta \rangle$ such that the index of an FEN at the level q in N_{fog} is smaller than that of a FEN at a level lower than q. That is, $f'_1 \sim f'_k$ $(k = 1, \ldots, \beta)$ are the children FENs of CDC in N_{fog} and f'_β is at the lowest level of N_{fog}. We define the coverage matrix V_{fog} representing N_{fog} such that the element of V_{fog}, v_{kj}, is set to 1 if the end device u_k is covered by the FEN f'_j. With given the service demand matrix D, $W = [w_{ij}] = D \times V_{fog}$ is the $\alpha \times \beta$ matrix whose element w_{ij} indicates the total demands on s_i from the end devices covered by f'_j. The notations for the SP mechanisms are listed in Table 2.

Table 2. Notations for the SP mechanisms.

Notation	Description
N_{fog}	The logical fog network
$\check{F}$	The ordered list of the FENs in F from the highest to the lowest level of N_{fog}; $\check{F} = \langle f'_1, \ldots, f'_\beta \rangle$
V_{fog}	The coverage matrix of N_{fog} consisting of v_{kj}'s
v_{kj}	The element of V_{fog} indicating whether the end device u_k is covered by the FEN f'_j or not; $v_{kj} = \begin{cases} 1, \text{ if } u_k \text{ is covered by } f'_j \\ 0, \text{ otherwise} \end{cases}$
W	The total demand matrix consisting of w_{ij}'s
w_{ij}	The element of W indicating the total demands on the service s_i from the end devices covered by the FEN f'_j
W_j	The jth column of W; $W_j = \langle W_j(1), \ldots, W_j(\alpha) \rangle$, where $W_j(q) = w_{qj}$ for $q \in \{1, \ldots, \alpha\}$
$\check{W}_j$	The sorted list of W_j in the decreasing order of the amount of demands; $\check{W}_j = \langle \check{W}_j(1), \ldots, \check{W}_j(\alpha) \rangle$, $\check{W}_j(q) \geq \check{W}_j(\check{q})$ if $q < \check{q}$ for $q, \check{q} \in \{1, \ldots, \alpha\}$
Y_j	The corresponding service list of W_j; $Y_j = \langle Y_j(1), \ldots, Y_j(\alpha) \rangle$, $Y_j(q) \in S$ for $q \in \{1, \ldots, \alpha\}$

The procedure of the MC-SP mechanism with given N_{fog}, $\check{F}$ and W is described in Figure 2.

```
[M1]  For j = 1 to β {
[M2]      W̆_j = Sort(W_j)
[M3]      Y_j = Service(W̆_j)
[M4]      For i = 1 to α {
[M5]          If (W̆_j(i)) > 0) {
[M6]              If (ChkCapacity(f'_j, Y_j(i)) and ChkDelay(f'_j, Y_j(i))) {
[M7]                  Provision(f'_j, Y_j(i))
[M8]                  Modify(N_fog, W, f'_j, Y_j(i))
[M9]              }
[M10]         }
[M11]     }
[M12] }
```

Figure 2. The procedure of the maximal coverage-SP (MC-SP) Mechanism.

In Figure 2, the jth column of W is denoted as W_j and the ith element of W_j as $W_j(i)$. $\check{W}_j$ is the sorted list of W_j, in the decreasing order of the amount of demands, which is obtained by carrying out the function $\overline{\text{Sort}}(W_j)$ (see **[M2]**). Y_j is the corresponding service list of $\check{W}_j$ obtained from the function Service($\check{W}_j$) (see **[M3]**). That is, $Y_j(i)$ has the information of the service whose demands imposed on f'_j is the ith largest among the services in W_j. The outer for-loop checks for each FEN from the highest to the lowest level of N_{fog} and the inne for-loop checks for each service from the list in Y_j. That is, for f'_j, the larger the demand on a service, the earlier the service is checked for its provisioning in f'_j. The condition i **[M5]** is for excluding those services with no demands from being provisioned in f'_j. I **[M6]**~**[M7]**, the provision of the service $Y_j(i)$ in f'_j is performed if the provision satisfie the Condition-C and the Condition-D of $Y_j(i)$. The functions ChkCapacity(f'_j, $Y_j(i)$) an ChkDelay(f'_j, $Y_j(i)$) check Condition-C and Condition-D, respectively, and return the tru or false value according to whether the corresponding requirement is satisfied or not. Th function Provision(f'_j, $Y_j(i)$) in **[M7]** provisions the FEN f'_j with the service image of th service $Y_j(i)$. The function Modify(N_{fog}, W, f'_j, $Y_j(i)$) in **[M8]** removes the demands o the service $Y_j(i)$ from the FENs which are the descendants of f'_j because all the servic demands are resolved by f'_j.

Figure 3 is a simple example showing N_{fog} with the FENs $f_1, \ldots, f_6$ and the en devices $u_1, \ldots, u_8$ and the services $s_1, \ldots, s_6$. The service demands of each end device ar listed in its corresponding box and the capacity of f_j is indicated by c_j. In this exampl to focus more on SP from the aspect of Condition-C, all the FENs are assumed to satisf Condition-D of s_i for all $i \in \{1, \ldots, 6\}$. For simplicity, f_0 is also assumed to satisfy th Condition-Ds of all services. In the figure, we can see the result of applying the MC-S mechanism, and the service placement matrix X = $[x_{ij}]$, indicated by a black solid line bo where $x_{ij} = 1$ implies that s_i is provisioned on f_j. Since all FENs satisfy Condition-D c each service, services are placed from f_0 to f_6. In MC-SP, the service with more demand has a higher chance to be placed on. In this example, f_0 can accommodate at most thre services and s_1, s_5, s_6 have demands of 4, 3, 3, respectively, so s_1, s_5, s_6 are provisioned o f_0. After that, the demands on s_1, s_5, s_6 are removed from the other FENs. Then, service s_2, s_4 are placed on f_1 and s_3 is placed on f_2.

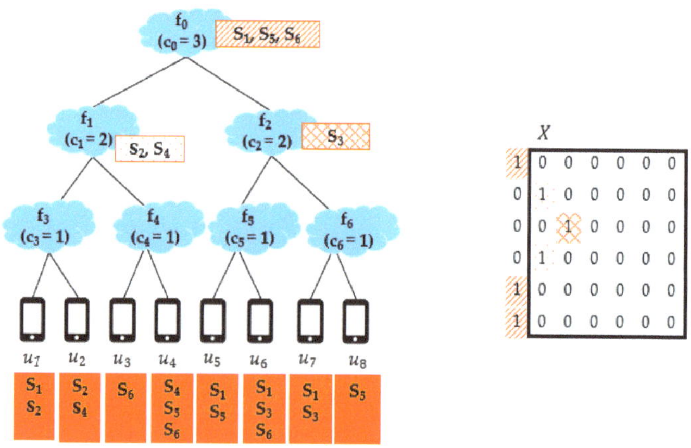

Figure 3. The service provisioning result of the MC-SP mechanism.

3.3.2. The Flexible Coverage-Service Provisioning Mechanism

The MC-SP mechanism takes into consideration only the hierarchical topology of N_{fog} in the provisioning service images on FENs. This mechanism is the lack of the awareness of the patterns of service demands which may depend on some geographical, industrial, o social characteristics. Thus, in this subsection, we propose the FC-SP mechanism, which adaptively places service images on FENs according to the pattern of service demands This is well suited for the situation with uneven service demands in a limited area. For this situation, an FEN near to the area is best placed to resolve the demands. Thus, we adopt the lowest-level common ancestor (LCA) FEN, in N_{fog}, of the end devices demanding a service as the candidate to be provisioned with the corresponding service image. If the LCA FEN f of a service s is installed with the service image of s, then those end devices having f as the LCA FEN demanding s can be commonly supported by f. In the FC-SP mechanism, the pattern of service demands is estimated by using the Shannon entropy [21]. The preference of a service image to be provisioned is determined based on the total amount of demands imposed on the FEN which is currently considered for the service provisioning. The additional notations for the FC-SP mechanism are listed in Table 3.

Table 3. Additional notations for the flexible coverage-SP (FC-SP) mechanism.

Notation	Description
Z	The ordered list of z_i's; Z = <$z_1, \ldots, z_\alpha$>, where z_i is the total demands on s_i, $z_i = \sum_{j \in \{1,\ldots,\beta\}} w_{ij}$
$\breve{Z}$	The sorted list of Z in the decreasing order of the total demands on services
$\breve{S}$	The corresponding service list of $\breve{Z}$; $\breve{S} = \breve{s}_1, \ldots, \breve{s}_\alpha$, where $\breve{s}_i$ is the service on which the total demands of $\breve{z}_i$ are imposed
A^{srv}	The ordered list of LCA FENs satisfying Condition-D of each service in $\breve{S}$
A^{fog}	The ordered list of the services with the same LCA FENs; the ith element, $A^{fog}(i)$, is the ordered list of the services, in the decreasing order of the total demands imposed on each service, whose LCA FEN is f'_i
σ	The current service to be provisioned
f^{min}	The minimum entropy service which was already provisioned on the current FEN
f^p	The parent node of the current FEN
f^c	The child node of the current FEN
p^{fog}	The ordered list of visited ancestor FENs starting from f'_i

The procedure of the FC-SP mechanism with given N_{fog} and W is described in detail in Figure 4. Here, we assume that, in N_{fog}, there exists at least one FEN satisfying Condition-C and Condition-D of each service in S. The pseudocode description in Figure 4b is somewhat lengthy and complicated because lots of notations were used, so we provided a simpler form of description, flowcharts, in Figure 4a.

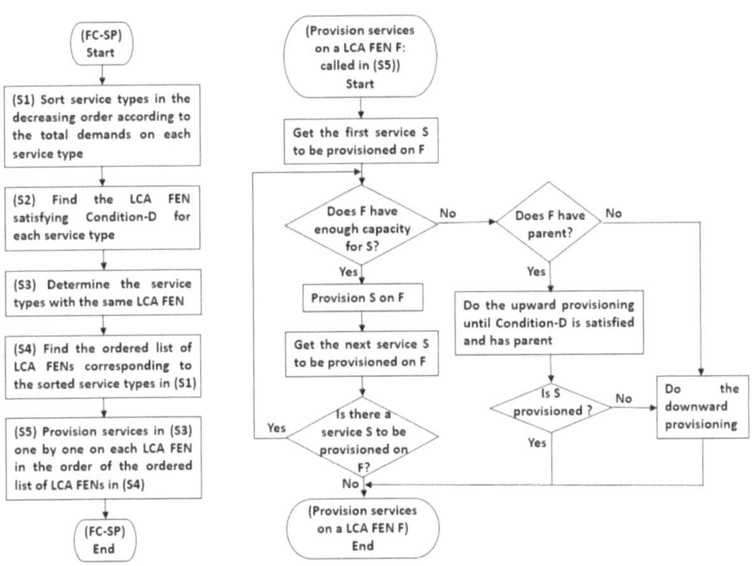

(a)

Figure 4. *Cont.*

```
[F1]  A^srv = LCAncestor(N_fog, W) /* the LCA FEN satisfying Condition-D is found for each service
[F2]  A^fog = FogLCAncestor(A^srv) /* the services with the same LCA FENs are determined
[F3]  For i = 1 to β {
[F4]    σ = GetFogService(A^fog(i)) /* σ is the service to be provisioned on f_i'
[F5]    If (σ ≠ NULL) {
[F6]      If (ChkCapacity(f_i', σ)) { /* f_i' has enough capacity for σ
[F7]        Do {
[F8]          Provision(f_i', σ) /* σ is provisioned on f_i'
[F9]          σ = GetFogService(A^fog(i))
[F10]       } While ((σ ≠ NULL) and ChkCapacity(f_i', σ)) /* repeat while Condition-C is met
[F11]     }
[F12]     If (σ ≠ NULL) { /* f_i' has no available capacity for σ
[F13]       If (Parent(f_i', N_fog) == NULL) { /*if f_i' is at the highest-level, downward provisioning is done
[F14]         Do {
[F15]           f^min = MinEntropyService(f_i') /* f^min is the min entropy service provisioned on f_i'
[F16]           If (Entropy(f^min) < Entropy(σ)) { /* if entropy of f^min < entropy of σ, σ replaces f^min
[F17]             ReleaseProvision(f^min) /* f^min is no more provisioned on f_i'
[F18]             SplitProvision(f^min, f_i') /* f^min is provisioned on a lower-level FEN
[F19]             Provision(f_i', σ) /* σ is provisioned on f_i'
[F20]           }
[F21]           Else { /* if entropy of σ is less than f_i^min, σ is provisioned on a lower-level FEN
[F22]             SplitProvision(σ, f_i')
[F23]           }
[F24]           σ = GetFogService(A^fog(i))
[F25]         } While (σ ≠ NULL) /* repeat until all the services in A^fog(i) are provisioned
[F26]       }
[F27]       Else { /* if f_i' has the parent node, upward provisioning is done and, then, downward if necessary
[F28]         f^p = f_i'
[F29]         p^fog =< f_i' > /* p^fog is the ordered list of visited ancestor FENs starting from f_i'
[F30]         Do { /* upward provisioning starts from f^p
[F31]           f^p = Parent(f^p, N_fog) /* the parent FEN of f^p is found
[F32]           While (ChkDelay(f^p, σ) and (f^p ≠ NULL)) {
[F33]             If (ChkCapacity(f^p, σ)) {
                    /* if f^p satisfies Condition-C and Condition-D, σ is provisioned on f^p
[F34]               Provision(f^p, σ)
[F35]               Break
[F36]             }
[F37]             p^fog = p^fog ∨ f^p /* f^p is appended to p^fog
[F38]             f^p = Parent(f^p, N_fog)
[F39]           }
[F40]           If (!ChkDelay(f^p, σ) or (f^p == NULL)) { /* if f^p does not satisfy Condition-D or
                                               /* has no parent, downward provisioning starts from f^p
[F41]             f^c = Child(f^p) /* Child(f^p) returns the child FEN, f^c, of f^p in p^fog
[F42]             f^min = MinEntropyService(f^c) /* f^min is the min entropy service provisioned on f^c
[F43]             If (Entropy(f^min) < Entropy(σ)) { /* if entropy of f^min is less than σ, σ replaces f^min
[F44]               ReleaseProvision(f^min) /* f^min is no more provisioned on f^c
[F45]               SplitProvision(f^min, f^c) /* f^min is provisioned on a lower-level FEN
[F46]               Provision(f^c, σ) /* σ is provisioned on f^c
[F47]             }
[F48]             Else { /* if entropy of σ is less than f^min, σ is provisioned on a lower-level FEN
[F49]               SplitProvision(σ, f^c)
[F50]             }
[F51]           }
[F52]           σ = GetFogService(A^fog(i))
[F53]         } While (σ ≠ NULL) /* repeat until all the services in A^fog(i) are provisioned
[F54]       }
[F55]     }
[F56]   }
[F57] }
```

(b)

Figure 4. The procedure of the FC-SP mechanism: (**a**) described in flowcharts; and (**b**) described in pseudocode.

The total demands on s_i is $z_i = \sum_{j \in \{1,\ldots,\beta\}} w_{ij}$, $z_i \in$ Z. The elements in Z are sorted in the decreasing order, resulting in the sorted list $\check{Z}$. The corresponding service list of $\check{Z}$ is $\check{S} = \check{s}_1, \ldots, \check{s}_\alpha$. For each service $\check{s}_i$ in $\check{S}$, the LCA FEN satisfying Condition-D is determined by the function LCAncestor$\left(N_{fog}, \text{W}\right)$ and the ordered list of the LCA FENs for $\check{S}$ is A^{srv}. Then, the function FogLCAncestor(A^{srv}) returns A^{fog} whose ith element, $A^{fog}(i)$, is the ordered list of the services, in the decreasing order, whose LCA FEN is f_i'. GetFogService$\left(A^{fog}(i)\right)$ in **[F9]** returns a service (the current to-be-provisioned service σ)

one by one from $A^{fog}(i)$ per call. In **[F7]**$\sim$**[F10]**, the services in $A^{fog}(i)$ are provisioned on f_i' while the Condition-C of f_i' is satisfied. If the condition in **[F12]** is met, the service σ cannot be provisioned on f_i' because of the capacity shortage. Then, according to the status of f_i', upward provisioning (see **[F31]**$\sim$**[F39]**) and/or downward provisioning (see **[F22]**, **[F41]**$\sim$**[F50]**) is performed. In the case that σ cannot be provisioned on f_i' and f_i' is located at the highest level of N_{fog}, σ may replace the already provisioned service with the minimum entropy, f^{min} found by calling EntropyService(f_i'), if σ's entropy is larger than the minimum entropy, i.e., f^{min}'s entropy (see **[F16]**$\sim$**[F20]**), or be provisioned on a lower-level FEN (by calling SplitProvision(σ, f_i'), i.e., the downward provisioning of **[F22]**). Or, if σ cannot be provisioned on f_i' and f_i' is not located at the highest level of N_{fog}, the upward provisioning of **[F31]**$\sim$**[F39]** is performed first and if the upward provisioning fails (that is, if all the ancestors of f_i' cannot provision σ), the downward provisioning of **[F41]**$\sim$**[F50]** is performed. This process is repeated until all the service images demanded by end devices are provisioned on FENs. Here, P^{fog} is the set of FENs fully provisioned with services.

In the FC-SP mechanism, the Shannon entropy of a service is used as a metric for the provisioning priority of the service. In Figure 4, the function Entropy(σ) is called when σ cannot be provisioned on f_i' because of Condition-C being not satisfied at f_i'. Entropy(σ) returns the Shannon entropy of σ by using the following Equation (8) which reflects the degree of the distribution of the end devices requesting σ:

$$\text{Entropy}(\sigma) = -\sum_{f' \in F'}^{|F'|} p_{f'}^{\sigma} \log_2\left(p_{f'}^{\sigma}\right), \text{ where } F' \subseteq F \tag{8}$$

In Equation (8), first, F' is determined for σ, which is the set of the descendent LCA FENs of the LCA FEN currently failed in provisioning σ. $p_{f'}^{\sigma}$ is obtained by dividing the total demands on σ at f' by the total demands on σ at the FENs in F'.

Figure 5 shows an example of applying the FC-SP mechanism to N_{fog} of Figure 1b with the result of the service placement matrix X. Here, we also assume that all FENs including f_0 satisfy Condition-Ds of all services. On f_0, services s_3 and s_4 are provisioned first since each of them has demands of 8 which is the largest. After that, because s_1 and s_2 have the same amount of total demands of 5, respectively, the Shannon entropy at f_0 is calculated for s_1 and s_2. For that, F' is determined for s_1 and s_2, respectively. F' of s_1 is $\{f_1, f_2\}$ and F' of s_2 is $\{f_1, f_5\}$. Then, the entropy of s_1 is "$-(3/5 \log_2(3/5) + 2/5 \log_2(2/5)$ (about 0.9709506)" because the total demands on s_1 at f_1 is 3 and the total demands on s_1 at f_2 is 2. The entropy of s_2 is "$-(4/5\, log_2(4/5) + 1/5\, log_2(1/5)$ (about 0.7219281)" because the total demands on s_2 at f_1 is 4 and the total demands on s_2 at f_5 is 1. Because s_1 has a higher entropy than s_2, s_1 is provisioned on f_0 and s_2 is provisioned on f_1 and f_5 by the downward provisioning of **[F22]**. The higher the entropy is, the more the service demands are distributed. Thus, by provisioning more distributed services on FECs at higher levels of N_{fog}, the capability of covering service demands can be enhanced.

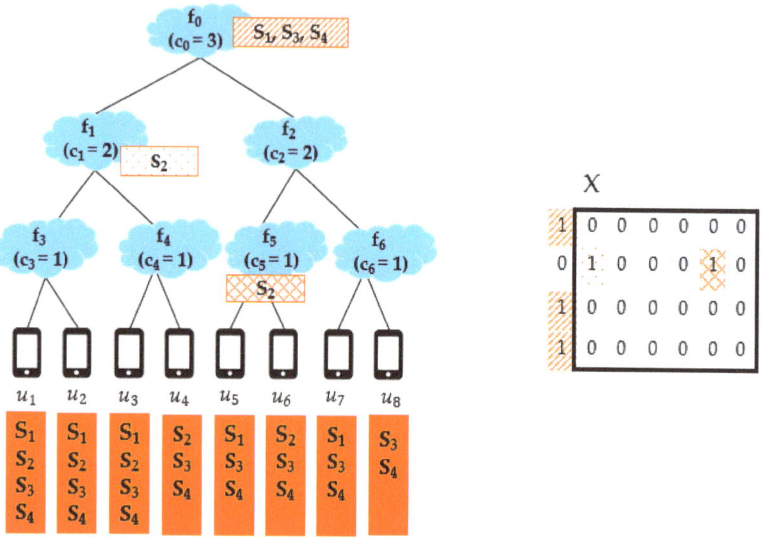

Figure 5. The service provisioning result of the FC-SP mechanism.

4. Performance Evaluation

Simulations were performed by using the NetworkX package [22] and Python. The simulation network environment is similar to that of [12]. The simulation network is obtained by designing an example of physical fog/edge network based on the administrative/population information obtained from Seoul open data center [23], South Korea, as shown in Figure 6a. The physical fog/edge network is built by locating one CDC (the largest dot in the figure) at the geographic center of Seoul and by locating 449 FENs according to the two-level administrative districts of Seoul. The capacity of a FEN is determined as proportional to the number of households of the corresponding administrative district. In the figure, the dots except for the largest dot (i.e., CDC) are FENs and the size of a dot implies the capacity of the corresponding FEN. The weight of an edge between two FENs is determined proportional to the physical distance between them. The corresponding logical fog network, shown in Figure 6b, was constructed based on the logical fog network construction mechanism described in Section 3.2.

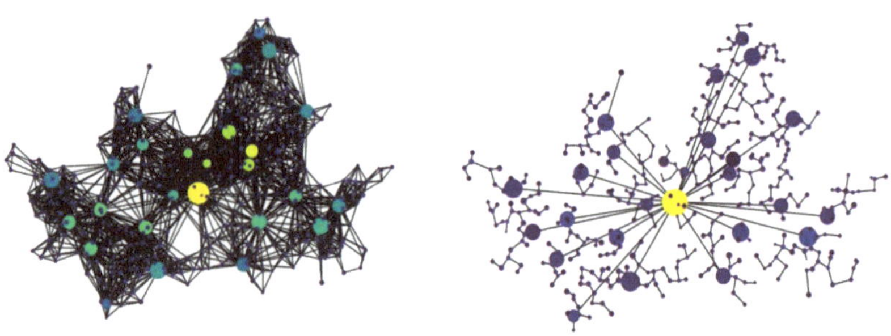

(a) (b)

Figure 6. An example fog network for simulations: (**a**) a physical network with FENs in a mesh topology; and (**b**) a logical representation of the physical network of (**a**) in a tree topology [12].

For the performance comparisons of the proposed MC-SP and FC-SP mechanisms, we designed and implemented a mechanism, called the on-demand mechanism, that dynamically places the corresponding service image upon a request based on the logical fog network. In the on-demand mechanism, if an end device receives a request on a service, it checks whether it has any FENs installed with the corresponding service image within 2-hop in the logical fog network. If there is none, it checks whether any of its ancestor FENs have the service image. If none of the ancestor FENs have the service image, it places the corresponding service image on itself. In the case when it is the lack of the resources, it places the service image on the 2-hop FEN which is the closest to itself. If it is not possible, it places the service image on one of its ancestor FENs which is the closest to itself. By comparing our proposed mechanisms with the on-demand mechanism operating on the basis of per-service request, we measure the performance of ours in terms of the resource utilization of FENs.

The number of service types, the distribution of service demands, the maximum capacity of CDC, and the maximum capacity of each FEN are the factors affecting performance. The number of service types requested by end devices is set to 1000 and 2000 service types and the number of end devices is 450. For the generation of service demands, we adopt the long-tailed distribution [24] of service demands in which a small set of service types is heavily requested and the rest of the service types are not frequently requested. The long-tailed distribution is adopted because a set of popular services are heavily used in the real world. For the simulations, the long-tailed distribution of service demands (we call this the L distribution case) is used, where about 10% of 1000 service types are heavily requested by the end devices, as shown in Figure 7 [12], which is obtained by using the zeta distribution [25]. This is achieved by using the function np.random.zipf (1.6, 1000) [26] where 1.6 is the value of the distribution parameter and 1000 is the number of service types. The function returns a value (we call this an L value) for each service type and the L value is used in determining the end devices with a service request for the service type. If the L value of a service type is greater than or equal to 100, all the end devices are assigned with a service request for the service type. Otherwise, the L value is used as the probability of assigning a service request for the service type to each end device. That is, for a service type, a higher L value implies a higher possibility of assigning a corresponding service request to each end device. If we sum up the service requests for all the service types in Figure 7, it becomes 42,210 in total.

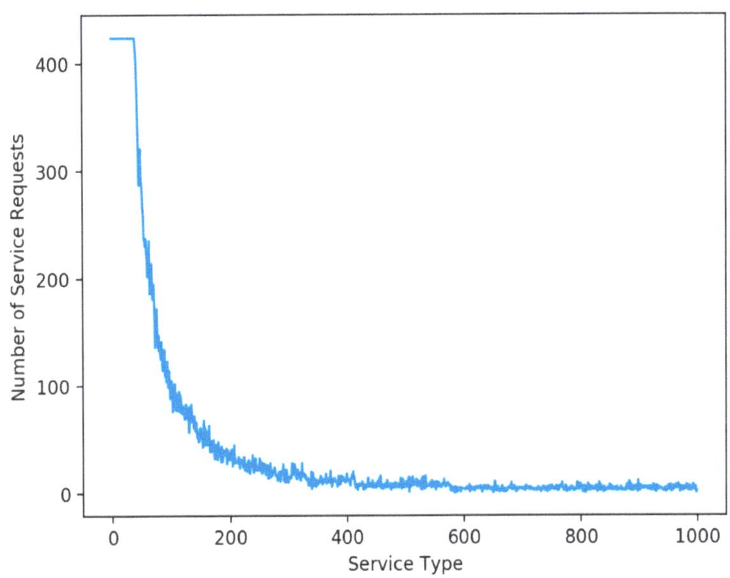

Figure 7. The L distribution case with 42,210 service demands for 1000 service types [12].

The measured performance factors are the number of service images placed, the number of non-accommodated service requests, and the average network cost per service request. The average network cost per service request is the average one-way physical distance required for providing a service which is assumed to be proportional to the delay. For the calculation of the average network cost per service request, the physical distance between the end device with a service request and the FEN accommodating the corresponding service request is measured on the logical fog network. The physical distance is measured for each service request and then, all the measured physical distances are summed up. The average network cost per service request is obtained by dividing the total physical distance by the number of service requests. The maximum capacity of CDC is set to 300, 500 and 750 resource units, and one service image placement is assumed to require one unit of resources.

Figures 8–11 show the performances of the proposed MC-SP and FC-SP mechanisms compared with the on-demand mechanism for various maximum CDC capacities of 300, 500 and 750 with 1000 services types for the L distribution case. Figure 8 depicts the graph showing the performance in terms of the number of service images placed on FENs. We can observe that, as the maximum capacity of CDC increases, more service images are placed on CDC, resulting in less service images placed on FENs in all the mechanisms. The on-demand mechanism places significantly more service images than our proposed mechanisms and even with increased CDC capacity, the on-demand mechanism slightly reduces the number of service images. The reason for this is that the on-demand mechanism places a service image near the end device which has requested the service, resulting in many service images near to end devices. On the other hand, the MC-SP and the FC-SP mechanisms reduce the number of service images placed significantly. Hence, we can assert that the FC-SP mechanism considering the service pattern performs the best in the resource utilization of FENs.

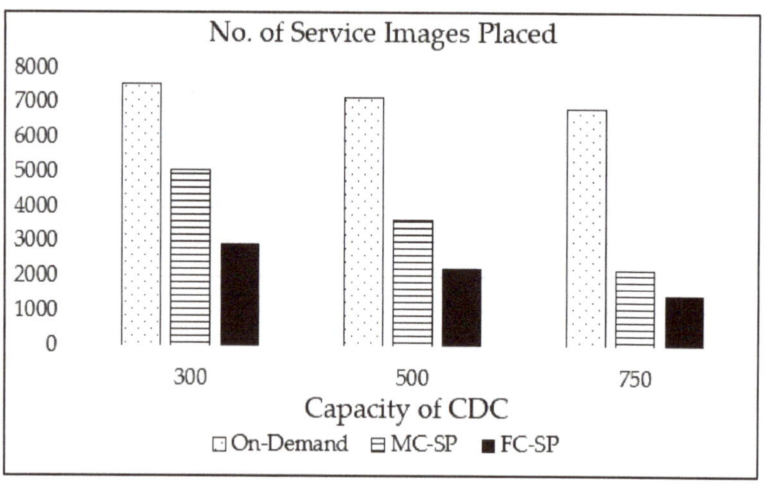

Figure 8. The number of service image placements for various maximum cloud data centers (CDCs) capacities with 1000 service types for the L distribution case.

Figure 9 is the graph showing the number of service requests for which the corresponding service images are not placed on any FENs for the L distribution case. Thanks to the well-utilized FEN resources, the FC-SP significantly outperforms the MC-SP mechanism and the on demand mechanism, especially for larger CDC capacities. As we expected, the number of non-accommodated service requests decreases as the maximum CDC capacity increases for all the mechanisms.

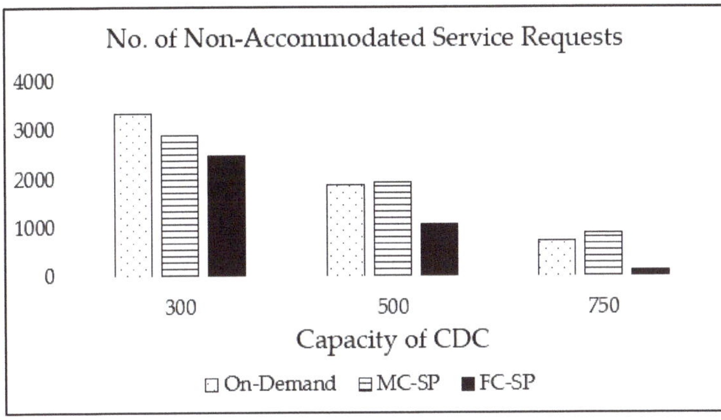

Figure 9. The number of non-accommodated service requests for various maximum CDC capacities with 1000 service types for the L distribution case.

In Figure 10, the average network cost of handling a service request is shown for various maximum CDC capacities for the L distribution case. As the maximum CDC capacity increases, the network cost also increases because more service images are placed on the CDC. Both of the proposed mechanisms perform much worse than the on-demand mechanism because service images are placed near the end devices in the on-demand mechanism with less service images placed. As for the proposed mechanisms, the FC-SP mechanism requires less network cost than the MC-SP mechanism. The reason is that the MC-SP mechanism does not consider the pattern of service demands and places service images on FENs at higher levels of the logical fog network for larger coverage.

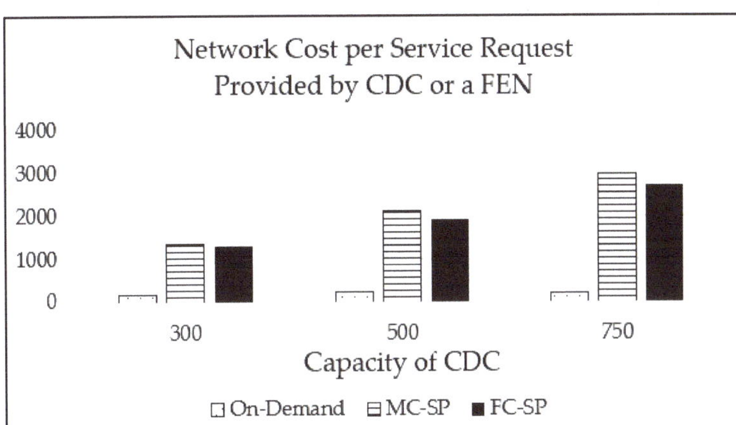

Figure 10. The network cost per service request for various maximum CDC capacities with 1000 service types for the L distribution case.

The performance of the proposed mechanisms is shown in Figure 11 with a varying FEN capacity and number of service types for the L distribution case. We simulated two cases of the FEN capacity, the basic case and the 10 times case. The basic case is the case of the FEN capacity in Figure 6b and the 10 times case is the FEN capacity of 10 times that of Figure 6b. The number of service types is set to 1000 and 2000.

Figure 11a shows the performance in terms of the number of service images placed on FENs, and Figure 11b shows the performance in terms of the number of non-accommodated service requests for the L distribution case. For the larger FEN capacity (i.e., the 10 time case), both MC-SP and FC-SP mechanisms place more service images, but the FC-S mechanism decreases the number of non-accommodated service requests more significantl than the MC-SP. This indicates that the FC-SP mechanism fully utilizes the advantage c the increased FEN capacity. That is, the FC-SP mechanism accommodates more servic requests than the MC-SP mechanism by provisioning more service images on FENs. I addition, as the number of service types increases, more service images are placed an more service requests are accommodated in both of the mechanisms. It is intuitive tha more service images are placed for more service types, but the MC-SP mechanism show a more noticeable increase in the number of service images placed with an insignifican decrease in the number of non-accommodated service requests. This implies that th MC-SP does not perform well in the situation of changed FEN capacities.

Figure 11c shows the graphs depicting the average network cost of a service reques for the L distribution case. As the FEN capacity increases, the network cost decreases i both of the mechanisms. The FC-SP mechanism shows a more significant decrease in th network cost compared to the MC-SP mechanism, especially for more service types. Thi indicates that the FC-SP mechanism performs better than the MC-SP mechanism even fo the case of more service types. Overall, the FC-SP mechanism is better than the MC-SP i utilizing the FEN resources and in adapting to changing environments such as the change in CDC capacity, FEN capacity, and the number of service types to support.

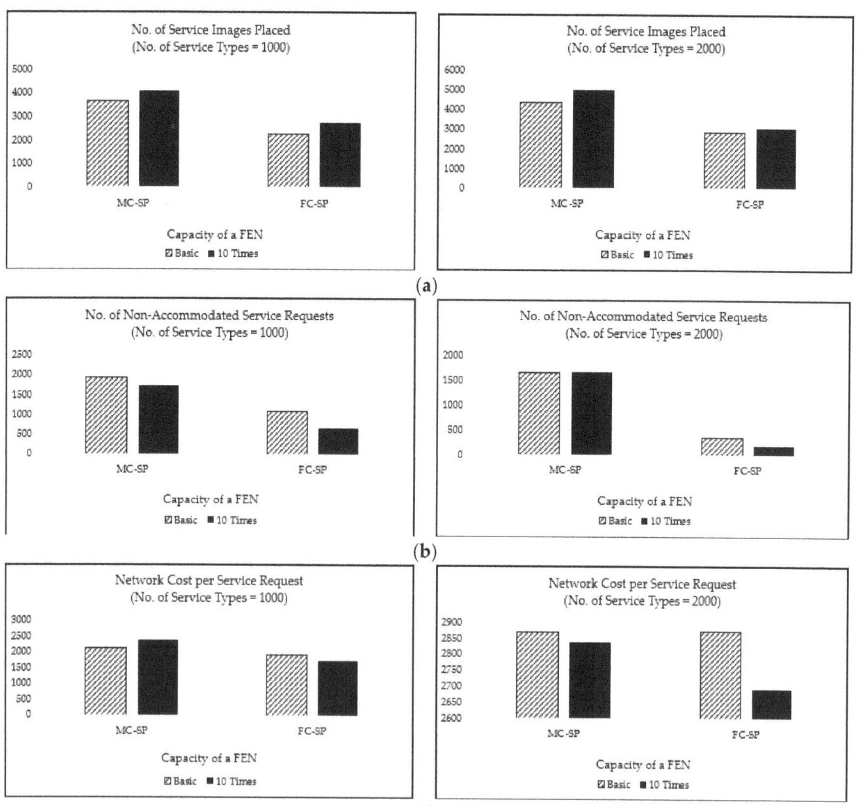

Figure 11. The performance according to the FEN capacity and the number of service types for the L distribution case: (**a**) the number of service images placed; (**b**) the number of non-accommodated service requests; and (**c**) the network cost per service request.

For a more concrete evaluation of the performance, we performed simulations for another service demand distribution shown in Figure 12 (we call this the U distribution case) which is obtained by using the uniform distribution. This is achieved by using the function np.random.uniform (0, 100, 1000) [27] where 0 and 100 are the lower and the upper boundary of the output interval, respectively, and 1000 is the number of service types. This function returns a value (we call this a U value) in the half-open output interval [0, 100) for each service type. The U value is used as the probability of assigning a service request for the service type to each end device. That is, the higher the U value of a service type is, the higher the possibility of assigning a service request of the service type to each end device. The U distribution case tends to assign service requests more evenly on service types than the L distribution case. The U distribution case generates 140,525 service requests for the performance evaluation of our mechanisms in a situation with heavy service requests.

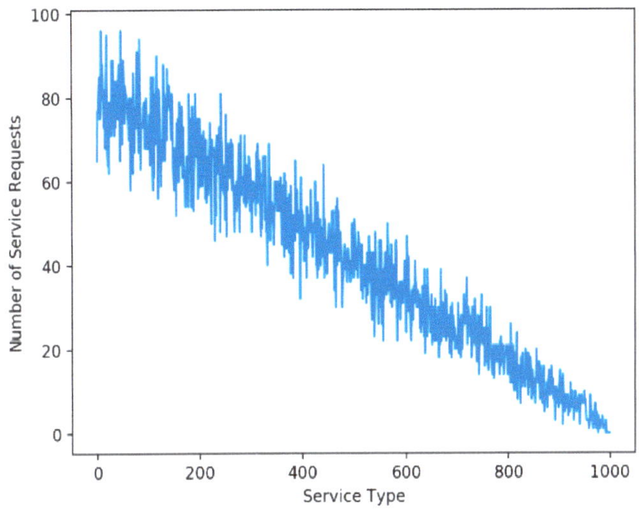

Figure 12. The U distribution case with 140,525 service demands for 1000 service types.

The performance of the proposed mechanisms for the U distribution case is depicted in Figure 13. We can easily see that the FC-SP mechanism outperforms the MC-SP mechanism in the aspect of all the performance factors. This indicates that the FC-SP mechanism performs better than the MC-SP mechanism even for the case with service requests relatively less biased over the service types (i.e., the U distribution case).

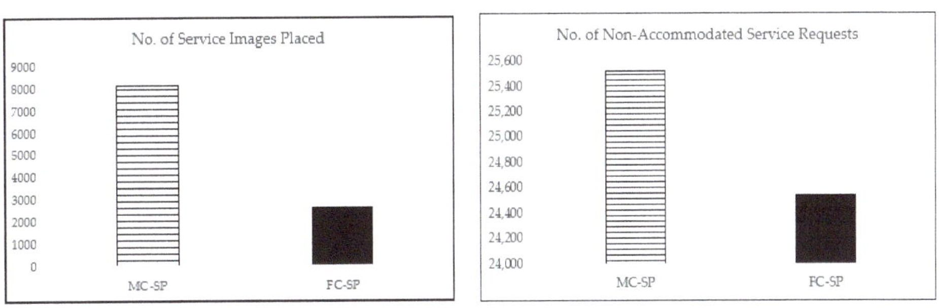

(a) (b)

Figure 13. *Cont.*

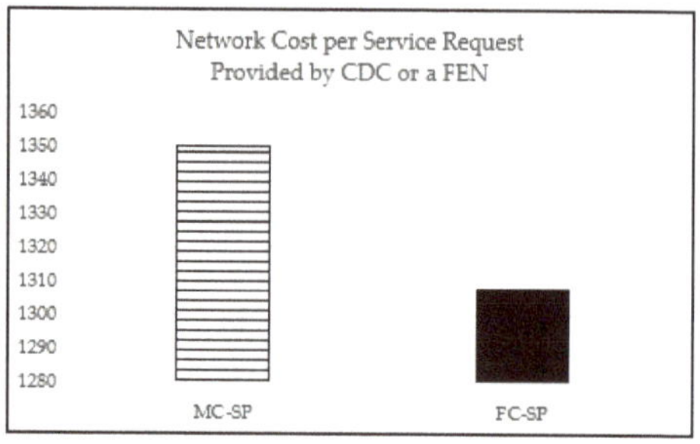

(**c**)

Figure 13. The performance with the maximum CDC capacity of 500 and 1000 service types for the U distribution case: (**a**) the number of service images placed; (**b**) the number of non-accommodated service requests; and (**c**) the network cost per service request provided by CDC or a FEN.

From Figure 14, we can observe the performance comparison of our proposed SP mechanisms for the L and the U distribution cases. In Figure 14a,b, it is clearly shown that our mechanisms perform better for the L distribution case than for the U distribution case in terms of the number of service images placed and the percentage of non-accommodated service requests. This can be intuitively expected because the U distribution case has about 3.33 times more service requests than the L distribution case. In addition, we can see that the FC-SP mechanism performs very effectively in the sense that it accommodates more service requests than the MC-SP mechanism with many less service images placed, even in the stressful situation (i.e., the U distribution case). Figure 14c shows that the average network cost per service request for the U distribution case is lower than that for the L distribution case. The reason for this is that, for the L distribution case, the majority of the service requests are likely covered by higher-level FENs because most of the service requests are biased on a few specific service types, resulting in higher average network cost per service request than the U distribution case.

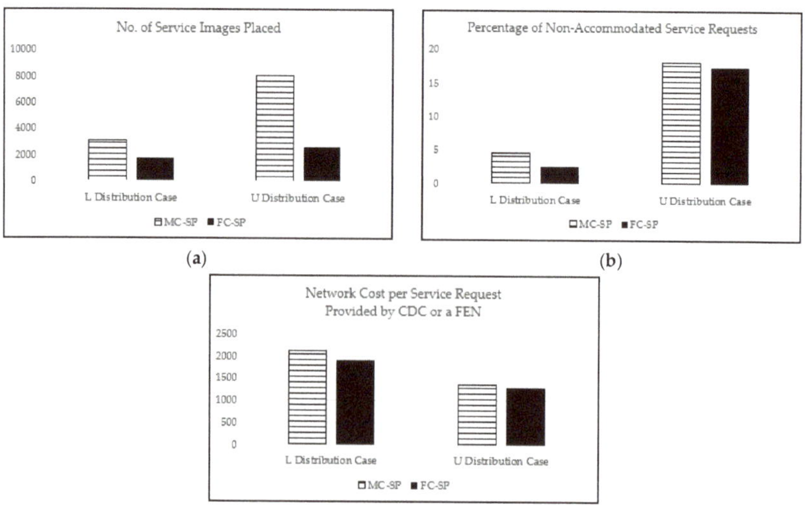

(**c**)

Figure 14. The performance comparision of the L distribution case and the U distribution case with the maximum CDC capacity of 500 and 1000 service types: (**a**) the number of service images placed; (**b**) the number of non-accommodated service requests; (**c**) the network cost per service request provided by CDC or a FEN.

5. Conclusions

Because of the limited resources of FENs, it is not possible to place all the services on an FEN. Moreover, if service images are placed on FENs in a per service request-based way, the resources of FENs may not be efficiently utilized due to the duplicate placement of service images, resulting in less accommodation of service requests. Therefore, in this paper, we proposed two SP mechanisms, the MC-SP mechanism and the FC-SP mechanism, for provisioning service images on FENs on the basis of logical fog network considering the pre-obtained service demands. The MC-SP mechanism provisions service images on FENs based on the number of service requests from end devices and the FC-SP mechanism based on the locations of end devices requesting services. The performance of the proposed mechanisms was evaluated by carrying out through simulations. According to the simulation results, we observed that both of our mechanisms perform better than the On-Demand mechanism which was designed for performance comparison and operates in a simple manner of placing a service image near to the end device requesting the service upon each service request. Therefore, we can say that our mechanisms are scalable because they can save the FEN resources by effectively placing service images on FENs and are good for the environment with plenty of IoT devices deployed.

Author Contributions: Conceptualization, J.C. and S.A.; methodology, J.C. and S.A.; software, J.C.; validation, J.C. and S.A.; formal analysis, J.C. and S.A.; investigation, J.C. and S.A.; resources, J.C. and S.A.; data curation, J.C. and S.A.; writing—original draft preparation, J.C. and S.A.; writing—review and editing, S.A.; visualization, J.C.; supervision, S.A.; project administration, S.A.; funding acquisition, S.A. All authors have read and agreed to the published version of the manuscript.

Funding: This work was supported by the National Research Foundation of Korea (NRF) grant funded by the South Korean government (MSIT) (No. NRF-2018R1D1A1B07047339). This research was supported by the MSIT (Ministry of Science and ICT), Korea, under the ITRC (Information Technology Research Center) support program (IITP-2020-2015-0-00363) supervised by the IITP (Institute for Information & Communications Technology Planning & Evaluation).

Conflicts of Interest: The authors declare no conflict of interest. The funders had no role in the design of the study; in the collection, analyses, or interpretation of data; in the writing of the manuscript, or in the decision to publish the results.

References

1. Internet of Things At-A-Glance. 2016. Available online: https://www.cisco.com/c/dam/en/us/products/collateral/se/internet-of-things/at-a-glance-c45-731471.pdf (accessed on 2 November 2019).
2. Vaquero, L.M.; Rodero-Merino, L.; Caceres, J.; Lindner, M. A Break in the Clouds: Towards a Cloud Definition. *ACM SIGCOMM Comput. Commun. Rev.* **2008**, *39*, 50–55. [CrossRef]
3. Mell, P.M.; Grance, T. *The NIST Definition of Cloud Computing*; NIST Special Publication 800–145; National Institute of Standards and Technology: Gaithersburg, MA, USA, 2011; Volume 800, pp. 1–7. [CrossRef]
4. Dillon, T.; Wu, C.; Chang, E. Cloud Computing: Issues and Challenges. In Proceedings of the 2010 24th IEEE International Conference on Advanced Information Networking and Applications, Perth, Australia, 20–23 April 2010; pp. 27–33.
5. Bonomi, F.; Milito, R.; Zhu, J.; Addepalli, S. Fog Computing and its Role in the Internet of Things. *MCC Workshop Mob. Cloud Comput.* **2012**, 13–16. [CrossRef]
6. Yi, S.; Li, C.; Li, Q. A Survey of Fog Computing: Concepts, Applications and Issues. In Proceedings of the 2015 Workshop on Mobile Big Data, Hangzhou, China, 22–25 June 2015; pp. 37–42.
7. Trakadas, P.; Nomikos, N.; Michailidis, E.T.; Zahariadis, T.; Facca, F.M.; Breitgand, D.; Rizou, S.; Masip, X.; Gkonis, P. Hybrid Clouds for Data-Intensive, 5G-Enabled IoT Applications: An Overview, Key Issues and Relevant Architecture. *Sensors* **2019**, *19*, 3591. [CrossRef] [PubMed]
8. Goudarzi, M.; Wu, H.; Palaniswami, M.S.; Buyya, R. An Application Placement Technique for Concurrent IoT Applications in Edge and Fog Computing Environments. *IEEE Trans. Mob. Comput.* **2020**. [CrossRef]
9. Salaht, F.A.; Desprez, F.; Lebre, A. An Overview of Service Placement Problem in Fog and Edge Computing. *ACM Comput. Surv.* **2020**, *53*, 1–35. [CrossRef]
10. Souza, V.B.C.; Ramirez, W.W.; Masip-Bruin, X.; Marin-Tordera, E.; Ren, G.; Tashakor, G. Handling Service Allocation in Combined Fog-Cloud Scenarios. In Proceedings of the IEEE International Conference on Communications (ICC), Kuala Lumpur, Malaysia, 22–27 May 2016; pp. 1–5.

11. Brogi, A.; Forti, S. QoS-Aware Deployment of IoT Applications Through the Fog. *IEEE Internet Things J.* **2017**, *4*, 1185–1192. [CrossRef]
12. Choi, J.; Ahn, S. Scalable Service Placement in the Fog Computing Environment for the IoT-based Smart City. *J. Inf. Process. Syst.* **2019**, *15*, 440–448.
13. Deng, R.; Lu, R.; Lai, C.; Luan, T.H. Towards Power Consumption-Delay Tradeoff by Workload Allocation in Cloud-Fog Computing. In Proceedings of the 2015 IEEE International Conference on Communications (ICC), London, UK, 8–12 June 2015; pp. 3909–3914. [CrossRef]
14. Skarlat, O.; Schulte, S.; Borkowski, M.; Leitner, P. Resource Provisioning for IoT Services in the Fog. In Proceedings of the IEEE 9th International Conference on Service-Oriented Computing and Applications (SOCA), Macau, China, 4–6 November 2016. [CrossRef]
15. Skarlat, O.; Nardelli, M.; Schulte, S.; Dustdar, S. Towards QoS-aware Fog Service Placement. In Proceedings of the IEEE International Conference on Fog and Edge Computing (ICFEC), Madrid, Spain, 14–15 May 2017.
16. Faticanti, F.; Pellegrini, F.D.; Siracusa, D.; Santoro, D.; Cretti, S. Cutting Throughput with the Edge: App-aware Placement in Fog Computing. In Proceedings of the IEEE International Conference on Edge Computing and Scalable Cloud (EdgeCom), Paris, France, 21–23 June 2019.
17. Saurez, E.; Hong, K.; Lillethun, D.; Ramachandran, U.; Ottenwälder, B. Incremental Deployment and Migration of Geo-Distributed Situation Awareness Applications in the Fog. In Proceedings of the 10th ACM International Conference on Distributed and Event-based Systems, Association for Computing Machinery (ACM), Irvine, CA, USA, 20–24 June 2016; pp. 258–269.
18. Yu, R.; Xue, G.; Zhang, X. Application Provisioning in FOG Computing-enabled Internet-of-Things: A Network Perspective. In Proceedings of the IEEE Infocom 2018 IEEE Conference on Computer Communications, Honolulu, HI, USA, 16–19 April 2018; pp. 783–791. [CrossRef]
19. Souza, V.; Masip-Bruin, X.; Marín-Tordera, E.; Sànchez-López, S.; Garcia, J.; Ren, G.; Jukan, A.; Ferrer, A.J. Towards a Proper Service Placement in Combined Fog-to-Cloud (F2C) Architectures. *Future Gener. Comput. Syst.* **2018**, *87*, 1–15. [CrossRef]
20. Eppstein, D. Subgraph Isomorphism in Planar Graphs and Related Problems. *J. Graph. Algorithms Appl.* **1999**, *3*, 1–27. [CrossRef]
21. Shannon, C.E. A Mathematical Theory of Communication. *Bell Syst. Tech. J.* **1948**, *27*, 379–423. [CrossRef]
22. NetworkX—NetworkX Document. Available online: https://networkx.org/ (accessed on 2 November 2019).
23. Seoul Open Data Plaza. Available online: https://data.seoul.go.kr/ (accessed on 2 November 2019).
24. Long Tail—Wikipedia. Available online: https://en.wikipedia.org/wiki/Long_tail (accessed on 2 November 2019).
25. Zeta Distribution—Wikipedia. Available online: https://en.wikipedia.org/wiki/Zeta_distribution (accessed on 2 November 2019).
26. numpy.random.zipf-NumPy v1.20 Manual. Available online: https://numpy.org/doc/stable/reference/random/generated/numpy.random.zipf.html (accessed on 2 November 2019).
27. numpy.random.uniform-NumPy v1.20 Manual. Available online: https://numpy.org/doc/stable/reference/random/generated/numpy.random.uniform.html (accessed on 2 November 2019).

sensors

MDPI

Article

Fuzzy Decision-Based Efficient Task Offloading Management Scheme in Multi-Tier MEC-Enabled Networks

Md Delowar Hossain, Tangina Sultana, Md Alamgir Hossain, Md Imtiaz Hossain, Luan N. T. Huynh, Junyoung Park and Eui-Nam Huh *

Global Campus, Department of Computer Science and Engineering, Kyung Hee University, Yongin-si 17104, Korea; delowar@khu.ac.kr (M.D.H.); tangina@khu.ac.kr (T.S.); alamgir@khu.ac.kr (M.A.H.); hossain.imtiaz@khu.ac.kr (M.I.H.); luanhnt@khu.ac.kr (L.N.T.H.); parkhans@khu.ac.kr (J.P.)
* Correspondence: johnhuh@khu.ac.kr; Tel.: +82-10-9582-9789

Abstract: Multi-access edge computing (MEC) is a new leading technology for meeting the demands of key performance indicators (KPIs) in 5G networks. However, in a rapidly changing dynamic environment, it is hard to find the optimal target server for processing offloaded tasks because we do not know the end users' demands in advance. Therefore, quality of service (QoS) deteriorates because of increasing task failures and long execution latency from congestion. To reduce latency and avoid task failures from resource-constrained edge servers, vertical offloading between mobile devices with local-edge collaboration or with local edge-remote cloud collaboration have been proposed in previous studies. However, they ignored the nearby edge server in the same tier that has excess computing resources. Therefore, this paper introduces a fuzzy decision-based cloud-MEC collaborative task offloading management system called FTOM, which takes advantage of powerful remote cloud-computing capabilities and utilizes neighboring edge servers. The main objective of the FTOM scheme is to select the optimal target node for task offloading based on server capacity, latency sensitivity, and the network's condition. Our proposed scheme can make dynamic decisions where local or nearby MEC servers are preferred for offloading delay-sensitive tasks, and delay-tolerant high resource-demand tasks are offloaded to a remote cloud server. Simulation results affirm that our proposed FTOM scheme significantly improves the rate of successfully executing offloaded tasks by approximately 68.5%, and reduces task completion time by 66.6%, when compared with a local edge offloading (LEO) scheme. The improved and reduced rates are 32.4% and 61.5%, respectively, when compared with a two-tier edge orchestration-based offloading (TTEO) scheme. They are 8.9% and 47.9%, respectively, when compared with a fuzzy orchestration-based load balancing (FOLB) scheme, approximately 3.2% and 49.8%, respectively, when compared with a fuzzy workload orchestration-based task offloading (WOTO) scheme, and approximately 38.6%% and 55%, respectively, when compared with a fuzzy edge-orchestration based collaborative task offloading (FCTO) scheme.

Keywords: multi-access edge computing; orchestrator; task offloading; fuzzy logic; 5G

Citation: Hossain, M.D.; Sultana, T.; Hossain, M.A.; Hossain,M.I.; Huynh, L.N.T.; Park, J.; Huh, E.-N. Fuzzy Decision-Based Efficient Task Offloading Management Scheme in Multi-Tier MEC-Enabled Networks. *Sensors* **2021**, *21*, 1484. https://doi.org/10.3390/s21041484

Academic Editor: Taehong Kim
Received: 17 January 2021
Accepted: 17 February 2021
Published: 20 February 2021

1. Introduction

Nowadays, with the rapid evolution of communication technology and the enormous popularity of high-demand applications (e.g., the Internet of vehicles, mobile augmented reality, map navigation, face/fingerprint/iris recognition, mobile healthcare, web browsing, cloud gaming, image identification), a huge number of devices are attached to the Internet of Things (IoT) infrastructure [1–5]. In conventional networking infrastructures, the demand poses an enormous burden due to the generation of huge volumes of data from using these devices. Moreover, storage capacity and computing capabilities in user devices is restricted. Due to these constraints, user devices cannot handle massive numbers of tasks, and it affects both quality of service (QoS) and performance. Therefore, these devices tend to offload their tasks to more powerful computing devices [6]. To resolve the above limitations, the mobile cloud computing (MCC) approach was introduced [7]. Thus, the workload of user devices and the processing latency are significantly reduced from offloading computation tasks to the MCC server. However, the location of the MCC server is on the core network, far from the user devices. Therefore, when a user wants to offload a task to the MCC server, the data must travel through the entire access network. The same scenario has to be followed when the processed results return. As a result, the MCC-based approach suffers from high transmission delays, data leakage, and compromised privacy due to the long-distance routing [8]. To reduce this network latency is very difficult if using the existing infrastructure. Therefore, for applications that need low latency in a real-time service environment, the MCC-based solution is not suitable. To cope with these challenges, ETSI proposed in December 2014 an emerging technology named mobile edge computing. In September 2017, ETSI removed the word mobile from multi-access edge computing (MEC) and officially renamed it multi-access edge computing [9,10]. MEC is an innovative network paradigm that brings the storage and computing resources to the network edge. As a result, it can overcome long transmission latency and the deficiencies from network congestion in the MCC system. Since the location of MEC servers is very close to the user terminals, end-to-end latency between the edge server and user device is significantly shortened. Therefore, the user can receive feedback immediately after processing, and this significantly improves QoS. Table 1 compares MCC and MEC [11,12].

Table 1. Comparison between mobile cloud computing (MCC) and multi-access edge computing (MEC) computing architectures.

Technical Aspect	MCC	MEC
Deployment	Centralized	Dense and distributed
Architectural style	Client-server	Peer-to-peer
Computing capabilities	Higher	Lower
Network access	Multi-hop	Single-hop
Support for client mobility	Limited	Supported
Support for server mobility	Not supported	Supported
Number of nodes	Small (100–1000)	Large (billions)
Heterogeneity	Limited support	Full support
Latency	High	Very Low
Storage capacity	Ample	Limited
Location	Large data center	With network ingress
Hierarchy	2 tiers	3 tiers

MEC is one of the premier ideas for rapidly computing user tasks offloaded to the edge server. The advantage of this technology is that users get the needed computing resources with only one-hop wireless transmission. Compared to MCC, it does not need to go through the core network to transmit the task to MEC servers. This reduces the delay and satisfies the low-latency requirements of different applications. In addition, the task's processed results return directly from the MEC server, which can alleviate the risk to privacy and helps to protect sensitive data. Moreover, the edge server (as well as user devices) can themselves collaboratively process the service workloads. As a result, it can save bandwidth, because most of the task is processed locally by the user device and the edge server, without sending the task to the cloud. Therefore, to handle context-aware and latency-sensitive applications, some researchers have proposed a framework for collaboration between the edge server and user devices to complete computed tasks [13]. Despite the multi-dimensional benefits of MEC, it faces challenges owing to finite storage capacity and limited computation resources. With the increase in high-demand applications and the popularity of smart mobile devices, the distinct edge server cannot efficiently handle multiple offload requests. To utilize adequate computing resources of remote cloud servers and benefit from using a MEC server, a collaborative cloud-MEC-based task offloading approach was proposed recently [14,15]. In collaborative offloading, there are still some challenges, such as how to decide where to offload the task (to either a MEC server or a cloud server). Therefore, the collaborative approach is more complicated in a dynamic environment. To exploit the advantages of unlimited storage space and powerful computing capabilities of a cloud server, and to utilize nearby MEC servers, a collaborative cloud-MEC-based FTOM scheme is introduced in this study. The novelty of our work is to improve the rate of successfully executing offloaded tasks and to reduce completed-task latency by utilizing the computing resources of nearby MEC servers that have excess computing resources. The key contributions of this paper are as follows:

- We investigate a low-complexity cloud-MEC-based offloading scheme to ensure QoS and accommodate more workload in the multi-tier MEC-enabled network.
- We develop a fuzzy decision-based, efficient task offloading management scheme by considering a vertical (local MEC with remote cloud) as well as horizontal (peer offloading among nearby MEC servers) task offloading scheme to meet the diverse needs of users.
- Based on the the states of server utilization, the delay sensitivity of the task, and the network conditions, the FTOM scheme can make a dynamic decision on where to offload the incoming task: local MEC, nearby MEC, or a cloud server.
- To improve resource utilization efficiency and the rate of successfully executed offloaded tasks, our system prefers to offload latency-sensitive tasks to local or neighboring MEC servers, whereas delay-tolerant, high resource-demand tasks go to a remote server.
- Performance evaluation demonstrates the effectiveness of our proposed FTOM scheme, compared to its competitors, for three different types of application: infotainment (I), augmented reality (AR), and health monitoring (HM).

The remainder of this paper is structured as follows. The related works on task offloading in the MEC-enabled networks are illustrated briefly in Section 2. Afterwards, the problem scenario and our proposed model are described in Section 3. Our introduced FTOM scheme for efficient task offloading management is presented in Section 4. Performance evaluations are illustrated in Section 5 and results summary of different evaluation metrics are presented in Section 6. The paper is finally concluded and future research suggested in Section 7.

2. Related Work

Task offloading and allocation of resources are primary key points of MEC-enabled networks. Based on previous research, these are divided into three main categories: binary or full offloading (the task cannot partition during processing) [16–18], partial offloading (the task is decomposed into several parts at the same time for local computing or for offloading) [19–21], and collaborative task offloading (integration between the edge and the cloud) [22–25]. For binary computation offloading (BCO), the tasks can be processed by the user devices themselves or by offloading them to the nearest edge servers. This scheme is mostly an NP-hard problem. To solve the problem of having multi-user participation and restrictive objectives, game theory is extensively used. Bi and Zhang [16] proposed a BCO policy to process a task with either user devices or by offloading it to an edge server for a multi-user MEC system. By using wireless power transfer (WPT), the users are wirelessly powered from the base station. Wang et al. [17] proposed a three-layer traffic system based on queueing theory for moving vehicle-based edge nodes to minimize the offloaded response time. Messous et al. [18] introduced a game theory-based strategy for solving offloaded problems with heavy task computations in unmanned aerial vehicles (UAVs). Recently, partial computational offloading (PCO) has gained widespread attention from many researchers into MEC-enabled networks. In this offloading model, a task is partitioned into some parts that are executed locally by mobile devices and other parts that are offloaded to, and processed by, MEC servers. The authors of [19] proposed a PCO approach for a single-user MEC system to minimize energy consumption and the task execution latency based on Lagrangian dual decomposition. To solve single-user PCO problems in latency-constrained networks, Ning et al. [20] apply a branch-and-bound algorithm. On the other hand, for multi-user PCO problems, a heuristic iterative algorithm was proposed for making offloading decisions and allocation of resources dynamically. To reduce latency for all user devices, Ren et al. [21] proposed a strategy named optimal closed-form data segmentation in partial computation offloading schemes for time-division multiple access based multi-user MEC systems.

However, due to the resource restrictions and limited storage capacity of the MEC server, researchers have proposed cloud-MEC-based collaborative integration to reap the benefits of both technologies. Most of the previous researchers proposed two-tier cloud MEC-based vertical offloading [22–25] and ignored horizontal offloading among nearby MEC servers in the same tier. Deng et al. [22] introduced a cloud-edge computing system to reduce power consumption as well as delay by formulating a mixed-integer nonlinear programming (MINLP) problem. By adopting the fiber-wireless (FiWi) access network Guo and Liu [23] proposed a collaborative cloud-MEC-based task offloading scheme. To obtain better offloading performance, a game-theory-based algorithm was used. To reduce the cost of the capacity, Lin et al. [24] constructed a three-tier cloud-edge system by using an iterative optimization algorithm. To minimize the network transmission load, Huang et al. [25] introduced a service orchestration scheme based on software-defined networking (SDN) technology. Furthermore, a heuristic algorithm was adopted to make the offloading decision between the cloud and the edge system. On the other hand, to utilize nearby MEC servers, some researchers focused on horizontal offloading between local MEC and nearby MEC servers in the same tier. To minimize the transmission distance and increase the capacity of edge caching systems, Yuan et al. [26] proposed a cooperation approach among edge clouds. Hossain et al. [27] used a collaboration approach based on fuzzy logic among MEC servers to reduce the task failure rate and service time. Moreover, Fan et al. [28] used a cooperation approach between different servers for balancing the computation workload.

Generally, the edge computing environment is dynamic and uncertain. On the other hand, fuzzy logic is one of the best-employed methods for rapidly changing uncertain systems. Therefore, for efficient task offloading management, the FTOM scheme is proposed. The main advantages of using fuzzy logic are that its complexity is low, compared with other decision-making algorithms [29–31], and it is significantly applied to workload management, vehicle routing, task scheduling, and network congestion-mitigation problems [32–34]. To satisfy the various security requirements in real time for mobile users, Li et al. [35] introduced a security service-chaining approach based on fuzzy logic for mobile edge computing. Nguyen et al. [36] proposed a fuzzy decision-based flexible task-offloading scheme for IoT applications. To minimize latency and the task failure rate, a fuzzy-based mobile edge orchestrator policy is used as a controller for application placement. Soleymani et al. [37] used fuzzy logic for the trust management system in a VANET. The proposed trust model executes a sequence of security checks to ensure vehicles are authorized. On the other hand, to determine the target server for task offloading Sonmez et al. [38] used two stages of fuzzy operation. The best-candidate edge server is found in the first stage from among all the edge servers. The target server is selected by comparing the candidate edge server with the cloud in the second stage. Our proposed system, however, uses only a single stage of fuzzy logic operation to select the optimal target server. In [27], Hossain et al. considered a collaborative approach for task offloading based on fuzzy logic. In this paper, authors considered local MEC and neighboring MEC servers to select the target server for task offloading decisions. To calculate the center of gravity (COG) value for choosing the target server to offload the task, authors did not consider the remote cloud server. Moreover, for performance evaluation, authors considered latency-sensitive AR application. However, for offloading tasks, our proposed FTOM scheme selects the optimal server from among local MEC, nearby MEC, or remote cloud servers. That is why we have considered a new input variable, named WAN bandwidth. The important role of WAN bandwidth is in making the decision about offloading the task to the remote cloud or not. Moreover, in FTOM scheme, for performance evaluation, we have considered latency-sensitive AR and HM applications and delay-tolerant infotainment application. The two key activities of the FTOM scheme are monitoring the continuously changing network conditions and finding the optimal target server for task offloading. As far as we know, an FTOM scheme for MEC-enabled networks has not been evaluated yet in this domain.

3. Problem Scenario and System Model

3.1. Problem Scenario

In MEC-enabled networks, task offloading is one of the challenging issues because of the delay constraint and limited computing resources. Moreover, congestion is caused from offloading multiple tasks from various users to the same edge server. Therefore, many users' processing tasks on the MEC server are left waiting in the queue. As a result, the processing delay is longer for all tasks because of the overload. Figures 1 and 2 show such scenarios, where some edge nodes are lightly loaded and some nodes are overloaded from too many user requests. Therefore, it is not always a better decision to offload a computing task to the closest edge server. From Figure 1, we can see that edge node-1 is already overloaded due to heavy user requests. In this situation, the overload tasks are forwarded to the remote cloud for processing. However, the nearby edge node-2 is lightly loaded and has more resources available to process computing tasks. This node can undoubtedly overcome the overload problem for edge node-1 without sending tasks to the remote cloud. In the ongoing 5G network, multiple edge servers are deployed near user devices within range of mobile communication. Therefore, users have multiple options for offloading tasks to nearby edge servers in order to receive services. On the other hand, when there are multiple edge servers available in MEC networks, it becomes a challenging issue to decide which edge server is best for task offloading. Thus, the design of an efficient task offloading mechanism is important, because QoS varies based on the task offloading decisions. Figure 1 shows the following two significant challenges faced when offloading tasks in MEC networks:

1. Should the edge server or the remote server be used to offload the computing task?
2. Which edge server is preferred for offloading the task?

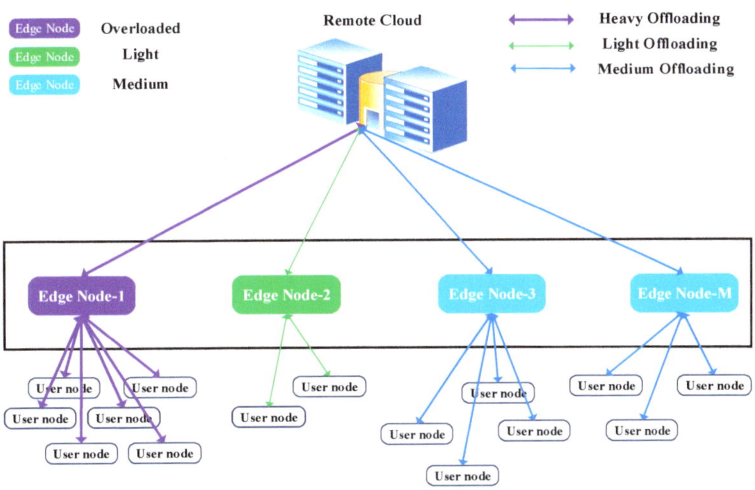

Figure 1. The overloaded problem in a multi-user MEC network.

To clearly understand the offloading problem, Figure 2 shows a multi-user MEC network scenario in detail. This network consists of $M = \{1, 2, 3, \ldots, M\}$ small base stations (SBSs), and a single MEC server is deployed in each SBS. There are $N = \{1, 2, 3, \ldots, N\}$ user devices and $T = \{1, 2, 3, \ldots, T\}$ independent tasks from each user. We denote the computing capacity of the edge server as τ^{mec}, and this server receives its mobile workload from N users, $\phi_1, \phi_2, ..., \phi_n$. Based on the user device capacity, some tasks are executed locally by the device, and the rest of the tasks are offloaded to a local MEC server. If the received workload exceeds the capacity of the edge server (i.e., $\sum \phi > \tau^{mec}$), it is hard to execute another task on this server. Therefore, due to the excessive workload, task 2 fails, as shown in Figure 2. To explore the neighboring SBSs and the remote cloud, we observed the following:

1. To overcome the local MEC server overload problem and utilize the neighboring MEC servers with the remote cloud, we can add a orchestrator management layer for efficient task offloading among MEC servers within the cloud.
2. Based on the task size, network condition, and delay sensitivity of the task, we can decide whether task offloading is more efficient if done by a local MEC server, a neighboring MEC server, or the remote cloud.
3. The rate of successfully executed tasks can improve, and task completion time can be significantly reduced, by offloading the task collaboratively among the MEC servers and the remote cloud server.

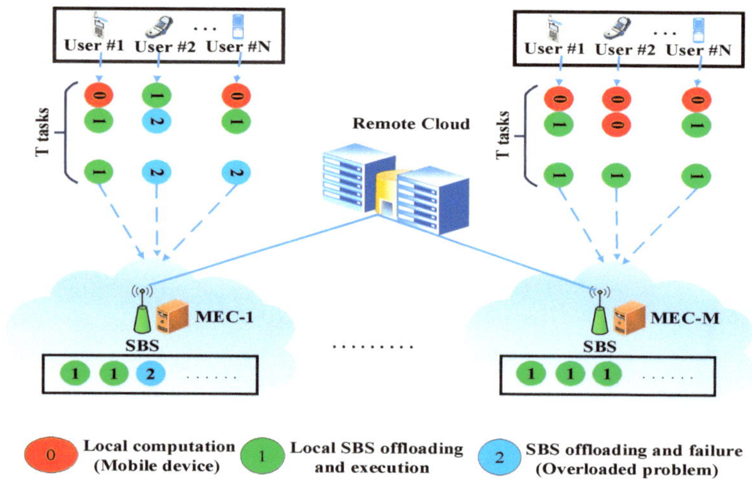

Figure 2. Multi-server multi-user MEC network.

3.2. *The Role of an Orchestrator Management Scheme*

To solve the overload problem in a distinct edge sever, we include a management layer for task orchestration among the MEC servers and the cloud in a multi-tier MEC-enabled network, which is shown in Figure 3.

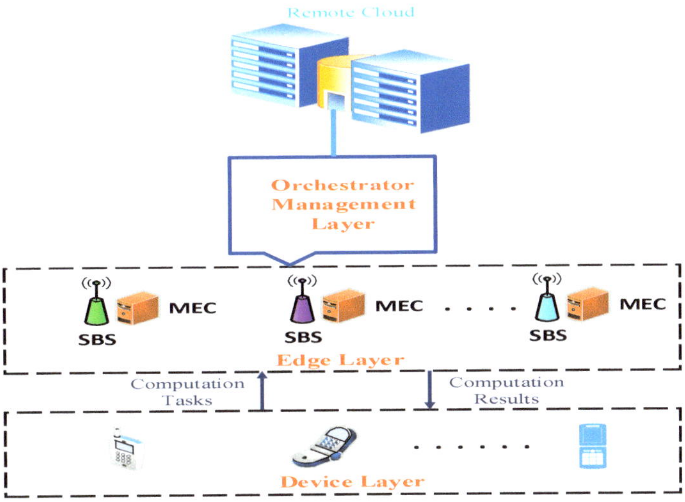

Figure 3. Orchestrator management scheme.

Without an orchestrator, all incoming user requests are offloaded and executed by the local MEC server. Therefore, it faces heavy congestion because of the numerous user requests, and sometimes, resources are not utilized efficiently. As depicted in Figure 3, we incorporate the orchestrator management layer between the edge layer and the remote cloud. Numerous devices, such as smartphones and sensing devices, are deployed on the device layer of the network and want to offload their computing tasks to an edge server or a remote server. The edge layer consists of multiple SBSs where a single MEC server is equipped at each SBS. The orchestrator management layer is responsible for collecting all information, including the computation resources of the MEC servers, the network information, and the input task sizes. Based on this information, it selects the optimal target node for task computing to ensure a sophisticated computation balance. Figure 4 describes the role of the orchestrator and the task offloading process. The user node selects the local SBS for task requesting. We assume that the task is already offloaded from the end-user device to the local edge node, and that each task is independent. There are six steps required to execute the process. (1) SBS integrates the edge node with task offloading information together and transmits the corresponding task offloading request to the orchestrator along with its requirements. (2) The orchestrator acts as the decision-maker of the system and, based on fuzzy rules, it decides where (i.e., in which resource) the tasks will be executed. When an edge node is connected to the network, the orchestrator links that node to the system. Then, during the offloading process, the orchestrator finds the best offload destination (i.e., the node that will execute the offloaded task) in the system. (3) The system sends the task to the optimal edge node based on fuzzy rules. (4) The selected edge node executes the task. (5) After executing the task, the result is returned to the orchestrator. (6) The orchestrator forwards the result to the corresponding edge node.

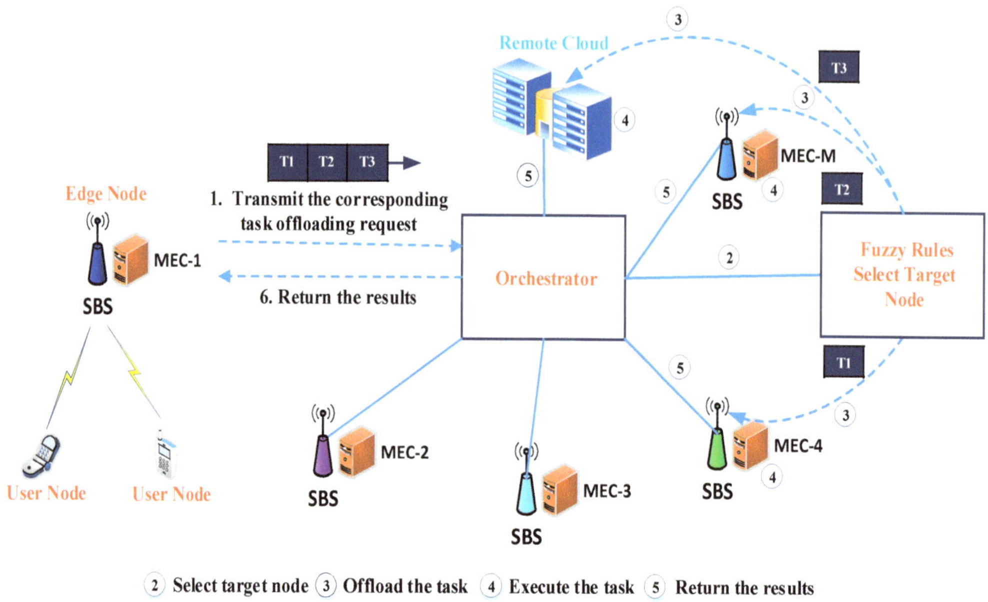

Figure 4. The role of the orchestrator and the flow of the task offloading process.

3.3. System Model

The proposed model is an integration framework with one centralized cloud server, M access points (AP), and many user devices which are all shown in Figure 5. There is a single MEC server in each AP which has limited storage and computing resources for processing tasks. The combination of the AP and its associated MEC server is considered an edge node. On the other hand, a centralized cloud server has a huge amount of storage capacity and powerful computing resources. Mobile users utilize wireless local area network to access edge resources, whereas wide area network connections are used if devices offload their tasks to a remote cloud server. We assume there are N user devices (UDs) where each user has T independent tasks. We denote the set of UDs as $U, U = \{U_i | i = 1, 2, 3, \ldots, N\}, |U| = N$, and the set of tasks that need to be executed for each user in the network is $T = \{T_i | i = 1, 2, 3, \ldots, T\}$. Each computation task is described by the following $T_i = \{\tau_i, \psi_i, \eth^i_{max}\}$. For task T_i, τ_i denotes the size of the task that needs to be offloaded for computation; ψ_i represents the required CPU cycle for task processing, which varies for various applications; and $\eth^i_{max}$ indicates the maximum tolerable latency of T_i. Moreover, we define the set of servers as $M = \{1, 2, 3, \ldots, M, M+1\}$, where $\{MEC_i | i = 1, \ldots M\}$ denotes the MEC servers and server $M+1$ represents the remote cloud server. For each MEC server $MEC_i = \{\mathfrak{r}^i_{max}, \mathfrak{s}^i_{max}\}$, where $\mathfrak{r}^i_{max}$ is the maximum resource capability of MEC_i, and $\mathfrak{s}^i_{max}$ is the local storage capacity of MEC_i. We assume that each MEC_i server has one host that operates four VMs. The resource capacity of each VM is 10 GIPS. If the required amount of resources is less than or equal to $\mathfrak{r}^i_{max}$, then the task will be executed only by MEC_i. On the other hand, the VMs running on the global cloud server are tens of times more powerful than the edge server in our scenario. The main aim of this study is to design an efficient cloud-MEC-based task offloading management approach to ensure satisfactory service requirements and reduce the overall latency.

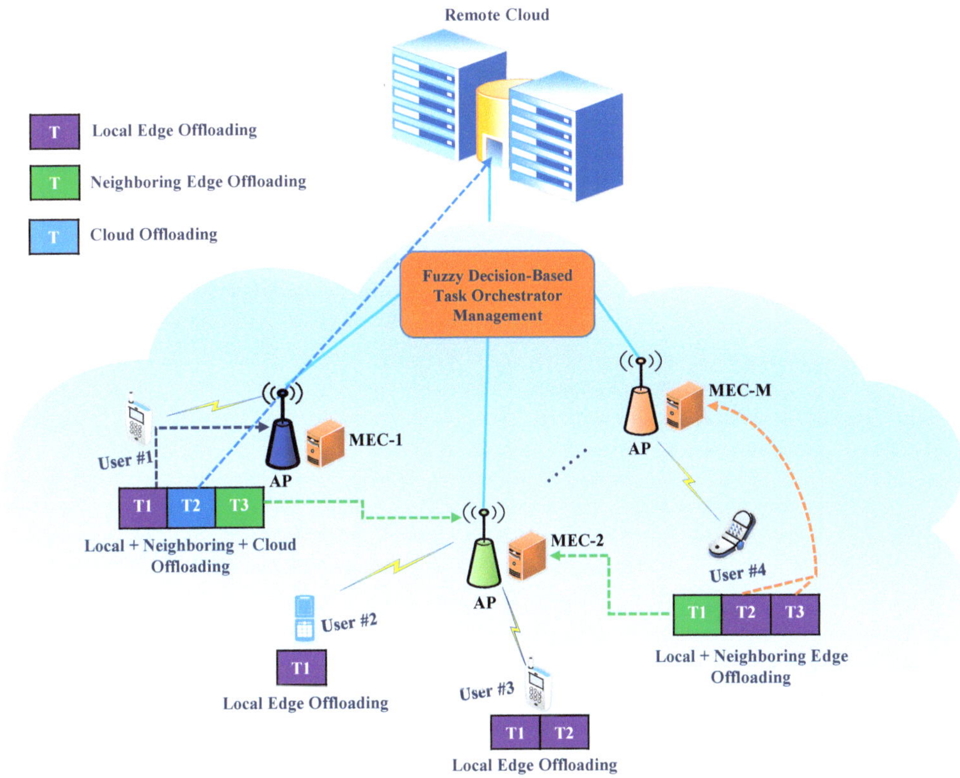

Figure 5. Proposed multi-tier MEC system architecture.

For each task, we consider the task offloading decision among the MEC servers and the cloud to be represented by

$$\mathfrak{D}_{ntm} \in \{0,1\} \tag{1}$$

where the t-th task of user device n is allocated to server m. Here, $n \in N, t \in T, m \in M$. When $\mathfrak{D}_{ntm} = 0$, the user device n will decide to offload its t-th task to the cloud server. Then, we have $\mathfrak{D}_{ntm} = 1, \forall m \in M \backslash \{M+1\}$. In this scenario, every task must use one of those servers for processing. Mathematically, it can be represented as follows:

$$\sum_{m=1}^{M+1} \mathfrak{D}_{ntm} = 1 \tag{2}$$

So, we can write the different computing modes mathematically, as given in Equation (3).

$$\begin{cases} \sum_{m=1}^{M} \mathfrak{D}_{ntm} = 1, & MEC\ computing \\ \mathfrak{D}_{ntm+1} = 1, & Cloud\ computing \end{cases} \tag{3}$$

for any $n \in N$ and $t \in T$.

In our proposed architecture, the following three cases may occur during task offloading.

1. Case 1: In this scenario, we consider the task to be offloaded and processed only by the local MEC server. For example, in Figure 5, we can see that User #2 has only one

task ($T1$) and User #3 has two tasks ($T1$ and $T2$). Because the local MEC server has enough capacity, both User #2 and User #3 process their computing tasks fully at the local MEC server.

2. Case 2: In this scenario, the offloaded task is executed by computation peer offloading between the local and nearby MEC servers. In describing Case 2, consider User #4 as having three tasks ($T1$, $T2$, and $T3$). Based on our proposed FTOM scheme, tasks $T2$ and $T3$ are processed locally because of the capabilities of the local MEC server, and task $T1$ is processed by a nearby MEC server.
3. Case 3: In this scenario, the offloaded task is executed through collaboration among a local MEC server, a neighboring MEC server, and the remote cloud. User #1 describes Case 3 and has three tasks ($T1$, $T2$, and $T3$). Based on the task orchestration management decision, $T1$ is processed by the local MEC server, $T2$ is handled by the remote cloud server, and $T3$ is executed by the nearby MEC server.

4. Fuzzy Decision-Based Task Offloading Management

For the efficient task offloading management of multi-tier MEC-enabled networks, we propose a fuzzy decision-based scheme for a multitude of reasons. The environment of edge computing is dynamic, and the stages of resources continuously change based on the offload requests. Due to this uncertainty, it is difficult to make a decision as to where a task should execute because we do not know the number of incoming user requests in advance. Moreover, task offloading management is basically online and considered an NP-hard problem. Therefore, we cannot apply conventional offline optimization techniques [36,38]. To handle these unpredictable environments, we need a low-complexity problem-solving technique. In addition, there are many input and output parameters involved in the MEC-enabled network environment, and these parameters are a part of the environmental behavior. This approach is inherently fuzzy. In this respect, fuzzy logic is one of the best alternatives to deal with the above-mentioned rapidly changing uncertain system. The advantage of fuzzy logic is that its complexity is very low, which is basically a very important criterion for an online algorithm [30]. Figure 6 shows the fuzzy logic architecture used in our proposed model. The main objective of our proposed fuzzy decision-based scheme is to identify a target server for the offloaded task by monitoring different factors including the incoming task's size, the network's condition, and the resources already utilized in the servers. The three main steps of the fuzzy reasoning mechanism are described as follows.

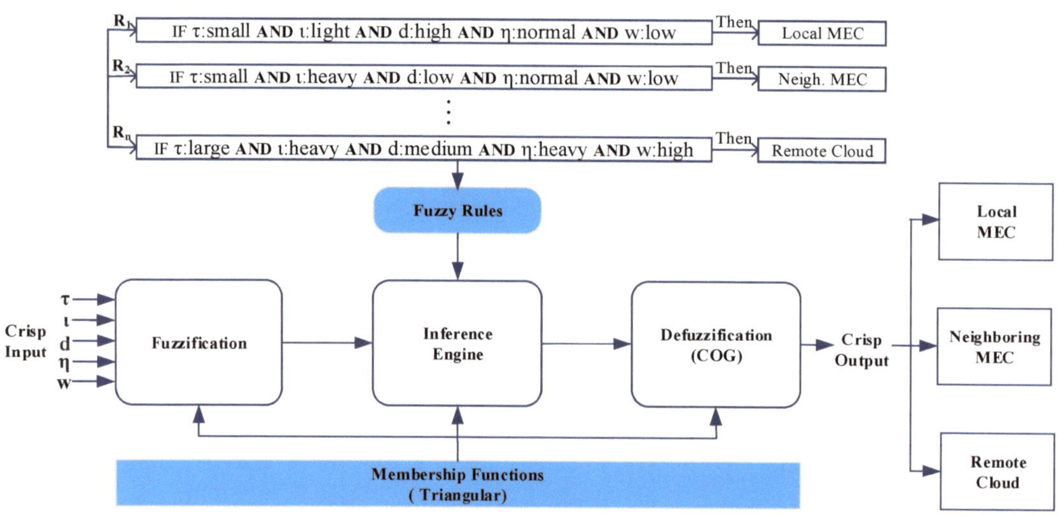

Figure 6. The proposed fuzzy logic architecture.

4.1. Fuzzification

During fuzzification, a crisp value is transformed into a fuzzy value by using membership functions (MFs). The crisp set of input parameters, which are described in Table 2, is the input for the fuzzy logic engine. It basically determines the degree of input data having the appropriate fuzzy sets by using the MFs. For efficient task offloading management, we define five significant fuzzy input variables: task size, local MEC VM utilization, network delay, neighboring MEC VM utilization, and WAN bandwidth. We represent these input variables mathematically as follows:

$$\Omega = [\tau,\ \iota,\ d,\ \eta,\ w] \tag{4}$$

where τ indicates the length of the incoming task in order to determine the task execution time; ι and η, respectively, represent the status of local MEC server and neighboring MEC server computational resources; d denotes the network delay; and w represents the WAN bandwidth. If the local MEC server is heavily congested and the latency of the network is very low, it will be advantageous to compute the incoming task by the neighboring MEC server. On the other hand, due to the heavily loaded neighboring MEC server and high network delay for handling large incoming requests, it is better to process the task in a local MEC server. The role of w is in making the decision about offloading the task to the remote cloud or not. If the local and neighboring servers are heavily loaded and WAN bandwidth is high, then it is appropriate to execute the incoming task to the remote cloud.

Table 2. Fuzzification input variables [36,38].

Input Variables	Notation	Fuzzy Set	Range
Task size (GI)	τ	Small Medium Large	0–8 6–18 16–50
Local MEC VM utilization (%)	ι	Light Normal Heavy	0–40 30–70 60–100
Network delay (ms)	d	Low Medium High	0–4 2–12 10–100
Neighboring MEC VM utilization(%)	η	Light Normal Heavy	0–40 30–70 60–100
WAN bandwidth (Mbps)	w	Low Medium High	0–4 3–7 6–21

Generally, a fuzzy logic system (FLS) uses non-numerical linguistic variables, such as Small, Medium, and Heavy, which come from natural language. Our FTOM scheme uses different linguistic variables to indicate the input parameters. Every base variable is represented by a linguistic variable, where the values are real numbers within a specific range. On the other hand, a linguistic variable is defined by using different terms that are the approximate value of a base variable. In Figure 7, we use a linguistic variable to represent WAN bandwidth. Based on the different bandwidths, the linguistic values for WAN bandwidth are Low, Medium, and High. For example, when the WAN bandwidth is up to 4 Mbps, we consider the bandwidth to be low. Moreover, we consider WAN bandwidth to be medium when the bandwidth range is between 3 Mbps and 7 Mbps. Furthermore, if the bandwidth range is between 6 Mbps and 21 Mbps, we consider the bandwidth to be high. A linguistic variable can be defined by using triplets (V, R, Ω_V), where V represents a fuzzy input variable such as network delay or WAN bandwidth, R denotes range of the variable, and Ω_V defines the set of linguistic terms for the fuzzy variable [39]. A linguistic variable for WAN bandwidth can be represented, based on Figure 7, as follows:

$$Linguistic\ variable, w = \begin{cases} w = WAN\ Bandwidth \\ R = \Re^{+} \\ \Omega_w = (Low, Medium, High) \end{cases} \quad (5)$$

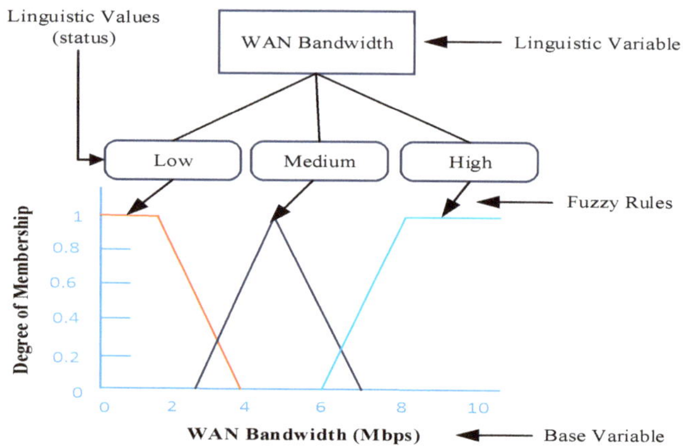

Figure 7. Example of linguistic variables for WAN bandwidth.

In this paper, we use three different linguistic terms such as Small (S), Medium (M), and Large (L) to represent the linguistic variable having task size τ. For network delay d and WAN bandwidth w, we use the linguistic terms Low (L), Medium (M), and High (H). Furthermore, the other two linguistic variables, ι and η, are Light (L), Normal (N), and Heavy (H). Mathematically, each of the above-mentioned input linguistic variables and their different terms are represented as follows:

$$\begin{cases} \Omega_\tau(x) = [\mu_\tau^S(x),\ \mu_\tau^M(x),\ \mu_\tau^L(x)] \quad \text{and} \\ \Omega_i(x) = [\mu_i^L(x),\ \mu_i^M(x),\ \mu_i^H(x)],\ \text{where } j \in \{d, w\}, \text{and} \\ \Omega_j(x) = [\mu_j^L(x),\ \mu_j^N(x),\ \mu_j^H(x)],\ \text{where } i \in \{\iota, \eta\} \end{cases} \quad (6)$$

4.1.1. Membership Functions

MFs play an important role in the performance of FLS. We use MFs for mapping the input variables to a membership value. It returns a value in the range [0, 1], which indicates the membership degree. For each fuzzy variable, we define a set of MFs. Mathematically it can be characterized by using Equation (7).

$$A_{Fuzzy} = \{(x,\ \mu_A(x)) : x \in X, \mu_A(x) \in [0,1]\} \quad (7)$$

Here, $\mu_A(x)$ represents the membership function of A. It quantifies the degree to which x belongs to A. The range of membership values is from 0 to 1, i.e., $\mu_A(x) \in [0,1]$ where x represents the element in a fuzzy set. According to Ω in Equation (4), we have used five fuzzy input variables. Based on the fuzzy input variable, we use five MF sets and each set includes three different linguistic terms, which are used in the fuzzification steps. The MFs are represented in various forms such as Gaussian, sigmoid, singleton, trapezoidal, or triangular [40]. In this paper, we use the triangular MF form because of its low complexity. Mathematically, the triangular MF is represented in Equation (8), where A is the fuzzy set. The parameters m and n indicate the lower limit and upper limit, respectively; and p represents the modal value of the triangle:

$$\mu_A^{triangular}(x) = \begin{cases} 0 & ; \ if \ x \leqslant m \\ \frac{x-m}{p-m} & ; \ if \ m \leqslant x \leqslant p \\ \frac{n-x}{n-p} & ; \ if \ p \leqslant x \leqslant n \\ 0 & ; \ if \ x \geq n \end{cases} \tag{8}$$

Determining the values used in the membership functions is critical because it has a notable impact on the overall FLS performance. Similar to other existing studies, the degree of membership values and the range of the values for each fuzzy variables are used from [31,36,38,39] because of their novel contribution to the edge computing environment based on fuzzy. The representations of MFs for the above-mentioned input fuzzy variables are shown in Figure 8. For example, if the size of the task is 8 GI, the degrees of the MF value are zero for Small, 0.4 for Medium, and zero for Large. So $\Omega_\tau(8) = [0, 0.4, 0]$, which is shown in Figure 8a.

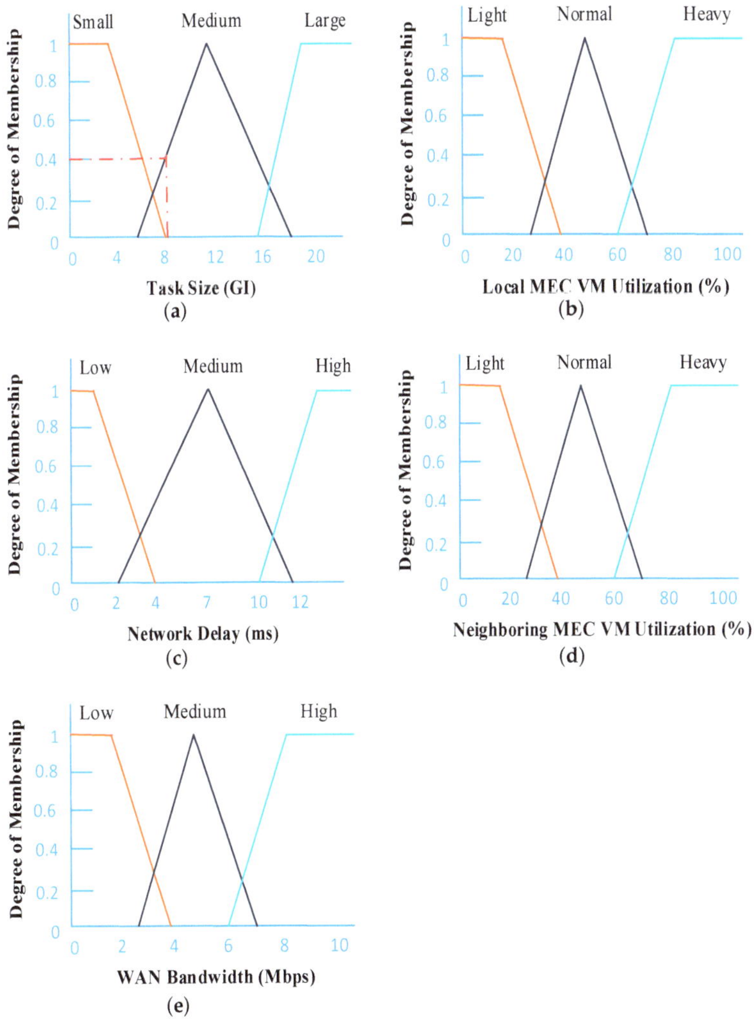

Figure 8. Membership functions (MFs) for the fuzzy input variables: (**a**) task size; (**b**) local MEC VM utilization; (**c**) network delay; (**d**) neighboring MEC VM utilization; (**e**) WAN bandwidth.

4.2. Fuzzy Inference Engine

It is the process of mapping the values of the given fuzzy input variables to an output using fuzzy logic. This step is the most crucial part of the FLS. For fuzzy inference inputs, different fuzzy sets (e.g., "Small", "Medium", "Large") have been considered as a confidence value. After evaluating and combining fuzzy rules, the output is generated A fuzzy rule is constructed by a series of simple IF-THEN rules and each rule defines a fuzzy implication between condition and conclusion. A fuzzy rule has the following form:

$$\begin{cases} If\ \ fvar_1 \in A\ and\ fvar_2 \in B\ ,\ \ldots,\ fvar_n \in N\ then\ f_{out} = OffDecision, \\ where\ A,\ B,\ ,\ \ldots,\ N\ are\ fuzzy\ sets, and \\ OffDecision \in \{localMEC,\ neighboringMEC,\ remotecloud\} \end{cases} \tag{9}$$

For fuzzification, we use five MFs sets, and we include three different linguistic terms in each set. Therefore, 243 fuzzy rules were used during the simulation. It is critical to define the fuzzy rules, because the overall performance of the system relies particularly on these rules. In this study, we use a better fuzzy rule set found empirically in [27,38]. Some examples of rules from our fuzzy rule set are given in Table 3. In each fuzzy rule, different linguistic variables are used. For example,

IF τ is Small
AND ι is Light
AND d is High
AND η is Normal
AND w is Low
THEN offload to the local MEC server.

Table 3. Example fuzzy rules.

Rule Index	Task Size (τ)	Local MEC VM Utilization (ι)	Network Delay (d)	Neighboring MEC VM Utilization (η)	WAN Bandwidth (w)	Offload Decision
R1	Small	Light	High	Normal	Low	Local MEC Server
R2	Medium	Heavy	Low	Light	Medium	Neighboring MEC Server
R3	Medium	Heavy	Medium	Heavy	High	Remote Cloud
R4	Small	Heavy	Low	Normal	Low	Neighboring MEC Server
R5	Large	Low	High	Heavy	Low	Local MEC Server
R6	Small	Normal	Low	Light	Medium	Neighboring MEC Server
R7	Large	Heavy	Medium	Heavy	High	Remote Cloud

Basically, there are three methods (aggregation, activation, and accumulation) that are used in the inference steps [36,38]. The aggregation method (also called the rule connection method) combines multiple rules within a rule set. The activation method explains the process of applying the evaluated result of the IF part of the rule to the THEN part. Based on the fuzzy rules (Table 3) and according to Equation (6), we can calculate the fuzzy value for selecting the target server from among the local MEC, neighboring MEC, and remote cloud as follows:

$$\begin{cases} \mu_{target} = max\{\mu^{R1}_{localMEC},\ \mu^{R2}_{neighboringMEC},\ \mu^{R3}_{cloud},\ \cdots,\mu^{Rn}_{cloud}\}, \\ where\ target \in \{localMEC,\ neighboringMEC,\ cloud\} \end{cases} \tag{10}$$

where $\mu^{R1}_{localMEC}$, $\mu^{R2}_{neighboringMEC}$, and μ^{R3}_{cloud} are represented as

$$\mu^{R1}_{localMEC} = [\mu^{R1}_{\tau}(\alpha),\ \mu^{R1}_{\iota}(\beta),\ \mu^{R1}_{d}(\gamma),\ \mu^{R1}_{\eta}(\delta),\ \mu^{R1}_{w}(\theta)] \tag{11}$$

$$\mu^{R2}_{neighboringMEC} = [\mu^{R2}_{\tau}(\alpha),\ \mu^{R2}_{\iota}(\beta),\ \mu^{R2}_{d}(\gamma),\ \mu^{R2}_{\eta}(\delta),\ \mu^{R2}_{w}(\theta)] \tag{12}$$

$$\mu^{R3}_{cloud} = [\mu^{R3}_{\tau}(\alpha),\ \mu^{R3}_{\iota}(\beta),\ \mu^{R3}_{d}(\gamma),\ \mu^{R3}_{\eta}(\delta),\ \mu^{R3}_{w}(\theta)] \tag{13}$$

where α, β, γ, δ, and θ represent the value of crisp input parameters τ, ι, d, η, and w respectively, in the fuzzy inference system. We can use a simple example to describe the inference process: 7 GI, 70%, 3 ms, 35%, and 4 Mbps are the values of α, β, γ, δ, and θ respectively. For the explanation, we considered only three rules (R1, R2, and R3) from Table 3. Then, we put these values into Equations (11)–(13). During our experiment, we considered the Minimum function in the activation phase, which is the most commonly used activation function. Therefore, we applied the aggregation and activation phases to rules R1, R2, and R3 to select the target server.

$$\mu^{R1}_{localMEC} = min[\mu^{R1}_{\tau}(7),\ \mu^{R1}_{\iota}(70),\ \mu^{R1}_{d}(3),\ \mu^{R1}_{\eta}(35),\ \mu^{R1}_{w}(4)] \tag{14}$$

$$\mu^{R2}_{neighboringMEC} = min[\mu^{R2}_{\tau}(7),\ \mu^{R2}_{\iota}(70),\ \mu^{R2}_{d}(3),\ \mu^{R2}_{\eta}(35),\ \mu^{R2}_{w}(4)] \tag{15}$$

$$\mu^{R3}_{cloud} = min[\mu^{R3}_{\tau}(7),\ \mu^{R3}_{\iota}(70),\ \mu^{R3}_{d}(3),\ \mu^{R3}_{\eta}(35),\ \mu^{R3}_{w}(4)] \tag{16}$$

Based on the fuzzification of input variables in Table 2, the fuzzy rules in Table 3, and MFs of the fuzzy input variables in Figure 8, we obtained fuzzy values for $\mu^{R1}_{localMEC}$, $\mu^{R2}_{neighboringMEC}$, and μ^{R3}_{cloud} are as follows:

$$\mu^{R1}_{localMEC} = min[0.2,\ 0,\ 0,\ 0.2,\ 0\,] = 0 \tag{17}$$

$$\mu^{R2}_{neighboringMEC} = min[0.2,\ 0.5,\ 0.3,\ 0.2,\ 0.5\,] = 0.2 \tag{18}$$

$$\mu^{R3}_{cloud} = min[0.2,\ 0.5,\ 0.25,\ 0,\ 0\,] = 0 \tag{19}$$

Finally, to determine the results from multiple rules, we considered the Maximum function as an accumulation method that can be represented as follows:

$$\mu_{target} = max[\mu^{R1}_{localMEC},\ \mu^{R2}_{neighboringMEC},\ \mu^{R3}_{cloud}] \tag{20}$$

After calculating the value of $\mu^{R1}_{localMEC}$, $\mu^{R2}_{neighboringMEC}$, and μ^{R3}_{cloud} from Equations (17)–(19), we can determine the value of the target server in the accumulation phase by using Equation (20), which is 0.2. Therefore, the target server is the neighboring edge server.

$$\mu_{target} = max[0,\ 0.2,\ 0] = 0.2 \tag{21}$$

4.3. Defuzzification

Defuzzification is the process of converting into a crisp value the output of the aggregated fuzzy set produced by the inference mechanism. It is an inverse transformation, compared with the fuzzification process, which is shown in Figure 9.

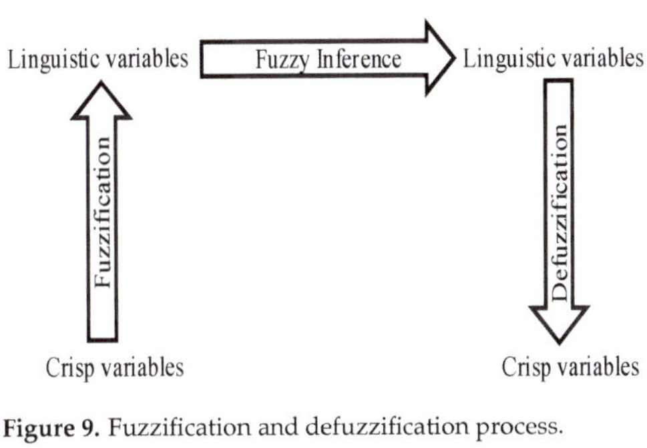

Figure 9. Fuzzification and defuzzification process.

The result of fuzzy inference is a linguistic value that translates into a numerical value in the defuzzification step. There are different methods for defuzzification, including fuzzy clustering defuzzification (FCD), weighted fuzzy mean (WFM), mean of maximum (MOM), and center of gravity (COG) [39]. The most popular and commonly used method is COG which is the defuzzification step in our proposed system. This method determines the value of the center of gravity under the curve and returns the corresponding crisp value. After implementing the COG method in our proposed system, we obtained the crisp value x^*, which is in the range [0, 100]. Based on the value of x^*, we defined the offloading decisions, all of which are shown in Table 4.

Table 4. Offloading Decisions.

Target Offloading Node	Range
Local MEC Server	0–40
Neighboring MEC Server	30–70
Remote Cloud Server	60–100

The centroid defuzzification process is shown in Figure 10. For example, if the value of $\mu_{localMEC}$, $\mu_{neighboringMEC}$, and μ_{cloud} are calculated as 0.2, 0.5, and 0.3 respectively, then the crisp result after the centroid defuzzfication process will be 53, as shown in Figure 10b. So, based on the crisp result, the task is offloaded to the neighboring edge server. Algorithm 1 is the FTOM algorithm. Mathematically, the COG method is represented as follows.

$$COG, \ x^* = \frac{\int x\mu(x)dx}{\int \mu(x)dx} \tag{22}$$

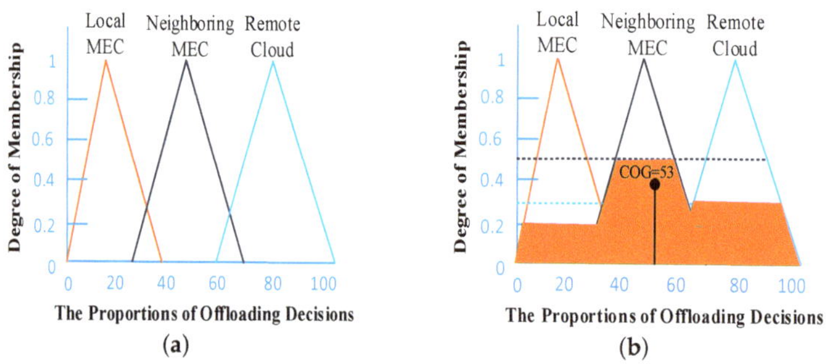

Figure 10. Defuzzification process: (**a**) output membership function; (**b**) the centroid defuzzification process.

Algorithm 1 Fuzzy Decision-Based Task Offloading Management (FTOM) Algorithm

Input: The incoming task, T

Output: Target offload node, O

1: Read the network topology;

2: Read the profile of incoming task T;

3: $f_v \leftarrow$ FuzzyLogic(τ, ι, d, η, w); // Output value that fuzzy logic returns

4: Calculate the center of gravity value for crisp output, COG $\leftarrow$ Equation (22);

5: Offloading decision, O $\leftarrow$ Table 4;

6: **return** O;

5. Performance Evaluation

In this section, we evaluate the effectiveness of our proposed FTOM scheme in terms of task failure rate, task processing latency, task completion time, and number of successfully executed tasks for different VM conditions in MEC-enabled networks with respect to various user devices through the EdgeCloudSim simulator [41]. To verify the performance, our proposed scheme was compared with five other benchmark task offloading schemes: local edge offloading (LEO), two-tier edge orchestration-based offloading (TTEO), fuzzy orchestration-based load balancing (FOLB), fuzzy workload orchestration-based task offloading (WOTO), and fuzzy edge-orchestration based collaborative task offloading (FCTO). In the LEO scheme, all users offload and execute their tasks by using the local MEC server. In the TTEO, FOLB, and WOTO schemes, all the neighboring edge servers and the remote cloud are connected to the orchestrator. The orchestrator distributes the incoming tasks and processes those tasks by using the edge servers and the cloud. On the other hand, orchestrator of the FCTO scheme distributes the incoming tasks among the edge servers. In order to present a realistic simulation for different real-life scenarios, we used three different applications during the experiments: an augmented reality (AR) application, an infotainment (I) application, and a health monitoring (HM) application [42–44]. Among them, the HM application is latency-sensitive, and the infotainment application is delay-tolerant. The AR application, however, is latency-sensitive as well as compute-intensive, requiring more CPU time. According to [36,38,41], Table 5 lists the key characteristic parameters of the AR, I, and HM applications, and the other simulation parameters used during the simulation are presented in Table 6.

Table 5. Applications used in the simulations [36,41].

	Augmented Reality (AR)	Infotainment (I)	Health Monitoring (HM)
Usage (%)	50	30	20
Interarrival time of tasks (sec)	2	5	10
Delay sensitivity (%)	0.9	0.3	0.7
Idle period (sec)	20	25	90
Active period (sec)	40	45	15
Upload data size (KB)	1500	25	1250
Download data size (KB)	25	750	250
Average task length (GI)	20	7.5	2.5
Task utilization of the VM (%)	10	5	2

Here, the tasks that are offloaded from the user device are represented as a set of predefined application categories, such as face recognition, infotainment services, and fall-risk detection. For example, in an AR application, a user wears smart glasses to upload images to the server for face identification. For a fall-risk detection service, the health monitoring application uses a foot-mounted inertial sensor that records the waking pattern of the user for a while; then, it sends the readings to a remote server for further processing. In Table 5, usages represent the percentage of mobile devices running for AR, I, and HM applications. In this study, we used 50%, 30%, and 20% for AR, I, and HM applications respectively. The task interarrival time depicts the frequency for transmitting the task to the orchestrator, which follows an exponential distribution. We considered the task interarrival time for AR, I, and HM applications were 2, 5, and 10 s respectively. We used a higher task interarrival time for HM application than others because we need to record the sensor data for a specific duration and send that collected data for further processing.

Table 6. Simulation parameters [36,38,41].

Parameter	Value
Number of mobile devices	500
Number of edge servers	14
Number of VMs per edge server	2∼8
Number of VMs in the cloud	4
VM processing speed per edge server	10 GIPS
VM processing speed in the cloud	100 GIPS
WAN/WLAN bandwidth	Empirical
MAN bandwidth	MMPP/M/1 model

To identify the sensitivity of the task (delay-sensitive or delay-tolerant), we used the delay sensitivity value in our simulation. The offloaded task is considered delay-tolerant if the delay sensitivity value is low. Because the infotainment application is delay-tolerant, we used a delay sensitivity value of 0.3 during the experiment. On the other hand, the AR and HM applications are delay-sensitive, and thus, 0.9 and 0.7, respectively, were the delay sensitivity values. The task is generated during the active period but stays idle in the waiting period. For example, in the AR I, and HM applications, we use 40, 45, and 15 s for active mode and 20, 25, and 90 s for idle mode, respectively. In the AR and HM applications, a user uploads a large amount of data for service and receives a comparatively lower amount of data in response. Therefore, during the simulation, we considered upload and download data sizes of <1.5 MB, 25 KB> for the AR application and <1.25 MB, 250 KB> for the HM application. Moreover, with the infotainment application, a user sends a very small amount of data with a service request and the corresponding service returns a large amount of data in response. Thus, we used an upload data size of 25 KB and the corresponding downloaded service was 750 KB in response. The task length defines the needed CPU resources for the corresponding task in the giga instructions (GI) unit. In the simulation analysis, we used 50 mobile devices in the lightly loaded scenario and 500 mobile devices in the heavily loaded scenario. Moreover, we used 14 APs, and each AP was equipped with a single MEC server.

To measure the efficiency of the proposed FTOM scheme, Figure 11a,b show the average processing time and the average task completion time (the y-axes), respectively, versus the number of mobile devices (the x-axes, varying from 50 to 500). From analyzing Figure 11a, the processing time tends to enhance in case of all scenarios to handle the excessive number of mobile devices, and the LEO scheme provides the worst performance than others. This is because the local MEC server experiences congestion due to its lower computing capabilities. On the other hand, the FOLB scheme provides better performance than the LEO scheme, since tasks are distributed between the MEC server and the remote cloud. Moreover, the FCTO scheme also provides better performance until 200 mobile devices than others except the FTOM scheme. In this scheme, tasks are easily distributed among the neighboring edge server. When it comes to the TTEO, WOTO, and our proposed FTOM scheme, they distribute the tasks among the MEC servers and the cloud. Therefore, for handling more mobile devices, the processing time does not increase, compared to the LEO scheme. However, when the number of mobile devices increases, for example, to 200, the average processing time for LEO, TTEO, FOLB, WOTO, FCTO, and our proposed FTOM scheme were 3.94, 1.94, 2.32, 2.19, 1.91, and 0.42 s, respectively. By comparing all schemes, the proposed FTOM scheme outperformed all the others as the load increased. In Figure 11b, the completion times for the above-mentioned task offloading schemes are given. The task completion time is derived by using the following formula: task completion time = processing time + network delay. Overall, the average task completion time tends to increase with increased numbers of mobile devices, and our proposed FTOM scheme showed the best performance, on average, because our proposed scheme can make dynamic decisions, and it efficiently balances both networking and edge computational resources, compared to the competitors. From the simulation results, we conclude that our proposed system can reduce the task completion time by approximately 66.6%, 61.5%, 47.9%, 49.8%, and 55% when compared to the LEO, TTEO, FOLB, WOTO, and FCTO schemes, respectively.

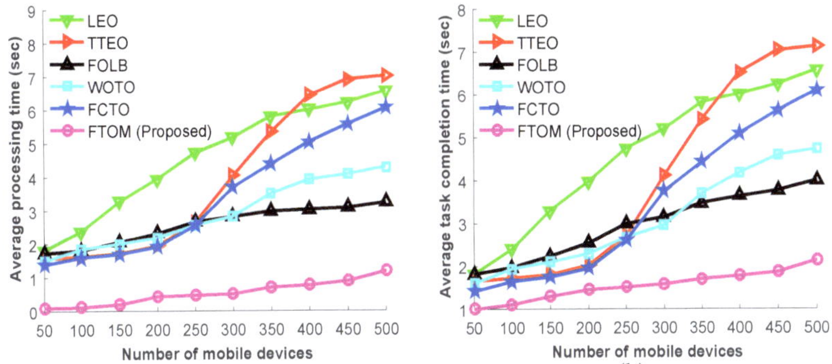

(a) (b)

Figure 11. Performance evaluations based on all application types: (**a**) average processing times; (**b**) average task completion times.

Moreover, to verify the necessity of the proposed FTOM scheme, Figure 12a,b show another experiment to investigate the task failure rate in terms of different numbers of mobile devices. The task failure rate indicates the percentage of task failures out of the total number of tasks. Figure 12a shows the task failure rate based on VM capacity. During the simulation, we used four VMs for each MEC server. Figure 12a shows that the LEO scheme starts to experience congestion after 100 mobile devices, the FOLB and FCTO schemes starts getting congested after 250 and 300 mobile devices, respectively. Due to its limited computing capacity, the LEO scheme faces an overload problem after 100 mobile devices and starts to congest. The FOLB scheme distributes the tasks between the local MEC server with the cloud and the FCTO scheme distributes the tasks among the neighboring MEC server. Therefore, the FOLB and FCTO schemes can easily handle 250 and 300 mobile devices respectively without congestion. After that, due to the WAN delay, the FOLB scheme faces congestion and due to the overloaded problem, the FCTO scheme faces congestion. On the other hand, the other three offloading schemes distribute the tasks among MEC servers and the cloud. Thus, the TTEO and WOTO schemes start to experience congestion after 350 and 400 mobile devices, respectively, and our proposed FTOM scheme can handle 500 devices without congestion. This is because our proposed system can utilize local and neighboring MEC servers more efficiently than its competitors in a dynamic environment. Similarly, Figure 12b shows the average task failure rate for the aforementioned task offloading schemes. There are three main factors contributing to task failure: server capacity, network delay, and mobility. In these experiments, we considered those three factors when calculating the average task failure rate. Analyzing Figure 12b, the task failure rate is approximately zero until there are 100 mobile devices. However, the situation changes as the number of devices increases. A heavily loaded system increases the task failure rate in all scenarios due to congestion. For example, the task failure rate rapidly increased from 1.3% at 100 devices to 43.7% at 500 devices in the LEO scheme; from 3.8% at 350 devices to 25.6% at 500 devices in the TTEO scheme; from 4% at 300 devices to 16.3% at 500 devices in the FOLB scheme; from 2.6% at 400 devices to 4.7% at 500 devices in the WOTO scheme; from 3% at 300 devices to 35.3% at 500 devices in the FCTO scheme; and from 0.82% at 400 devices to 0.98% at 500 devices in our proposed FTOM scheme. Comparing all the schemes, our proposed FTOM provided a lower task failure rate than the others because it makes better decisions about sending tasks to MEC servers and, based on the network condition, sending some tasks to the remote cloud.

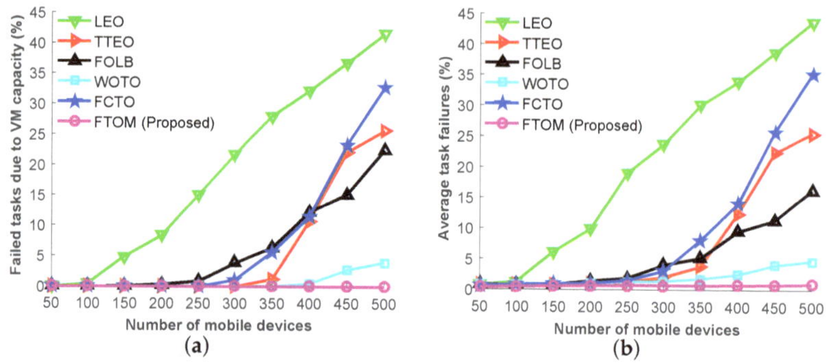

Figure 12. Performance analysis based on each application type: (**a**) failed tasks due to VM capacity; (**b**) average task failure rate.

By varying the ratio between the latency-sensitive AR application and the latency-tolerant infotainment application, Figure 13a,b show the task failure rate and the task completion time, respectively, for the aforementioned task offloading schemes. In these experiments, we considered the average task length of the AR application to be higher than the infotainment application, because the AR application is not only latency-sensitive but also compute-intensive. Initially, we considered the ratio between two applications to be 0:10, meaning all the offloaded tasks are latency-tolerant. Then, the task failure rate of the LEO, TTEO, FOLB, WOTO, and FTOM schemes were 0.43%, 0.36%, 0.25%, 0.25%, and 0.23%, respectively. The task failure rate is low at this ratio because all the tasks are latency tolerant. On the other hand, if we use all latency-sensitive applications, the ratio is 10:0. In this scenario, the task failure rate of the LEO, TTEO, FOLB, WOTO, FCTO, and FTOM schemes was 29.71%, 16.93%, 9.69%, 19.43%, 9.23%, and 8.73%, respectively. Therefore it is seen that, when we use all latency-sensitive applications, the FCTO scheme provides lower task failure rate than others except the FTOM scheme. From the above analysis in Figure 13a, we observe that when there are more latency-sensitive tasks compared to latency-tolerant tasks, the average task failure rate increased in all scenarios. But our proposed FTOM scheme reduced the average task failure rate, compared to the others, because the proposed system utilizes local and neighboring MEC servers for offloading latency-sensitive tasks and, based on the network condition, utilizes a remote server to offload latency-tolerant tasks. Similarly, Figure 13b shows the task completion times for the different ratios between latency-sensitive and latency-tolerant applications. In this experiment, latency-sensitive AR applications are relatively heavy, compared to latency-tolerant applications. Thus, the average task completion time of the latency-sensitive tasks is higher than the latency-tolerant tasks. Our proposed scheme reduces the task completion time in all scenarios, compared to the other schemes.

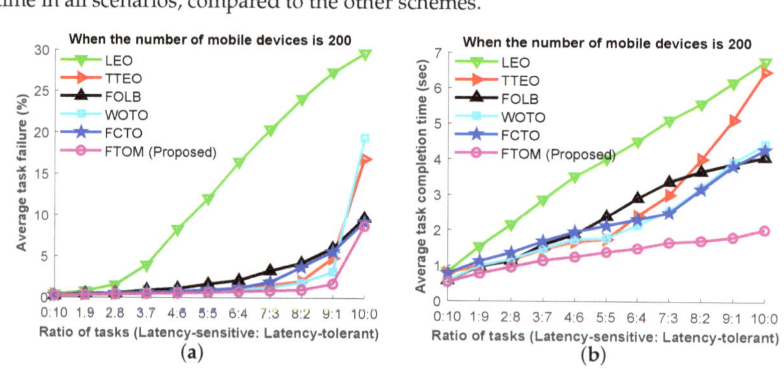

Figure 13. Performance analysis based on latency-sensitive to latency-tolerant task ratio: (**a**) average task failure rate ; (**b**) average task completion time.

Furthermore, Figure 14a,b, show the successfully executed offloaded tasks for two different MEC server capacities versus the number of mobile devices. From the simulation results, we observed that most of the offloaded tasks were executed successfully when the system was lightly loaded. However, this success rate decreased because of the growing number of devices. In Figure 14a,b, for two VMs and four VMs deployed, respectively, the number of successfully executed tasks dropped after 150 mobile devices had been added and after 250 mobile devices had been added for all schemes except LEO. At both capacities, the LEO scheme could not handle more tasks due to congestion in the VMs. Thus, with two VMs in each MEC server, the number of successfully executed tasks rapidly dropped from 94.3% at 50 devices to 31% at 500 devices. With four VMs in each MEC server, successfully completed tasks dropped from 99.2% at 50 devices to 56.2% at 500 devices under LEO. On the other hand, with two VMs in each MEC server, the number of successfully executed tasks dropped from 99% at 50 devices to 44.8% at 500 devices when using the TTEO scheme. For the FOLB scheme, completed tasks dropped from 98.8% at 50 devices to 75% at 500 devices, for the WOTO scheme, completed tasks dropped from 99.2% at 50 devices to 86.2% at 500 devices and for the FCTO scheme, completed tasks dropped from 99.15% at 50 devices to 40.6% at 500 devices. Our proposed FTOM scheme, however, saw the successful execution rate drop from 99.6% at 50 devices to 93.5% at 500 devices. Figure 14a,b, show that the rate of successfully executed tasks tends to increase with the increasing number of VM. When comparing the five schemes, our proposed FTOM approach outperformed the others because it can alleviate the load on the local edge server and efficiently distribute tasks to neighboring MEC servers and the remote cloud based on the network condition. After analyzing the simulation results, we can summarize that using our proposed FTOM scheme improves the successfully executed task rate by almost 68.5% compared with the LEO scheme, by 32.4% compared with the TTEO scheme, by 8.9% compared with FOLB, by 3.2% compared with WOTO, and by 38.6% compared with FCTO.

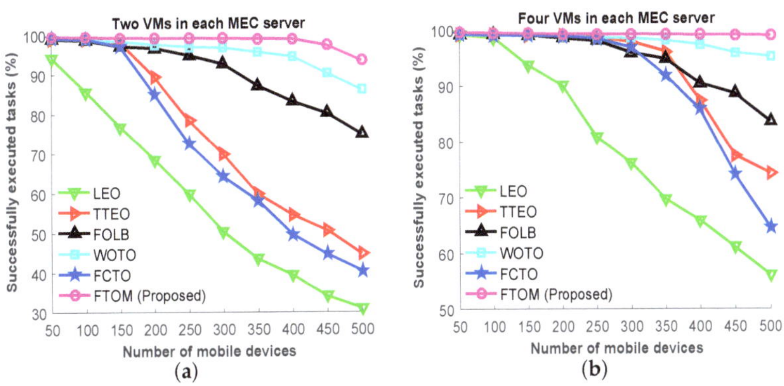

Figure 14. Successfully executed tasks versus the number of mobile devices: (**a**) with two VMs in each MEC server; (**b**) with four VMs in each MEC server.

Finally, in the last simulation result, the effect of different MEC server capacities in terms of the number of mobile devices was investigated, and the results are shown in Figure 15. In this experiment, we assigned three different numbers of VMs (eight, four, and two) to each MEC server. Figure 15 shows that the completion time with the LEO scheme was worse than the others in all scenarios. The main reason is that, for processing tasks on the local MEC server, many users wait a long time in the queue. For example, when the number of mobile devices is 100 and two VMs are deployed in each MEC server, the completion times for the LEO, TTEO, WOTO, and FTOM schemes were 3.95, 2.41, 2.3, and 1.21 s, respectively. However, the completion times for the LEO, TTEO, WOTO, and FTOM schemes were 1.95, 1.62, 1.75, and 1.1 s, respectively, when eight VMs were deployed in each MEC server. From the above analysis, we can say that each scheme can handle more user devices, as well as reduce the average task completion time, if the number of VMs is increased. However, our proposed FTOM scheme outperformed in all the scenarios, since it can avoid congestion and balances loads more efficiently among the MEC servers in the same tier.

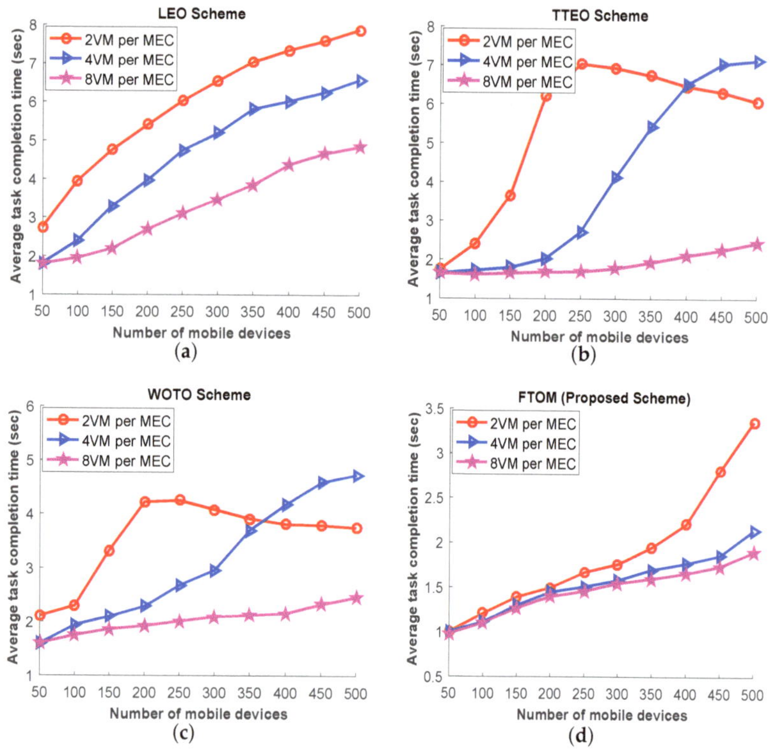

Figure 15. Performance evaluation based on each application types for different VM condition: (**a**) LEO scheme; (**b**) TTEO scheme; (**c**) WOTO scheme; (**d**) FTOM scheme.

6. Discussion

In this section, we have summarized the previously proposed various task offloading schemes in MEC-enabled networks and analyzed the performance evaluation results with respect to various evaluation metrics to show the effectiveness of our proposed FTOM scheme. The different task offloading schemes for various scenarios, including single/multiple users, single/multiple tasks, and different computing locations (local MEC/neighboring MEC/cloud server) are summarized in Table 7. The previous work mostly focused on vertical offloading between MEC and the cloud or on horizontal offloading among neighboring MEC servers. For example, Chen et al. [45] considered multi-user, single task, and local MEC computation offloading scheme. They ignored the neighboring MEC as well as the remote cloud server. Dinh et al. [46] considered single-user and multi-task offloading schemes. For processing the offloaded task, authors utilized local MEC as well as neighboring MEC servers. However, they ignored the remote cloud server which has powerful computing capabilities and did not consider the multi-user scenarios. On the other hand, Liu et al. [47] considered multi-user and single task offloading scheme. For task offloading, authors considered local MEC and remote cloud servers while they ignored neighboring MEC servers. Most of the previous work did not consider the collaborative integration between vertical and horizontal task offloading schemes. Therefore, to take the advantage of both task offloading schemes, in this paper, we propose an efficient fuzzy decision–based task offloading management (FTOM) scheme. Our proposed scheme used multi-user and multi-task offloading scenarios. Moreover, to utilize the neighboring MEC servers as well as a remote cloud server, our FTOM scheme considers vertical and horizontal task offloading schemes.

Table 7. Summary of different task offloading in MEC-enabled networks.

Publication	**User**		**Task**		**Computing Location**		**Cloud**
	Single	**Multiple**	**Single**	**Multiple**	**Local MEC**	**Neighboring MEC**	**Server**
Bi and Zhang [16]		✓	✓		✓		
Ning et al. [20]	✓	✓	✓		✓	✓	✓
Huang et al. [25]		✓		✓	✓		✓
Hossain et al. [27]		✓		✓	✓	✓	
Chen et al. [48]		✓		✓	✓	✓	
Huang et al. [49]		✓		✓	✓	✓	✓
Sonmez et al. [38]		✓		✓	✓	✓	✓
Li et al. [50]		✓	✓		✓	✓	
Chen et al. [45]		✓	✓		✓		
Dinh et al. [46]	✓			✓	✓	✓	
Liu et al. [47]		✓	✓		✓		✓
Wei et al. [51]	✓			✓	✓		
Our Work		✓		✓	✓	✓	✓

Table 8 is summarized from Figures 11–15, which shows the comparisons of our scheme with the existing task offloading approaches. We have used many key performance evaluation metrics in this study to analyze the effectiveness of our proposed FTOM scheme. From Table 8, we observe that our proposed FTOM scheme provides lower processing and task completion time compared with other task offloading schemes. It reduces task completion time by approximately 66.6%, 61.5%, 47.9%, 49.8%, and 55% when compared to the LEO, TTEO, FOLB, WOTO, and FCTO schemes, respectively. Moreover, due to the VM capacity and overloaded problem, LEO, TTEO, FOLB, WOTO, and FCTO schemes starts getting congested after 100, 350, 250, 400, and 300 mobile devices respectively. On the other hand, our proposed scheme can handle 500 mobile devices without any congestion. Furthermore, the average task failure rate and task completion time increase in all scenarios when there are more latency-sensitive tasks compared to latency-tolerant tasks, and our proposed scheme outperforms others. When the system was lightly loaded, most of the offloaded tasks were executed successfully in all task offloading schemes. However, our proposed FTOM scheme improves the successfully executed task rates by almost 68.5%, 32.4%, 8.9%, 3.2%, and 38.6% compared with LEO, TTEO, FOLB, WOTO, and FCTO schemes respectively. Therefore, after analyzing Table 8, we can conclude that our proposed system significantly improves the rate of successfully executing offloaded tasks compared to others.

Table 8. Results summary of different methods.

Evaluations Metrics	**Methods**					
	LEO	**TTEO**	**FOLB**	**WOTO**	**FCTO**	**FTOM**
Average processing time (sec)	4.58	3.91	2.57	2.87	3.38	0.53
Average task completion time (sec)	4.61	4.01	2.96	3.08	3.43	1.54
Failed tasks due to VM capacity (%)	18.93	5.98	6.12	0.72	7.4	0
Average task failure (%)	20.79	7.12	5.21	1.94	9.12	0.70
Average completion time for different ratio of tasks (sec)	3.92	2.63	2.41	2.19	2.28	1.35
Average task failure for different ratio of tasks(%)	13.2	2.75	2.76	2.73	2.3	1.44
Successfully executed tasks for 2VM MEC server (%)	58.43	74.4	90.47	95.46	71.04	98.49
Successfully executed tasks for 4VM MEC server (%)	79.21	92.88	94.78	98.06	90.85	99.29

7. Conclusions

Efficient task offloading management in a MEC-enabled network is an intrinsically difficult online problem because the environment of edge computing is extremely dynamic and the states of computing resources change rapidly based on the offload requests. On the other hand, without proper task offloading management, the distinct MEC server is not fully utilized or is sometimes overloaded by handling so many user requests. To handle this uncertainty and provide an automated management system, we proposed an efficient fuzzy decision-based task offloading management (FTOM) scheme. Our proposed approach makes dynamic decisions as to where to offload incoming tasks based on the states of server resources, the network conditions, and the latency sensitivity of the tasks. Moreover, our proposed system utilizes nearby MEC servers as well as the remote cloud to handle the overload problem and increase performance in a MEC server. To offload decisions, our system analyzes the computing resources to determine if they are already overloaded or underutilized. It can efficiently balance both networking and computational resources where small and latency-sensitive tasks are better offloaded to a local or nearby MEC server. To evaluate our FTOM scheme, we used infotainment, augmented reality, and health monitoring applications and compared the proposed scheme with five benchmark schemes. According to the evaluations, our proposal outperformed its competitors in terms of task failure rate, task completion latency, and number of successfully executed tasks in all scenarios. For future work, we will consider a machine learning approach to efficient task offloading in MEC-enabled networks

Author Contributions: Conceptualization, M.D.H.; Project administration, E.-N.H.; Software, M.D.H. Supervision, E.-N.H.; Writing—original draft, M.D.H.; Writing—review and editing, M.D.H., T.S. M.A.H., M.I.H., L.N.T.H., J.P., and E.-N.H. All authors have read and agreed to the published version of the manuscript.

Funding: This work was supported by Institute for Information & Communications Technology Planning & Evaluation (IITP) grant funded by the Korea government (MSIT) (No.2017-0-00294 Service mobility support distributed cloud technology). This work was also supported by the ICT Division, Ministry of Posts, Telecommunications and Information Technology, Bangladesh, through the research fellowship program.

Institutional Review Board Statement: Not applicable.

Informed Consent Statement: Not applicable.

Conflicts of Interest: The authors declare no conflict of interest.

References

1. Su, X.; Cao, J.; Hui, P. 5G edge enhanced mobile augmented reality. In Proceedings of the 26th Annual International Conference on Mobile Computing and Networking (MobiCom'20), London, UK, 21–25 September 2020; pp. 1–3.
2. Storck, C.R.; Duarte-Figueiredo, F. A Survey of 5G Technology Evolution, Standards, and Infrastructure Associated With Vehicle-to-Everything Communications by Internet of Vehicles. *IEEE Access* **2020**, *8*, 117593–117614.
3. Sigwele, T.; Hu, Y.F.; Ali, M.; Hou, J.; Susanto, M.; Fitriawan, H. Intelligent and Energy Efficient Mobile Smartphone Gateway for Healthcare Smart Devices Based on 5G. In Proceedings of the IEEE Global Communications Conference (GLOBECOM), Abu Dhabi, UAE, 9–13 December 2018; pp. 1–7.
4. Mao, Y.; You, C.; Zhang, J.; Huang, K.; Letaief, K.B. A survey on mobile edge computing: The communication perspective *IEEE Commun. Surv. Tutor.* **2017**, *19*, 2322–2358.
5. Taleb, T.; Ksentini, A.; Jantti, R. "Anything as a service" for 5G mobile systems. *IEEE Netw.* **2016**, *30*, 84–91.
6. Sabella, D.; Vaillant, A.; Kuure, P.; Rauschenbach, U.; Giust, F. Mobile-Edge Computing Architecture: The role of MEC in the Internet of Things. *IEEE Consum. Electron. Mag.* **2016**, *5*, 84–91.
7. Khan, A.U.R.; Othman, M.; Madani, S.A.; Khan, S.U. A Survey of Mobile Cloud Computing Application Models. *IEEE Commun. Surv. Tutor.* **2014**, *16*, 393–413.
8. Liu, Y.; Peng, M.; Shou, G.; Chen, Y.; Chen, S. Toward Edge Intelligence: Multiaccess Edge Computing for 5G and Internet of Things. *IEEE Internet Things J.* **2020**, *7*, 6722–6747.
9. Hu, Y.C.; Patel, M.; Sabella, D.; Sprecher, N.; Young, V. Mobile edge computing—A key technology towards 5G. *ETSI White Pap.* **2015**, *11*, 1–16.

10. Pham, Q.; Fang, F.; Ha, V.N.; Piran, M.J.; Le, M.; Le, L.B.; Hwang, W.; Ding, Z. A Survey of Multi-Access Edge Computing in 5G and Beyond: Fundamentals, Technology Integration, and State-of-the-Art. *IEEE Access* **2020**, *8*, 116974–117017.
11. Wang, S.; Zhang, X.; Zhang, Y.; Wang, L.; Yang, J.; Wang, W. A Survey on Mobile Edge Networks: Convergence of Computing Caching and Communications. *IEEE Access* **2017**, *5*, 6757–6779.
12. Mach, P.; Becvar, P. Mobile edge computing: A survey on architecture and computation offloading. *IEEE Commun. Surv. Tutor.* **2017**, *19*, 1628–1656.
13. Ridhawi, I.A.; Aloqaily, M.; Kotb, Y.; Ridhawi, Y.A.; Jararweh, Y. A collaborative mobile edge computing and user solution for service composition in 5G systems. *Trans. Emerg. Telecommun. Technol.* **2018**, *29*, 1–19.
14. Ren, J.; Yu, G.; He, Y.; Li, G.Y. A Collaborative Cloud and Edge Computing for Latency Minimization. *IEEE Trans. Veh. Technol.* **2019**, *68*, 5031–5044.
15. Hossain, M.D.; Sultana, T.; Hossain, M.A.; Lee, G.; Huh, E.-N. Efficient Load Management in Multi-Access Edge Computing Using Fuzzy Logic. *KIISE Trans. Comput. Pract.* **2020**, *26*, 482–492.
16. Bi, S.; Zhang, Y.J. Computation rate maximization for wireless powered mobile-edge computing with binary computation offloading. *IEEE Trans. Wirel. Commun.* **2018**, *17*, 4177–4190.
17. Wang, X.; Ning, Z.; Wang, L. Offloading in Internet of Vehicles: A Fog-Enabled Real-Time Traffic Management System. *IEEE Trans. Ind. Inform.* **2018**, *14*, 4568–4578.
18. Messous, M.-A.; Senouci, S.-M.; Sedjelmaci, H.; Cherkaoui, S. A game theory based efficient computation offloading in an UAV network. *IEEE Trans. Veh. Technol.* **2019**, *68*, 4964–4974.
19. Kuang, Z.; Li, L.; Gao, J.; Zhao, L.; Liu, A. Partial Offloading Scheduling and Power Allocation for Mobile Edge Computing Systems. *IEEE Internet Things J.* **2019**, *6*, 6774–6785.
20. Ning, Z.; Dong, P.; Kong, X.; Xia, F. A cooperative partial computation offloading scheme for mobile edge computing enabled Internet of Things. *IEEE Internet Things J.* **2019**, *6*, 4804–4814.
21. Ren, J.; Yu, G.; Cai, Y.; He, Y.; Qu, F. Partial offloading for latency minimization in mobile-edge computing. In Proceedings of the IEEE Global Communications Conference (GLOBECOM), Singapore, 4–8 December 2017; pp. 1–6.
22. Deng, R.; Lu, R.; Lai, C.; Luan, T.H.; Liang, H. Optimal Workload Allocation in Fog-Cloud Computing Toward Balanced Delay and Power Consumption. *IEEE Internet Things J.* **2016**, *3*, 1171–1181.
23. Guo, H.; Liu, J. Collaborative computation offloading for multi-access edge computing over fiber–wireless networks. *IEEE Trans. Veh. Technol.* **2018**, *67*, 4514–4526.
24. Lin, Y.; Lai, Y.; Huang, J.; Chien, H. Three-Tier Capacity and Traffic Allocation for Core, Edges, and Devices for Mobile Edge Computing. *IEEE Trans. Netw. Serv. Manag.* **2018**, *15*, 923–933.
25. Huang, M.; Liu, W.; Wang, T.; Liu, A.; Zhang, S. A cloud-MEC collaborative task offloading scheme with service orchestration. *IEEE Internet Things J.* **2020**, *7*, 5792–5805.
26. Yuan, P.; Cai, Y.; Huang, X.; Tang, S.; Zhao, X. Collaboration Improves the Capacity of Mobile Edge Computing. *IEEE Internet Things J.* **2019**, *6*, 10610–10619.
27. Hossain, M.D.; Sultana, T.; Nguyen, V.; Rahman, W.; Nguyen, T.D.T.; Huynh, L.N.T.; Huh, E.-N. Fuzzy Based Collaborative Task Offloading Scheme in the Densely Deployed Small-Cell Networks with Multi-Access Edge Computing. *Appl. Sci.* **2020**, *10*, 3115.
28. Fan, W.; Liu, Y.;Tang, B.; Wu, F.; Wang, Z. Computation Offloading Based on Cooperations of Mobile Edge Computing-Enabled Base Stations. *IEEE Access* **2018**, *6*, 22622–22633.
29. Dhanya, N.M.; Kousalya, G.; Balarksihnan, P.; Raj, P. Fuzzy-logic-based decision engine for offloading iot application using fog computing. In *Handbook of Research on Cloud and Fog Computing Infrastructures for Data Science*; IGI Global: Hershey, PA, USA, 2018; Chapter 9, pp. 175–194.
30. Abdullah, L. Fuzzy multi criteria decision-making and its applications: A brief review of category. *Procedia Soc. Behav. Sci.* **2013**, *97*, 131–136.
31. Mehamel, S.; Slimani, K.; Bouzefrane, S.; Daoui, M. Energy-efficient hardware caching decision using Fuzzy Logic in Mobile Edge Computing. In Proceedings of the 6th International Conference on Future Internet of Things and Cloud Workshops, Barcelona, Spain, 6–8 August 2018; pp. 237–242.
32. Rout, R.R.; Vemireddy, S.; Raul, S.K.; Somayajulu, D.V.L.N. Fuzzy logic-based emergency vehicle routing: An IoT system development for smart city applications. *Comput. Electr. Eng.* **2020**, *88*, 106839.
33. OmKumar, C.U.; Bhama, P.R.K.S. Fuzzy based energy efficient workload management system for flash crowd. *Comput. Commun.* **2020**, *147*, 225–234.
34. An, J.; Hu, M.; Fu, L.; Zhan, J. A novel fuzzy approach for combining uncertain conflict evidences in the Dempster-Shafer theory. *IEEE Access* **2019**, *7*, 7481–7501.
35. Li, G.; Zhou, H.; Feng, B.; Li, G.; Li, T.; Xu, Q.; Quan, W. Fuzzy theory based security service chaining for sustainable mobile-edge computing. *Mob. Inf. Syst.* **2017**, *2017*, 1–13.
36. Nguyen, V.D.; Khanh, T.T.; Nguyen, T.D.T.; Hong, C.S.; Huh, E.-N. Flexible computation offloading in a fuzzy-based mobile edge orchestrator for IoT applications. *J. Cloud Comput.* **2020**, *9*, 1–18.
37. Soleymani, S. A.; Abdullah, A.H.; Zareei, M.; Anisi, M.H.; Rosales, C.V.; Khan, M.K.; Goudarzi, S. A secure trust model based on fuzzy logic in vehicular ad hoc networks with fog computing. *IEEE Access* **2017**, *5*, 15619–15629.

38. Sonmez, C.; Ozgovde, A.; Ersoy, C. Fuzzy Workload Orchestration for Edge Computing. *IEEE Trans. Netw. Serv. Manag.* **2019** *16*, 769–782.
39. Dernoncourt, F. *Introduction to Fuzzy Logic*; Massachusetts Institute of Technology: Cambridge, MA, USA, 2013; pp. 1–21.
40. Mendel, J.M. Fuzzy logic systems for engineering: A tutorial. *Proc. IEEE* **1995**, *83*, 345–377.
41. Sonmez, C.; Ozgovde, A.; Ersoy, C. EdgeCloudSim: An environment for performance evaluation of Edge Computing systems *Trans. Emerg. Telecommun. Technol.* **2018**, *29*, 1–17.
42. Silva, M.; Freitas, D.; Neto, E.; Lins, C.; Teichrieb, V.; Teixeira, J.M. Glassist: Using Augmented Reality on Google Glass as an Aid to Classroom Management. In Proceedings of the 2014 XVI Symposium on Virtual and Augmented Reality, Piata Salvador, Brazil, 12–15 May 2014; pp. 37–44.
43. Guo, J.; Song, B.; He, Y.; Yu, F.R.; Sookhak, M. A Survey on Compressed Sensing in Vehicular Infotainment Systems *IEEE Commun. Surv. Tutor.* **2017**, *19*, 2662–2680.
44. Tunca, C.; Pehlivan, N.; Ak, N.; Arnrich, B.; Salur, G.; Ersoy, C. Inertial Sensor-Based Robust Gait Analysis in Non-Hospital Settings for Neurological Disorders. *Sensors* **2017**, *17*, 825.
45. Chen, X.; Jiao, L.; Li, W.; Fu, X. Efficient Multi-User Computation Offloading for Mobile-Edge Cloud Computing. *IEEE/ACM Trans. Netw.* **2016**, *24*, 2795–2808.
46. Dinh, T.Q.; Tang, J.; La, Q.D.; Quek, T.Q.S. Offloading in Mobile Edge Computing: Task Allocation and Computational Frequency Scaling. *IEEE Trans. Commun.* **2017**, *65*, 3571–3584.
47. Liu, F.; Huang, Z.; Wang, L. Energy-Efficient Collaborative Task Computation Offloading in Cloud-Assisted Edge Computing for IoT Sensors. *Sensors* **2019**, *19*, 1105.
48. Chen, M.; Hao, Y. Task offloading for mobile edge computing in software defined ultra-dense network. *IEEE J. Sel. Areas Commun.* **2018**, *36*, 587–597.
49. Huang, L.; Feng, X.; Zhang, L.; Qian, L.; Wu, Y. Multi-Server Multi-User Multi-Task Computation Offloading for Mobile Edge Computing Networks. *Sensors* **2019**, *19*, 1446.
50. Li, S.; Tao, Y.; Qin, X.; Liu, L.; Zhang, Z.; Zhang, P. Energy-Aware Mobile Edge Computation Offloading for IoT Over Heterogenous Networks. *IEEE Access* **2019**, *7*, 13092–13105.
51. Wei, X.; Wang, S.; Zhou, A.; Xu, J.; Su, S.; Kumar, S.; Yang, F. MVR: An Architecture for Computation Offloading in Mobile Edge Computing. In Proceedings of the 2017 IEEE International Conference on Edge Computing (EDGE), Honolulu, HI, USA, 25–30 June 2015; pp. 232–235.

sensors

MDPI

Article

Dynamically Controlling Offloading Thresholds in Fog Systems

Faten Alenizi * and Omer Rana

School of Computer Science and Informatics, Cardiff University, Cardiff CF24 3AA, UK; RanaOF@cardiff.ac.uk
* Correspondence: AleniziF@cardiff.ac.uk

Abstract: Fog computing is a potential solution to overcome the shortcomings of cloud-based processing of IoT tasks. These drawbacks can include high latency, location awareness, and security—attributed to the distance between IoT devices and cloud-hosted servers. Although fog computing has evolved as a solution to address these challenges, it is known for having limited resources that need to be effectively utilized, or its advantages could be lost. Computational offloading and resource management are critical to be able to benefit from fog computing systems. We introduce a dynamic, online, offloading scheme that involves the execution of delay-sensitive tasks. This paper proposes an architecture of a fog node able to adjust its offloading threshold dynamically (i.e., the criteria by which a fog node decides whether tasks should be offloaded rather than executed locally) using two algorithms: dynamic task scheduling (DTS) and dynamic energy control (DEC). These algorithms seek to minimize overall delay, maximize throughput, and minimize energy consumption at the fog layer. Compared to other benchmarks, our approach could reduce latency by up to 95%, improve throughput by 71%, and reduce energy consumption by up to 67% in fog nodes.

Keywords: fog computing; computational offloading; dynamic offloading threshold; resource management; minimizing delay; minimizing energy consumption; maximizing throughputs

Citation: Alenizi, F.; Rana, O. Dynamically Controlling Offloading Thresholds in Fog Systems. *Sensors* **2021**, *21*, 2512. https://doi.org/10.3390/s21072512

Academic Editors: Marco Picone and Taehong Kim

Received: 11 February 2021
Accepted: 31 March 2021
Published: 3 April 2021

1. Introduction

The number of IoT devices and their generated tasks are constantly growing, imposing a burden on cloud infrastructure, in particular if processing of these tasks must take place within Quality of Services (QoS) constraints [1,2]. The processing of these tasks in the cloud can trigger systems to suffer high communication latency, security issues, and network congestion [3]. This is due to the distance between IoT devices and cloud-hosted servers [4,5]. Fog computing has emerged to address limitation of processing IoT tasks at the cloud and ensure the processing of these tasks takes place within pre-defined time periods [6]. Fog computing is an intermediate layer situated between cloud and IoT devices that brings location awareness, low latency, and wide-spread geographical distribution for IoT devices [7,8]. It consists of limited-resource devices called fog nodes, providing storage, processing, and networking resources close to IoT devices where tasks are produced [8,9]. Fog computing was introduced by Cisco in 2012 [4,10]. With limited-resource devices used in fog systems, poor utilization of these resources would limit their benefit.

Computational offloading enables workload/computational tasks to be shared between IoT devices, fog nodes, and cloud servers [11–14]. When computational offloading occurs between fog nodes, this is called "fog cooperation" [15], in which overloaded fog nodes send part of their workload to other underloaded fog nodes to meet their QoS requirements [16,17]. Resource management can involve multiple factors, saving energy consumption in the fog environment is one of these factors, and is considered in this work. Integrating computational offloading and resource management is essential to effectively utilize fog resources [1].

In online dynamic fog systems, where uncertainties are arising due to multiple factors, with no prior awareness of task arrival rate, the number of connected IoT devices, and computational capacity of fog nodes, addressing computational offloading and resource management is challenging to obtain optimum outcomes [1]. Computational offloading has mostly been explored in offline fog systems, where all system data are known beforehand, and limited work has been carried out in online dynamic fog systems. There is also limited work on understanding the impact of dynamically changing the offloading threshold which is a factor that determines when a fog node begins sharing its workload with other neighboring fog nodes within its proximity.

1.1. Contributions

Specifically, this work provides the following contributions:

- We propose a fog node architecture that dynamically decides whether to process the received tasks locally or offloads them to other neighbors. This is based on a dynamic threshold that considers the queuing delay of the primary fog node and the availability (i.e., the queuing delay) of its neighbors.
- Computational offloading and the associated computational resource management was investigated using an online dynamic system with the aim to solve the multi-objective problem that aims to minimize delay, minimize energy consumption, and maximize throughput.
- We conducted extensive experiments to evaluate the performance of our proposed scheme and compare our proposed algorithm to various benchmarks.
- This paper extends our previous work [1] by introducing a dynamic offloading threshold, made use of in an online model for evaluating service delay.

1.2. Paper Organization

The remainder of this paper is organized as followed. Related work is provided in Section 2, followed by the system model and associated constraints in Section 3. In Section 4, we decompose the multi-objective problem into two sub-problems: delay minimization and energy saving, followed by a description of our solution in Section 5. In Section 6, we compare the performance of our proposed scheme against other benchmarks, followed by conclusions in Section 7.

2. Related Work

This section is divided into three main parts. The first focuses on computational offloading between entities within a specific system; the second addresses the impact of dynamically managing servers to enhance power efficiency. Finally, a comparison of state-of-the-art of related approaches in fog computing is provided and summarized in Table 1.

Table 1. Computational Offloading State-Of-Art Comparison.

Ref	Offloading Deployment	Architecture Model				Fog Cooperation	Offloading Threshold	Communication Type		Objectives						Evaluation Tool
										Delay	Energy				Throughput	
		IoT-Fog	IoT-Fog-Cloud	Fog-Cloud	Fog			Vertical	Horizontal		Yes	Relation with Delay	Which Energy	Which Device		
Tang et al. [18]	Offline	✓	-	-	-	X	static	✓	X	✓	✓	Energy constraint	Energy for data processing and data transmission	IoT	X	MATLAB
Wang and Chen [19]	Offline	✓	-	-	-	X	static	✓	X	✓	✓	Energy constraint	Energy for local processing, processing at fog nodes, transmitting tasks	IoT devices and Fog nodes	X	Simulation
Liu et al. [11]	Offline	-	✓	-	-	X	static	✓	X	✓	✓	Trade-off	Energy spent by local processing and transmitting tasks	Mobile devices	X	Simulation
Mukherjee et al. [20]	Offline	-	✓	-	-	✓	static	✓	✓	✓	X	X	X	X	X	Monte Carlo simulations
Zhu et al. [13]	Offline	-	✓	-	-	X	static	✓	X	✓	✓	Offloading policy is designed to minimize task execution delay and to save energy	The energy spent by uploading and receiving tasks	Mobile devices	X	Simulation (MATLAB)
Mukherjee et al. [21]	Offline	-	✓	-	-	✓	static	✓	✓	✓	✓	Minimize total systems cost which includes total energy consumption & total processing delay	Energy consumed by local computing and uploading tasks to fog nodes	End user devices	X	MATLAB
Chen and Hao [14]	Offline	✓	-	-	-	X	static	✓	X	✓	✓	Battery capacity	Energy for transmitting tasks to edge devices and for local processing	End user devices	X	Simulation

Table 1. *Cont.*

Ref	Offloading Deployment	Architecture Model				Fog Cooperation	Offloading Threshold	Communication Type		Objectives						Evaluation Tool
										Delay	Energy				Throughput	
		IoT-Fog	IoT-Fog-Cloud	Fog-Cloud	Fog			Vertical	Horizontal		Yes	Relation with Delay	Which Energy	Which Device		
Sun et al. [22]	Online	-	✓	-	-	X	static	✓	X	✓	✓	Selects the lowest overhead cost which involves total computational time plus total energy spent by either processing task at IoT devices or transmission tasks to fog or cloud	Processing and transmission power	IoT, fog nodes, cloud servers	X	iFogSim
Zhao et al. [12]		-	✓	-	-	X	static	✓	X	✓	✓	Minimizes the system cost which is the total offloading latency and the total energy consumption	Transmission and processing energy	Whole System: end user devices, fog nodes, and cloud servers	X	Simulation
Meng et al. [23]		-	✓	-	-	X	static	✓	X	✓	✓	Minimizing energy given delay constraint	Transmission + computational energy	Mobile terminal, fog nodes, and cloud servers	X	Simulation
Xiao and Krunz [8]		-	-	✓	-	✓	static	✓	✓	✓	✓	Trade-off	Transmission energy	Fog nodes	X	Simulation
Yousefpour et al. [16]		-	✓	-	-	✓	static	✓	✓	✓	X	X	X	X	X	Simulation
Yin et al. [24]		-	-	✓	-	X	static	✓	X	✓	X	X	X	X	✓	Simulation
Al-Khafajiy et al. [15]		-	✓	-	-	✓	static	✓	✓	✓	X	X	X	X	X	MATLAB-based simulation
Gao et al. [17]		-	-	✓	-	✓	static	✓	X	✓	✓	Trade-off	Processing and transmitting tasks between fog nodes	Fog nodes	X	Simulation

Table 1. *Cont.*

Ref	Offloading Deployment	Architecture Model				Fog Cooperation	Offloading Threshold	Communication Type			Objectives					Evaluation Tool
											Energy					
		IoT-Fog	IoT-Fog-Cloud	Fog-Cloud	Fog			Vertical	Horizontal	Delay	Yes	Relation with Delay	Which Energy	Which Device	Throughput	
Mukherjee et al. [25]		-	-	-	✓	✓	static	X	✓	✓	X	X	X	X	X	Simulation
Alenizi and Rana [1]		-	-	✓	-	✓	static	✓	✓	✓	✓	X	POWERING ON and processing tasks	Fog nodes	✓	iFogSim
The proposed approach		-	-	✓	-	✓	Dynamic	✓	✓	✓	✓	X	POWERING ON and processing tasks	Fog nodes	✓	iFogSim

2.1. Computational Offloading

Computational offloading can be implemented offline or online. In offline implementation, all system information needed to make the offloading decision is previously known and is based on historical or predictive knowledge, such as the computational capabilities of fog nodes, the total number of IoT devices, and their total workload (number of requests). This is applied during the system design stage. In online deployment the computational offloading decision takes place at run-time and considers the current system status and process characteristics, such as the current waiting time and the current available computational resources, without prior knowledge of system inputs considered in the offline deployment. Several studies investigate computational offloading in offline deployment [8,11–14,18–23]. In [19], Wang et al. investigated the optimized offloading problem to minimize task completion time given tolerable delay and energy constraints The optimization problem was formulated as a mixed integer nonlinear programming problem that jointly optimizes the local computation capability for IoT devices, the computing resource allocation of fog nodes and the offloading decision. Wang et al. [19] decomposed it into two independent sub problems to find the optimal amount of workload that should be processed locally at IoT devices and at fog nodes. A hybrid genetic-simulated annealing algorithm was developed to optimize the offloading decision. Tang et al. [18] aimed to increase the total number of executed tasks on IoT devices and fog nodes under deadline and energy constraints. The authors in [18] considered this as a decentralized, partially observable offloading optimization problem in which end users are partially aware of their local system status, including the current number of remaining tasks, the current battery power, and the nearest available fog node. Such parameters are used to assess if tasks should be processed locally or offloaded to the nearest fog node. Their approach enables IoT devices to make an appropriate decision based on its locally observed system.

Liu et al. [11] addressed a multi-objective optimization offloading problem in a fog environment with the aim of minimizing execution delay, energy consumed at mobile devices, and offloading payment cost for using fog/cloud resources. The multi-objective problem was formulated into a single problem using scalarization method [11]. The proposed solution found the optimal offloading probability that accomplishes the stated objectives. Mukherjee et al. [20] designed an offloading technique focusing on jointly optimizing the computing and communication resources at fog systems to reduce end-to-end latency. Their technique considers the trade-off between transmission delay and task execution delay when making the offloading decision, in which a fog node can seek additional computational resources from either one of its neighbors, or the cloud data center, to reduce task execution delay at the expense of the transmission delay. The optimization problem was transformed into convex quadratically constraint quadratic programming and solved using CVX toolbar, which is a MATLAB-based modelling system for convex optimization. Their simulation results demonstrated that their proposed solution offers minimal end-to-end latency in comparison to executing all tasks at end-user devices and executing all tasks at the primary fog nodes.

Zhu et al. [13] proposed a task offloading policy based on execution time and energy consumption. This approach helps mobile devices to make an appropriate decision on whether to process their tasks locally or offload them to a fog node, or the cloud. During the decision-making procedure, mobile devices calculate both the execution time and the energy consumed when executing the task on the local device and compare this with the execution time and the energy consumed when offloading and receiving the processed task on a fog node; the energy consumed when executing the tasks on fog nodes are not considered. Based on this comparison, the IoT device makes a decision with the least cost (execution time plus energy consumption). Comparing their scheme to Random, no offloading, and only offloading when considering only execution time, their simulation results showed an optimization of the execution time of tasks and energy consumption of mobile devices. Mukherjee et al. [21] formulated the offloading problem as an optimization problem with the goal to minimize the total system cost, which is the sum of the total delay of end-users' tasks and the total energy consumed at end-users' devices due to local processing of tasks and uploading tasks to the fog environment for processing. Under delay and energy constraint, the optimization problem was transformed into a quadratically constraint quadratic programming problem and solved by semidefinite relaxation method. Within a heterogeneous environment where fog nodes have different computational resources, the proposed solution enables the optimal amount of workload to be identified that should be processed at end-user devices, primary and neighboring fog nodes, and cloud servers. The decision on when to offload depends entirely on the availability of computational resources. The authors stated that having higher computational resources at fog nodes helps to reduce the system cost. Increasing number of end-users leads to greater congestion at fog nodes, leading to fog nodes preferring to send their workload to the cloud server for processing rather than other neighboring fog nodes.

Chen and Hao [14] studied offloading problem in dense software-defined networks, formulating this as a mixed-integer nonlinear problem that is decomposed into: (i) deciding whether the task is processed locally at the end-user device or offloaded to the edge device; (ii) determining the computational resources that are dedicated to each task. Chen and Hao [14] developed an efficient software-defined task offloading scheme to solve these sub-problems. The results of their proposed scheme demonstrated the superiority of their approach at decreasing end user device energy consumption and overall task execution latency. In IoT-Fog-Cloud architecture, Sun et al. [22] presented the "ETCORA" algorithm, which consists of two parts. The first part aims to find the optimal offloading decision based on minimizing time and energy, and the second part optimizes resource allocation in terms of transmission power allocation. Their proposed solution helps to minimize energy consumption and completion time of tasks compared to other schemes. Zhao et al. [12] investigated the computational offloading problem in the context of radio access networks to reduce the weighted sum of total offloading latency plus total energy consumption. To improve the offloading decision and enhance the allocation of computation and radio resources, the authors formulated the problem as a non-linear, non-convex joint optimization problem. Their proposed solution was more effective than mobile cloud computing (MCC), which processes all end-user tasks on a cloud server, and mobile edge computing (MEC), which processes all end-user tasks in the edge computing system. The reason their approach was more effective compared to MCC and MEC is that it made use of a combination of available resources at the cloud and fog nodes, compared to *cloud only* as in MCC, and edge only as in MEC.

The hybrid-computational offloading optimization problem was investigated by Meng et al. [23], where two types of models were considered; namely cloud computational offloading and fog computational offloading. The authors aimed to minimize the consumption of energy caused by transmitting and processing tasks at mobile terminals, fog, and cloud servers under deadline constraints. Meng et al. [23] introduced a new concept called *computation energy efficiency* that is defined as "the number of computation tasks that are offloaded by consuming a unit of energy", to solve the optimization problem. Based on the proposed solution that considers offloading tasks to fog and cloud servers for execution, simulation results show the effectiveness of the solution compared to only offloading tasks to either cloud only or fog only resources. Xiao and Krunz [8] proposed a workload scheduling method that ensures user response time is minimized under available power constraints. In their study, the energy spent while processing tasks was ignored and only the energy consumed for offloading each unit of received workload was considered. Cooperation between fog nodes to offload workload by an agreement between neighboring nodes, the workload arrival rates, and the workload processing capabilities determines the amount of offloading carried out. Their experimental results indicated that the average response time decreased due to allowing cooperation between fog nodes. Additionally, a crucial trade-off between the fog node's power efficiency and the average response time was observed. Xiao and Krunz [8] proposed that the response time of end-user tasks should be set to its highest tolerable point to optimize energy consumption at fog computing systems. This enables most of the tasks to be processed at end-user devices, avoiding any offloading.

Regarding online deployment of computational offloading, few studies have addressed this, such as [15–17,24–26]. Yousefpour et al. [16] suggested a delay-minimization approach to reduce overall service delay. In their approach, the estimated queueing delay, which is utilized as the offloading threshold, determines whether a fog node processes its incoming task(s), or offloads these to one of its neighbors or the cloud server. If the offloading threshold has been reached, then the best neighboring fog node in its domain is selected to offload its upcoming tasks. The best neighboring fog node is chosen based on having the minimum total of propagation delay and queuing delay. Compared to other models, their results achieved the minimum average service delay. Yin et al. [24] determined where to process end user tasks into task scheduling and resource allocation problems, where tasks are either processed locally at end-user devices or offloaded to fog nodes or cloud servers. In an intelligent manufacturing environment, the authors introduced fog computing and utilized the concept of the container within the fog system, intending to reduce overall delay and optimize the number of concurrent tasks for the fog node. In their online model, generated tasks by end-users are transmitted to the request evaluator, which is located at a fog node that decides whether to accept or reject the task based on its deadline requirement. If the task is accepted, then the task is transmitted to the task scheduler, which determines whether the task is processed at fog nodes or cloud servers based on the available resources and the execution time of this task, which involves computation and transmission time. Finally, the resource manager is responsible for reallocating the required resources to process the task at fog nodes. Experimental results showed the effectiveness of their approach compared to other benchmarks.

Al-Khafajiy et al. [15] proposed an offloading mechanism that allows fog-to-fog collaboration in heterogeneous fog systems, intending to minimize overall service latency. Their mechanism utilizes a FRAMES load balancing scheme that aims to detect congested fog devices, determine the amount of workload located at fog devices' queues that require offloading, based on their deadline requirement, and finally select the best fog node that provides the minimal service latency for the selected workload. They evaluated their proposed mechanism using a simulation. Their numerical results indicated the effectiveness of their proposed model in terms of minimizing overall latency in comparison with different algorithms. In a fog-cloud computing system, Gao et al. [17] investigated the issue of dynamic computational offloading and resource allocation. In order to reduce energy consumption and delay while having a stable queueing status, the authors formulated the problem as a stochastic network optimization problem. They provided a predictive approach to computational offloading and resource allocation that depended on the trade-off between delay and energy use. Their approach implied that a delay reduction can be induced by increasing the allocation of computational resources at fog nodes; however, because of the processing of more tasks, energy consumption increases, and vice versa. Compared to other systems, the authors showed the importance of their method. Mukherjee et al. [25] developed a scheduling strategy that managed to fulfil the deadline constraint of end-user tasks, taking into account computational resources. The deadline constraint of a given task and the availability of a neighbor, in their scheduling policy, help to decide on whether to place a given task in the fog node queue, e.g., in its high-priority queue or low-priority queue, or offload it to one of its neighboring fog nodes. Their findings illustrated the efficacy of their suggested strategy as opposed to the no offloading and random schemes. Table 1 presents a summary of relevant articles concerning computational offloading at fog computing systems and the forms in which these systems execute.

2.2. Dynamic Server Energy Management

Dynamic Server Energy Management has been used in the wireless local area network and the cloud, and it has proven to be efficient in terms of improving power quality.

Although up to the time of our study, this has not yet been implemented in the fog area. In WLANs, the energy efficiency was enhanced by placing access points (APs) in sleep mode or turning them off. In [27], Marsan and Meo observed that in a community of APs, getting one AP in each community to control the system and service the incoming clients when all others are turned off will minimize energy consumption by up to 40 percent. Furthermore, an additional 60% of consumed energy can be saved if all APs are turned off, especially during idle periods, e.g., at night. Li et al. [28] suggested an energy-saving method for state transformations in which APs are not only turned on and off based on consumer requirements, but there is also an intermediary stage that aims to reduce the frequency of switching. The authors stated that increasing the switching frequency will shorten AP's service life. In addition to that, the intermediary stage will help to avoid latency and energy overhead caused by switching on APs.

It was suggested that servers could be periodically switched off [29,30] or placed into sleep mode [31–33] in cloud computing systems to conserve energy resources. In [29–33], the authors examined the issue of the placement of virtual machines (VMs) to save resources concerning energy and yet retain QoS. When underutilized data centers are detected, all VMs are migrated to other active data centers, and these underutilized data centers are placed in sleep mode according to [31–33] or shutdown as per [29,30]. This is intended to reduce the consumption of energy at cloud computing systems and is called 'VM consolidation'. Numerous VM migration approaches were suggested to assess which virtual machines can be migrated from overloaded data centers. Moreover, in order to satisfy the QoS specifications of the system, a switched-off data center may also be activated to handle the migrated VMs. According to Mahadevamangalam [31], the energy demand for an idle data center is ~70% percent of the energy generated by a fully utilized data center. Thus, by switching off idle-mode data centers, up to 70% of the energy consumed can be saved in the cloud system.

2.3. Comparison of the State-of-the-Art

Table 1 provides a summary of related work in computational offloading in fog computing systems, highlighting the architecture model, e.g., IoT-Fog means that end-user tasks are processed locally at IoT devices or offloaded and processed at fog nodes, use of fog cooperation, communication, the stated objectives of the work, and evaluation tools arranged by offline or on-line offloading decisions. Offline deployment helps to predict the best output for the system at its design stage; and online deployment mimics various scenarios in real-world environments, involving uncertainty and unpredictable events, and helps the system to produce a better outcome. However, most of the literature is focused on offline deployment. Additionally, the problem of computational offloading is usually investigated with the aim of minimizing an overall delay in the system; managing system resources is sometimes included, especially minimizing energy consumption of IoT/end user devices.

Managing resources in the system is much easier within offline deployment than online deployment, especially when all the system data is known in advance. In offline deployment, most attention has been given to addressing the energy consumed at IoT devices compared to fog devices. Additionally, when considering energy consumed at fog nodes, often the trade-off between delay and energy has been investigated.

In this work, we consider online deployment of computational offloading and the potential for minimizing energy consumption at fog nodes (compared to energy consumption of networks or cloud servers). Computational offloading and resource management at fog environments has received limited attention so far. When considering computational offloading, existing efforts utilize a fixed threshold that determines when to start offloading; in the current work, a dynamic threshold is investigated to address its impact on the system.

3. System Modelling and Constraints

Based on the model in [1], we describe an extended fog node architecture in Section 3.1.3.

3.1. System Model

The proposed model consists of one cloud server, 'N' fog nodes that are located at roadside units (RSU), a fog controller, and M vehicle nodes. Each vehicle node connects to the associated fog node through a wireless local area network, and the connection to the remote cloud server is via a wide area network. A single task contains the following data T = {Type, S_m, D_m, $Task^{CPU}$, $Task^{Network}$}, where Type is the category of task being considered (e.g., urgent or non-urgent); S_m, D_m respectively represent the source application module (from where the task is emitted) and the destination application module (where the task is heading); $Task^{CPU}$ indicates the computational complexity of the tasks, captured in number of instructions (Million Instructions Per Second (MIPS)); $Task^{Network}$ represents the size of the encapsulated data in the task that needs to be transmitted across the network. In iFogSim, the simulator used to model the system, tasks are represented as tuples. A network diagram is presented in Section 3.1.1 and the associated application module is described in Section 3.1.2.

3.1.1. Network Diagram

Figure 1 shows an illustration of the fog computing architecture, which comprises of three layers:

- **The IoT devices layer:** This layer is composed of mobile vehicles—represented as vehicle nodes, containing an actuator and a collection of sensors. Each sensor produces a task, labelling it as "non-urgent" or "urgent". Non-urgent tasks include data such as current position, speed, and path. Urgent tasks require a quicker response and can have stringent Quality of Service (QoS) requirements. This task may contain a video stream around a moving vehicle, requiring short latency or processing. This is necessary, for instance, in self-driving vehicles.
- **Fog computing layer:** This layer is comprised of a series of fog nodes and a fog controller. Fog nodes are located in RSU that are installed alongside a road. If fog nodes are situated in communication proximity of each other, they can interact and share data with each other [34]. Hence, fog nodes form an ad hoc network to exchange and share data. All fog nodes are linked to the fog controller, which is responsible for managing fog resources and controlling fog nodes. Fog nodes process two different types of tasks, urgent tasks are given priority and their processing results are sent back to the vehicle. For non-urgent tasks, fog nodes process these tasks and transfer the findings to the cloud for further analysis and storage, e.g., for retrieval by traffic management organizations.
- **Cloud computing layer:** This layer is composed of a set of cloud servers, hosted within one or more data centers. This layer is able to aggregate traffic information across a geographical area over time.

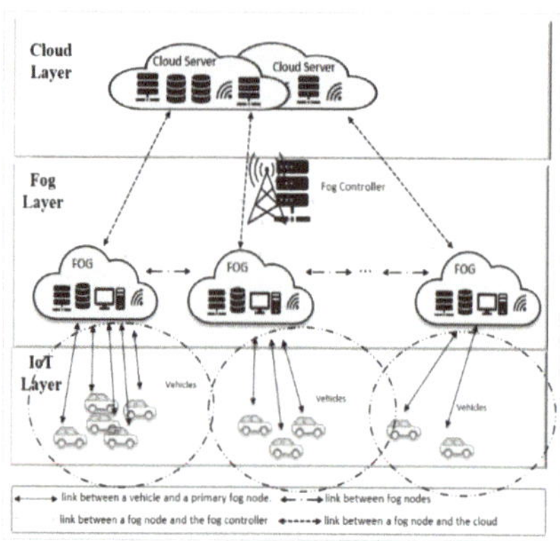

Figure 1. Fog Computing Model.

3.1.2. Application Module Description

Three modules: road control, global road monitor, and process priority tasks are part of the application model. The first two modules focus on traffic light management, while the last module is responsible for processing urgent tasks. The operations carried out by these modules are outlined below.

- **Road Monitor:** This module is placed at fog nodes. When a vehicle comes into communication proximity of a fog node, a sensor immediately sends data to the connected fog node for analysis. This data contains the current position, the speed of the vehicle, weather, and road conditions. After processing these data by the specified module, the results are transmitted to a cloud server for further processing.
- **Global Monitor:** this module is placed at a cloud data center, receiving data that has already been processed by the road monitor module.
- **Process Priority task:** this module is placed at fog nodes and is responsible for processing priority requests from a user. The results are then sent back to the user. An application in iFogSim is specified as a directed acyclic graph (DAG) = (M, E), where M represents the deployed application modules M = {m_1, m_2, m_3, ... , m_n}, e.g., process priority task, road monitor, and global road monitor modules. 'E' represents a set of edges describing data dependencies between application modules, as illustrated in Figure 2.

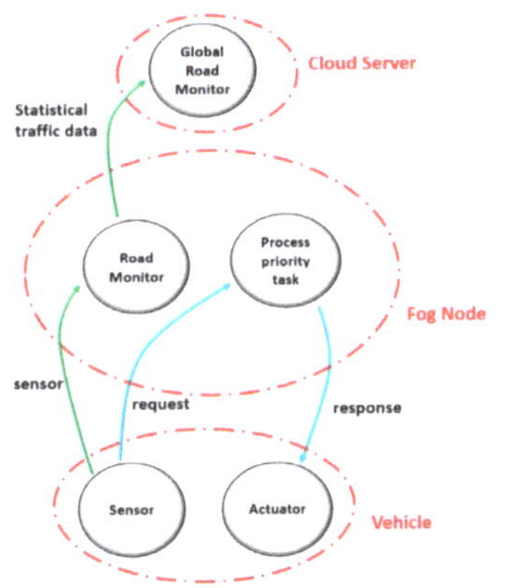

Figure 2. Directed Acyclic Graph (DAG) of the application model.

3.1.3. Fog Node Architecture

The proposed fog node architecture consists of a task scheduler, best neighbor selector, and threshold monitor (see Figure 3). Task scheduler receives tasks generated from IoT devices within the proximity of the primary fog node and from other neighboring fog nodes. If a fog node receives a task that is already offloaded from another neighbor, task scheduler immediately inserts this task in the processing queue. If the task is generated from other IoT devices, then task scheduler will check the offloading threshold and compare this to the queuing delay at the current node. If the queueing delay reaches the offloading threshold, then task scheduler sends this task to the best neighbor selection, which in turn decides the best neighbor node to offload this task to. The selection of the best neighbor is described in more detail in Section 3.2.2. Threshold monitor is responsible for dynamically increasing and decreasing the offloading threshold for both the primary fog node and all its neighbors, based on the workload and the availability of other neighbors: this is done from the perspective of the primary fog node. On the one hand, it is assumed that fog nodes are cooperative and accept tasks coming from their neighbor nodes, even if this exceeds their threshold. On the other hand, each neighbor has its own threshold monitor, and the primary fog node and all its neighbors may not have the same threshold value.

In Table 2, we can see that primary fog node A set its threshold to 9 ms for itself and all its neighbors. At the same time, primary fog node B in Table 3 set its threshold to 6 ms, even for its neighbor fog node A; therefore, it can be seen that fog node A is congested and will not be selected as the best neighbor for fog node B. Determining when to increase and decrease the offloading threshold is described in Section 5.

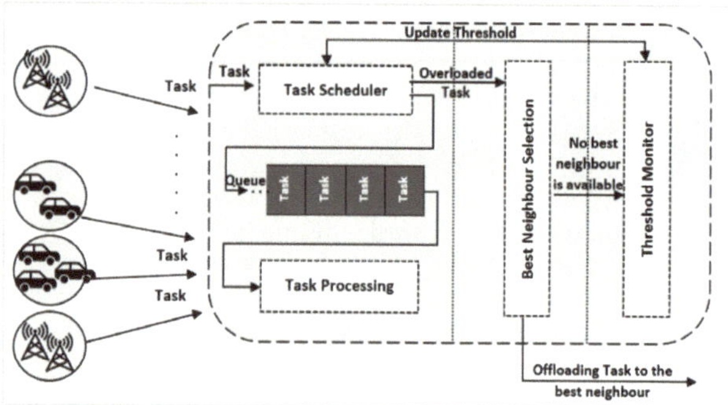

Figure 3. Fog Node Architecture Model.

Table 2. Example of Offloading Threshold set for Fog Node A and its neighbors.

Fog Node Type	Primary Fog Node	Neighboring Fog Nodes			
	Fog node A	Fog node B	Fog node C	Fog node D	Fog node E
Threshold	9 ms	9 ms	9 ms	9 ms	9 ms

Table 3. Example of Offloading Threshold set for Fog Node B and its neighbors.

Fog Node Type	Primary Fog Node	Neighboring Fog Nodes			
	Fog node B	Fog node A	Fog node F	Fog node G	Fog node H
Threshold	6 ms	6 ms	6 ms	6 ms	6 ms

3.2. *Types of Connections and Constraints*

This section explains the relations between a vehicle and a fog node, between fog nodes, and between fog nodes and cloud servers. Additionally, we also specify the restrictions that render these relations appropriate.

3.2.1. Connection between Vehicles and Fog Nodes

The interaction between a vehicle and a fog node is controlled by communication and processing restrictions.

- Communication Constraints

Each vehicle connects to a fog node if it is located within the communications coverage radius of that fog node, as specified in Constraint (1).

$$D_{v,f} \leq max\ Coverage_f;\ \forall\, v \in V,\ \forall\, f \in FN \tag{1}$$

where V represents all vehicles, v is a single vehicle, FN represents all fog nodes, and f is a single fog node. $D_{v,f}$ is the distance between a vehicle v and a fog node f, and calculated as:

$$D_{v,f} = \sqrt{\left(X_v - X_f\right)^2 + \left(Y_v - Y_f\right)^2};\quad \forall\, v \in V,\ \forall\, f \in FN \tag{2}$$

where (X_v, Y_v) and (X_f, Y_f) are the location of the coordinates of a vehicle v and a fog node f, respectively. When a vehicle is within a range of several fog nodes, it will connect to the nearest fog node. This is to decrease delay, as the propagation delay relies on the distance between the two connected nodes, propagation delay (PD) is calculated as

$$PD = \frac{D_{v,f}}{PS} \tag{3}$$

Signal propagation (PS) speed is assumed to be equivalent to the speed of light [35], i.e., $c = 3 \times 10^8$.

- Processing Constraints

To enable placement of application modules, fog nodes should have enough resources to meet the demands of these application modules.

$$\sum_{i=0}^{M} Required_{Capacity} m_i \leq \sum_{j=0}^{FN} Available_{Capacity} f_j;\ \forall\ m_i \in \mathrm{M},\ \forall\ f_j \in \mathrm{FN} \tag{4}$$

The required capacity of an application module and fog node is therefore captured using CPU, RAM, and bandwidth. Constraint (4) indicates that the overall needed capability of all application modules should not surpass the available capacity of the fog node on which they are installed. In the iFogSim simulator, if there is no available capacity at fog nodes, the application will be placed at cloud servers. The required CPU capacity to place an application module is calculated as followed:

$$CPU = NV \times (Rate \times TaskCPU) \tag{5}$$

where NV is the number of vehicles attached to the fog node, $TaskCPU$ is the number of instructions contained in each task, specified in Million Instructions Per Second (MIPS). Rate is calculated as:

$$Rate = \frac{1}{Transmission\ Time\ in\ ms} \tag{6}$$

In iFogSim, application module placement takes place at the design stage. Increasing the number of connected vehicles at a fog node will increase the required CPU requirement, to execute the required application modules. If the fog node does not have enough CPU capacity, these applications will be placed in the cloud. In this case the number of connected vehicles for each fog node is limited, as specified in Constraint (7).

$$\sum_{i=0}^{V} v_i f_j \leq\ MAX_{vehicles};\ \forall\ v_i \in V,\ \forall\ f_j \in\ FN \tag{7}$$

3.2.2. Connection between Fog Nodes

In this section, we describe the waiting queue for fog nodes, based on which an offloading decision is determined.

- Fog Node Waiting Queue.

All fog nodes contain a queue for arriving tasks, served on a sequential First In First Out (FIFO) basis. A queueing delay triggers the decision to begin offloading tasks from the arrival queue to neighboring fog nodes [16]. To begin offloading tasks, the queue waiting time should exceed the predetermined offloading threshold.

$$T^{Queue} > Offloading_{threshold}. \tag{8}$$

T^{Queue} is calculated as

$$T^{Queue} = \sum T_i \times\ T_i^{process} + \sum T_z \times\ T_z^{process};\ \forall\ i, z\ \ \in T \tag{9}$$

where T_i and T_z are the total number of tasks of the type i and z, urgent or non-urgent, respectively. T is all tasks and is the expected execution time of a specific task and calculated as

$$T^{process} = \frac{TaskCPU}{F\_MIPS \times N\,of\,PS} \tag{10}$$

where F_MIPS is the total computational capacity (measured in $MIPS$) available at a fog node, and 'N of PS' is the total number of processing units at that fog node.

- Coverage Method

  Fog nodes can have overlap in their coverage area [35], as illustrated in Figure 4.

Figure 4. Overlapping Fog Nodes.

- Selecting the Best Neighboring Fog Node

  Fog nodes form an ad hoc network between them to share and exchange data such as their queueing delay. Following [16], the best neighboring fog node is selected based on propagation delay plus queueing delay. The selection of the *best* neighboring fog node begins if the offloading threshold of a fog node is reached, determined by the waiting queue time. A fog node can communicate with neighbors in its coverage area, as specified in Constraint (11)

$$d_{ij} \leq Coverage_{radius};\ \forall\, i,\ j\ \in\ FN \tag{11}$$

where d_{ij} represents the distance between fog nodes i and j. In Figure 4, the neighboring fog nodes for FOG 1 are FOG 2 and FOG 3. Additionally, the neighboring fog nodes for FOG 3 are FOG 1, FOG 4, and FOG 5. The criterion for choosing the best neighboring fog node is based upon the coverage radius of the primary fog node, and the sum of queueing and propagation delay to the neighbor, where PD is calculated in (3) above.

$$Min\ \sum T^{Queue} + PD \tag{12}$$

3.2.3. Between Fog Nodes and the Cloud

We primarily focused on sharing workload with other neighboring fog nodes in preference to using cloud servers. This is attributed to the availability of other neighboring fog nodes as they overlap and to get maximum utilization of the available resources in the fog system. Task exchange between fog nodes and the cloud is only considered if the primary fog node and all its neighbors are congested, e.g., all their offloading thresholds reach the maximum threshold. We assume that cloud servers are much more efficient, and their queueing latency is ignored—i.e., tasks are processed immediately upon arrival at a cloud server [36–38]. The maximum offloading threshold is calculated as followed:

$$\text{Maximum threshold} = 2 \times \left(Transmission_{cloud}^{delay}\right) \tag{13}$$

4. Problem Formulation

The multi-objective problem of minimizing delay, maximizing throughput, and minimizing energy consumption is decomposed into two sub-problems [1]: (i) delay minimization and throughput maximization, and (ii) energy saving.

4.1. Delay Minimization and Throughput Maximization

The response time is the time required for sending the workload from a vehicle to the connected fog node and getting the results back. It consists of the transmission delay, propagation delay, queuing delay, and processing delay. If task processing takes place at the primary fog node, then the service latency is calculated as:

$$T = T^{sTv} + 2 \times (T_{vTf}^{Transmision} + PD_{vTf}) + T^{Queue} + T^{procss} + T^{vTa} \tag{14}$$

where T^{sTv}. and T^{vTa} are the latency time between a vehicle and its sensor, and between the vehicle and its actuator, respectively. $T_{vTf}^{Transmision}$ is transmission delay between the vehicle and its primary fog node. It is based on the network length of the task (i.e., its data size) and the available bandwidth, and is calculated as:

$$T^{Transmision} = \frac{Network\ Length\ of\ Task}{Bandwidth} \tag{15}$$

If a neighboring fog node is used to carry out processing of the received task, then latency is calculated as:

$$T = T^{sTv} + 2 \times\ T_{vTf}^{Transmsiion} + PD_{vTf} + 2 \times\ T_{fTf}^{Transmission}\ PD_{fTf}) + T^{Queue} + T^{Process}\ T^{vTa} \tag{16}$$

If the cloud is incorporated in the processing of the task, then the latency is calculated as:

$$T = T^{sTv} + 2 \times\ T_{vTf}^{Transmsiion} + PD_{vTf} + 2 \times\ T_{fTc}^{Transmission} + T^{Process}\ T^{vTa} \tag{17}$$

Throughput is measured as the percentage of the processed tasks as the following.

$$Throughputs = \frac{total\ number\ of\ processed\ tasks\ in\ the\ system}{total\ number\ of\ genertaed\ tasks\ in\ the\ system} \times 100 \tag{18}$$

4.2. Energy Saving

Minimizing the power consumption of fog nodes brings many advantages, including but not limited to decreasing the overall cost of electricity and reducing the environmental impact. Two power modes are presented for each fog node: idle and busy. In the idle mode, the fog node is not performing any processing, but the power is ON, and in the busy mode the fog node is processing tasks and power is ON. The energy consumed is determined by how much power the fog node consumes when processing workload and when the fog node is idle. The total energy consumption in iFogSim is calculated [39] as:

$$E = PR + (TN - LUT) \times LUP \tag{19}$$

where PR is the previously calculated total energy consumed at this fog node, TN is the time now, which is the time that the updateEnergyConsumption() is called when utilizing this fog node, updateEnergyConsumption() is a method located at Fog Device class in iFogSim, LUT is the last time this fog node has been utilized, and finally LUP is the last used power status (for either idle or busy period). The problem of minimizing delay and energy is formulated as followed:

$$\text{Min} \sum T \ \& \sum E,\ s.t.\ (1),\ (7)\ and\ (4)\ hold$$

$$T^{Queue} \leq Offloading_{threshold} \tag{20}$$

$$P_F + P_N = 1,\ P_F \,\&\, P_N = \{0, 1\} \tag{21}$$

Equation (1) ensures the connection between a fog node and a vehicle that is located within its coverage range. Equation (7) guarantees the number of vehicles connected to a fog node does not exceed the threshold number. Constraint (4) ensures the placement of the required application modules at fog nodes. Equation (20) ensures the stability of fog node queues so that to process incoming tasks, the waiting queue time should not exceed its threshold. In constraint (21), P_F (Primary Fog), and P_N (Primary Neighbor), i.e., $P_F = 1$ and $P_N = 0$ if the task is processed on the node where it is generated.

5. Proposed Algorithms

Two algorithms are proposed [1] called dynamic task allocation (DTA) and dynamic resource saving (DRS)—which need to be combined. Previously, these algorithms were applied to a static offloading threshold [1]. In the current work, we proposed a dynamic offloading threshold, in which the offloading threshold is adapted based on the workload and the availability of neighboring fog nodes, as described in Section 5.1.

5.1. Dynamic Offloading Threshold

The dynamic threshold is managed by the threshold monitor, which adjusts its value periodically according to the received workload and the availability of other neighbors, as described in Algorithm 1. The first part of the algorithm (Procedure 1) determines whether to increase the offloading threshold of the primary node and its neighbors. This runs each time a new task arrives at the primary fog node. It starts by checking if the current threshold exceeds the maximum offloading threshold calculated in Equation (13), if this occurs then the best decision for the arrival task is to be migrated to the cloud for processing. Otherwise, it checks whether the queuing delay of the primary fog node has reached its offloading threshold, i.e., to decide whether to process the task locally and add it to its queue or select the best neighbor with the least queueing delay as per lines 4–16. The current threshold is then updated using Equation (23) and Procedure 2 is called.

The second part of the algorithm (Procedure 2) determines whether the threshold should be decreased. This runs each time a new task is received, and when the fog node finishes the execution of a task. It starts by checking if the current threshold of the primary fog node is larger than a threshold, as per line 25. If this occurs, then the average queueing delay for all the neighbors is calculated as in (22) and the current threshold is updated, as per lines 26–27. The computational complexity of the proposed algorithm is O(n). Parameters used in this algorithm are fin Table 4.

$$VQ = \frac{\sum_{s=0}^{Ns} Qs}{Ns} \tag{22}$$

$$\delta^{n+1} = \begin{cases} \delta^n - p, & Q \geq \alpha,\ VQ < x \\ \delta^n, & Q \geq \alpha,\ VQ \geq x \\ \delta^n, & Q < \alpha \\ \delta^n + p, & Q \geq \delta^n,\ VQ \geq \delta^n \end{cases} \quad \forall\, x,\ \alpha,\ p > 0 \tag{23}$$

Table 4. Description of Parameters used for Dynamic Threshold Algorithm.

Symbol	Description
δ^n	Refers to the initial offloading threshold and the current threshold.
VQ	Average queueing delay of all the neighbors
δ^{n+1}	New offloading threshold.
x	$x = \delta^n/2$.
Ns	All neighboring fog nodes
Qs	Set of all queueing delay of all its neighbors
QNeighbours	Set of all neighbors and their queueing delay
α	When the queuing delay reaches this threshold, the fog node might consider decreasing its offloading threshold.
N	The best neighbor fog node
T	The arrival task
Q	Queuing delay in the primary fog node
p	A number bigger than zero that determines how much to modify the offloading threshold based on the current offloading threshold
Q_N	Queueing delay of one neighbor

Algorithm 1 Dynamic Offloading Threshold

Input: δ^n, $Q_{Neighbours}$=[(N_1,Q_1), (N_2,Q_2), …,(N_M,Q_M)], T, Q;
Initialization: δ^{n+1}= Ø, *max_threshold = 2*Latency;*
Result/s: δ^{n+1};

Procedure_1.
  IF (δ^n<*max_threshold*)
    IF (Q >= δ^n)
      Best ← N_1
      min ← Q_1
      For each $n \in Q_{Neighbours}$ **do**
        IF (Q_i < Q_1) **then**
          Best ← N_i
          min ← Q_i
        end if
      end for
      Offload (T,*Best*)
      Update(δ^n) using Equation (23)
    Else
      Queue.add(T)
      Update(δ^n) using Equation (23)
    end if
  else
    migrate to cloud
  end if
  return δ^{n+1}
  Call procesdure_2
end Procedure_1.
Procedure_2.
  IF (δ^n >= α) **then**
    Calculate *VQ* using Equation (22)
    Update(δ^n) using Equation (23)
  end if
  return δ^{n+1}

5.2. Dynamic Resource Saving

We used the algorithm from [1] to reduce energy consumption at fog nodes by dynamically switching ON/OFF fog nodes.

6. Experimental Results

In this work, iFogSim was used to simulate the environment. It was a toolkit developed by Gupta et al. [40], as an extension of the CloudSim simulator. It was a toolkit allowing the modelling and simulation of IoT and fog environments and can monitor various performance parameters, such as energy consumption, latency, response time, cost, etc Simulation settings values were used as in [1]. The simulation was run with one cloud server, seven fog nodes, a fog controller, and a total of 50 vehicles. Each vehicle transmitted two different tasks from two sensors every 3ms. In iFogSim, the workload was represented as tuples, generated from vehicle nodes, and the following main classes were considered FogDevice class was used to define the main characteristics of fog nodes and cloud servers, including RAM size, processor capacity in MIPS, uplink and downlink bandwidth, idle and busy power. Sensor class in the FogDevice class represented the attributes of a vehicle sensor, such as the vehicle node id to which the sensor is connected and the latency between them. Tuple class was used to represent computational tasks. The metrices used to measure the performance were:

- **Service Latency** is the average round trip time for all tasks processed in the fog environment. Two control loops were used in the simulation: **Control loop A:** *Sensor -> Process Priority Tasks -> Actuator*. This control loop represented the path of priority requests. **Control loop B:** *Sensor -> Road Monitor -> Global Road Monitor*. This control loop represented the path of non-priority requests.
- **Throughput**, which was measured as the percentage of processed tasks within a time window.
- **Total Energy Consumption** in fog environment caused by powering on fog nodes and processing tasks.

6.1. Performance Comparisons with Various Computation Offloading Schemes

To evaluate the effectiveness of our proposed algorithm, the comparisons with various computation offloading schemes were provided, where the number of vehicles was set to 50 and total number of fog nodes was set to 7. Although we implemented uncertainty within the system to mimic real world scenarios, we maintained the number of total generated tasks, the capacity of fog nodes, and the size of the generated tasks to be identical for fair comparison. In particular, the following four schemes were selected as benchmarks:

Benchmark 1: *No Offloading Scheme (NO):* in this scheme, each primary fog node processes all the tasks without cooperation with other neighboring fog nodes.

Benchmark 2: *Joint Task Offloading and Resource Allocation Scheme (JTORA)* [20]: in this scheme, if the primary fog node does not have enough computational resources that meet the delay requirement of a task, then the task will be offloaded to a neighboring fog node within the proximity of the primary fog node that has enough computational resources. Any underutilized neighbor is a candidate of processing the overload, ignoring the selection of the least utilized fog node. In this scheme, a static threshold is applied.

Benchmark 3: *Workload Offloading Scheme (WO)* [26]. In their work, end users offload their computational tasks to a broker node that manages the system, the broker node will send tasks to a fog node closest to end users (primary fog node). If the primary fog node is congested (e.g., its queueing delay reaches 50 ms), then the broker node will offload the task to any underutilized neighboring fog node. In this scheme, a static threshold is determined.

Benchmark 4: *Static Threshold 50ms Scheme (ST50)* [1]: where offloading threshold is set to 50 ms, upon which the primary fog node makes the decision on whether to process the task locally or offload it to the best neighboring fog node. The four benchmarks are compared to the proposed offloading policy called Dynamic Threshold (DT).

6.2. *Impact of the Proposed Scheme and Different Offloading Schemes on Delay and Throughputs*

In Figure 5, the impact of various offloading schemes on average latency is addressed. It can be seen in Figure 5 that delay was very high in the no offloading scheme; this was due to a long queueing delay as tasks were not shared by the primary node with other neighboring fog nodes, so they were waiting to be executed by the primary fog node. The impact of allowing cooperation between fog nodes in terms of sharing workload is shown, comparing other schemes to the no offloading scheme. Additionally, the impact of selecting which neighbor to share the workload with was very clear when comparing the WO, JTORA, ST50, and DT schemes. In the WO and JTORA schemes, when the primary fog node was congested (e.g., reaching its offloading threshold), it selected any underutilized neighbor to share the workload, rather than selecting the least utilized neighbor, as in ST50 and DT.

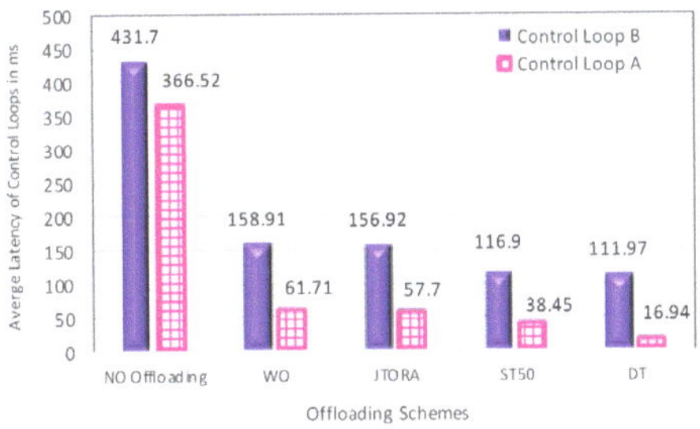

Figure 5. The Comparison of the Average Latency with Various Offloading Schemes.

In addition, the delay was higher in the WO scheme compared to JTORA; this was due to a communication overhead caused by sending tasks to a broker node first, which in turn decides whether to process these tasks at the primary fog node or any underutilized neighbor. The least delay was achieved for both control loops when applying our proposed algorithm, DT, compared other benchmarks.

The impact of various offloading schemes on throughput is shown in Figure 6. It can be seen that the lowest percentage of processed task was when no offloading was applied; this was obvious as most of the tasks were waiting in the queue to be executed by the primary fog nodes. The impact of sharing workload with any underutilized neighbors was very clear in WO and JTORA schemes, resulting in processing almost 90.88 and 91.06% of tasks, respectively, compared to 94.56 and 95.67% in ST50 and DT schemes, respectively.

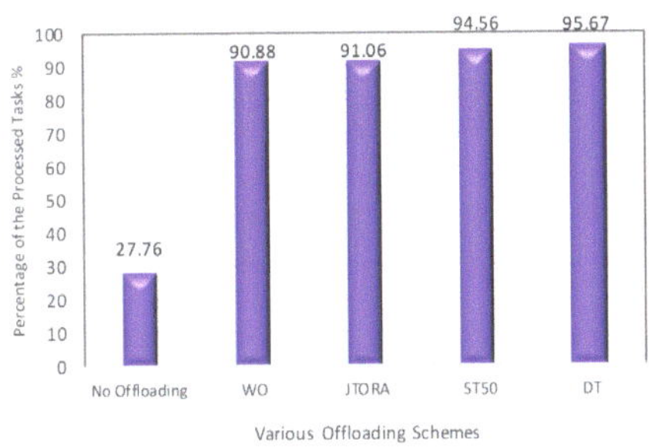

Figure 6. The Comparison of the Throughputs with Various Offloading Schemes.

6.3. *Impact of Increasing Number of Vehicles on Delay and Throughputs with Different Offloading Schemes*

The impact of increasing the number of vehicles was to investigate how the delay was maintained as we increased the workload in the online system. In this experiment, the total number of fog nodes was set to seven and the number of vehicles ranged from 4 to 48. In Figure 7, we can observe that when the number of vehicles was small, between 4 to 12 vehicles, the DT, ST50 and JTORA schemes exhibited an identical pattern. This was because the generated workloads were small, resulting in the primary fog nodes processing most of these workloads themselves. When the number of vehicles increased, all three approaches, ST50, JTORA, and WO, showed a dramatic increase in delay compared to DT, which displayed a stable pattern with a slight increase in delay that increased as the number of vehicles increased.

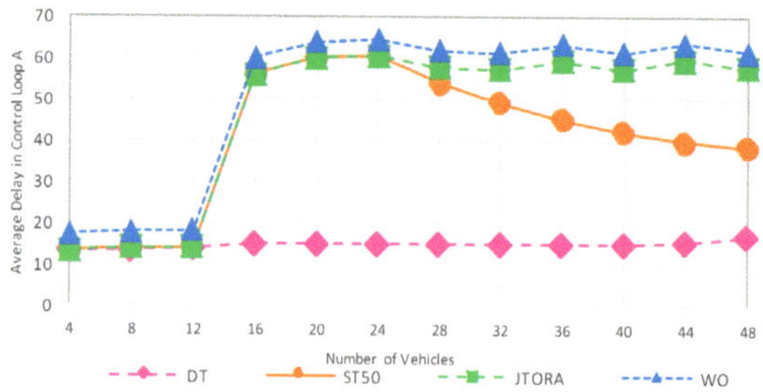

Figure 7. Impact of Increasing Number of Vehicles on Average Delay with Different Offloading Schemes.

The reason for the huge increase in delay for ST50, JTORA, and WO was that increasing the workload made the primary fog nodes almost reach their offloading threshold (e.g., 50 ms), but not always exceeding it, resulting in the primary nodes processing most of the workload with little help from neighboring nodes. The impact of selecting the best neighbor to share the workload becomes clear when the number of vehicles is high (i.e., 28 vehicles). The overall results show the effectiveness of the DT scheme even when increasing the number of vehicles.

The impact on throughput was also investigated while increasing the number of vehicles. When there was a small number of vehicles, ranging from 4 to 12, all the offloading schemes operated in a similar way; this was because the workload was minimal and can be processed at the primary fog nodes without using capacity of neighbors. When the number of vehicles was increased, DT achieved the highest throughput, with ~96% compared to other schemes, which accomplished 94.5, 91, and 90% for ST50, JTORA, and WO, respectively.

6.4. *Impact of Increasing Number of Neighbors on Delay, Throughputs, and Energy with Various Offloading Schemes*

The impact of increasing the number of neighbors was carried out to investigate its impact on overall system performance and to find the optimal number of neighbors that are required. From Figure 8, we observe that as the number of neighbors was increased, the delay decreased for both control loops. However, when a certain number of neighbors was reached (e.g., five neighbors), the delay remained almost stable despite adding further neighbors. This means that the optimal number that is required to achieve minimum delay was reached, and no additional neighbors were needed to save energy consumption of the fog paradigm. The reason for the stable pattern was attributed to the workload, as most of the generated tasks were processed.

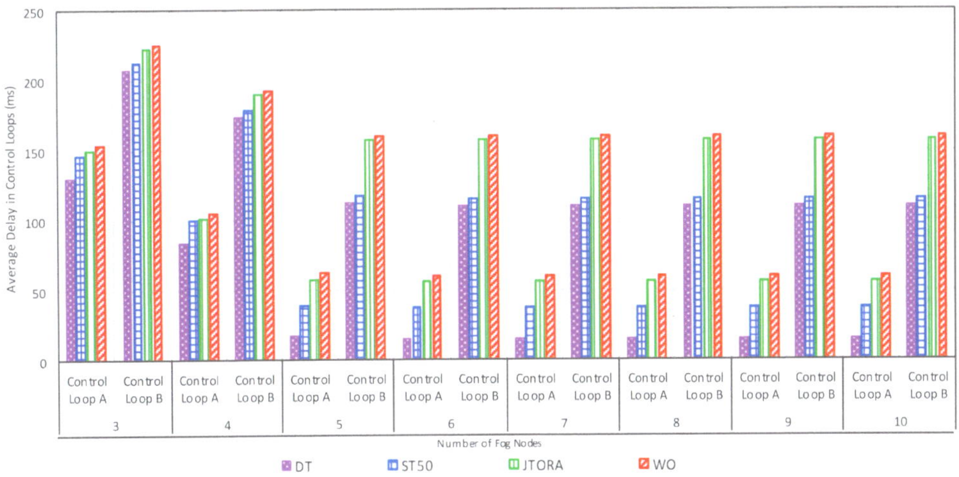

Figure 8. Impact of Increasing Number of Neighbors on Average Delay in Control Loops with Various Offloading Schemes.

We note that DT accomplished the least delay for both control loops as the number of neighbors was increased, compared to the other schemes: ST50, JTORA, and WO. With three neighbors, DT decreased delay by 10.80, 13.38, and 15.29% compared to ST50, JTORA, and WO, respectively. When the number of neighbors was five, the DT scheme reduced delay by 55.94, 70.64, and 72.55% in comparison to ST50, JTORA, and WO, respectively.

Increasing the number of neighbors on throughput showed a similar pattern as increasing the number of neighbors to decrease delay. As the number of neighbors increased, the percentage of processed tasks increased, until a certain number of neighbors was achieved (e.g., five neighbors), after which the pattern remained almost stable. The reason behind the stable pattern was due to the workload, as most of the generated tasks were processed. The reason why the percentage of processed tasks did not reach 100% was that this study implemented an online dynamic system, therefore vehicles were still generating tasks until the end of the simulation; 5% of the total generated tasks were not processed because they were newly generated.

In terms of the comparison with other schemes, DT improved throughputs by 0.10% when the number of neighbors was three, 1.16% when the number of neighbors was four, and 1.11% when the number of neighbors was five, six, seven, eight, nine, and ten, compared to ST50 scheme. When the optimal number of five neighboring fog nodes was reached, the DT processed 95.66% of the total generated tasks, while ST50, JTORA, and WO processed 4.55, 91.06, and 90.88%, respectively. The DT scheme improves throughput compared to other stated schemes as the number of neighboring fog nodes was increased.

The impact of increasing the number of neighbors on energy consumption was investigated with various offloading schemes, as shown in Figure 9. When increasing the number of neighbors, the energy consumption in the system was increased because of operating additional fog nodes. Addressing the impact of increasing the number of neighbors helped to find the optimal number of neighboring fog nodes that was necessary to achieve optimum results. When having five neighbors, the difference between the energy consumed with and without DEC was very low; then as we increased the number of neighbors, the difference started to increase. In the no offloading scheme, the impact of utilizing DEC can be observed, i.e., reducing the wastage of energy by 55.72% when the number of neighbors was three, and up to 80.74% when the number of neighbors was ten. This method can also be applied to ST50 and DT, as DEC saved up to 38.58 and 32.16% of energy for each scheme, respectively, when the number of neighbors was ten. When comparing ST50 to DT after applying DEC, more energy was consumed with DT. This was because of the nature of this scheme, as more tasks were processed in DT than ST50, so the energy consumed by processing these tasks caused an increase in overall energy consumption in the system.

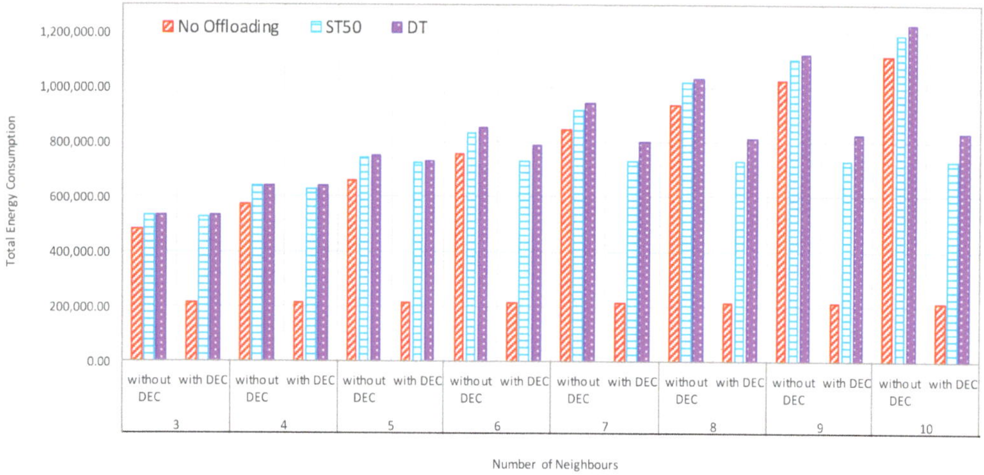

Figure 9. Impact of Increasing Number of Neighbors on Energy Consumption with Various Offloading Schemes.

7. Conclusions

In this paper, we studied the problem of computational offloading and resource management in online fog computing systems and proposed a dynamic offloading threshold that allows a fog node to adjust its threshold dynamically, with a combination of two efficient and effective algorithms: dynamic task scheduling (DTS) and dynamic energy control (DEC). Our proposed scheme exploited the available resources of nearby fog nodes and the remote cloud, selecting the best candidate to handle the overloaded tasks. Moreover, our proposed approach made dynamic decisions as to when to increase/decrease the offloading threshold, which in turn determined whether the incoming task was processed locally at the primary fog node or offload to the best neighbor, based on the states of the fog node's resources, and its neighbors. Therefore, once the primary fog node was considered congested (e.g., reaching its offloading threshold), it tended to migrate its workloads to the best neighbor.

The performance of the proposed approach was evaluated in terms of average round trip time, throughputs and total energy consumed at fog nodes. In addition to that performance comparisons with more recent offloading schemes were presented to validate the efficiency of the proposed solution. Furthermore, the effect of increasing the number of vehicles was addressed, this was to analyze the performance of the proposed algorithm in cases where traffic congestion occurred in a specific region. Along with that, the impact of the increasing number of neighbors was investigated to examine how the system would perform in situations where there were more available neighbors willing to help. Various numerical results were included, and the performance evaluations were presented to illustrate the effectiveness of the proposed scheme and demonstrate the superior performance over existing schemes.

For future work, one can consider the impact of latency and energy overhead caused by switching on/off fog nodes. Moreover, considering task offloading in an environment that takes user mobility into account. An interesting direction for future research is to examine how latency, energy consumption, security could be optimized simultaneously.

Author Contributions: Conceptualization, F.A.; methodology and software F.A.; writing, original draft, F.A.; review and proofread the manuscript, O.R.; supervision, O.R. All authors have read and agreed to the published version of the manuscript.

Funding: The Ministry of Education, Saudi Arabia.

Data Availability Statement: Data sharing not applicable.

Conflicts of Interest: The authors declare no conflict of interest.

References

1. Alenizi, F.; Rana, O. Minimising Delay and Energy in Online Dynamic Fog Systems. In Proceedings of the International Conference on Internet of Things and Embedded Systems (IoTE 2020), London, UK, 28–29 November 2020; Volume 14, pp. 139–158.
2. Arkian, H.R.; Diyanat, A.; Pourkhalili, A. MIST: Fog-based data analytics scheme with cost-efficient resource provisioning for IoT crowdsensing applications. *J. Netw. Comput. Appl.* **2017**, *82*, 152–165. [CrossRef]
3. Sarkar, S.; Misra, S. Theoretical modelling of fog computing: A green computing paradigm to support IoT applications. *IET Netw.* **2016**, *5*, 23–29. [CrossRef]
4. Mahmud, R.; Kotagiri, R.; Buyya, R. Fog Computing: A Taxonomy, Survey and Future Directions. In *Smart Sensors, Measurement and Instrumentation*; Springer Science and Business Media LLC: Berlin/Heidelberg, Germany, 2018; pp. 103–130.
5. Hu, P.; Dhelim, S.; Ning, H.; Qiu, T. Survey on fog computing: Architecture, key technologies, applications and open issues. *J. Netw. Comput. Appl.* **2017**, *98*, 27–42. [CrossRef]
6. Brasil. Decreto—Lei nº 227, de 28 de Fevereiro de 1967. Dá Nova Redação ao Decreto-lei nº 1.985, de 29 de Janeiro de 1940 (Código de Minas). Brasília. 1967. Available online: http://www.planalto.gov.br/ccivil_03/Decreto-Lei/Del0227.htm (accessed on 19 October 2020).
7. Ma, K.; Bagula, A.; Nyirenda, C.; Ajayi, O. An iot-based fog computing model. *Sensors* **2019**, *19*, 2783. [CrossRef]
8. Xiao, Y.; Krunz, M. QoE and power efficiency tradeoff for fog computing networks with fog node cooperation. In Proceedings of the IEEE INFOCOM 2017—IEEE Conference on Computer Communications, Atlanta, GA, USA, 1–4 May 2017; pp. 1–9.
9. Jamil, B.; Shojafar, M.; Ahmed, I.; Ullah, A.; Munir, K.; Ijaz, H. A job scheduling algorithm for delay and performance optimization in fog computing. *Concurr. Comput. Pr. Exp.* **2019**, *32*, 5581. [CrossRef]
10. Dastjerdi, A.V.; Gupta, H.; Calheiros, R.N.; Ghosh, S.K.; Buyya, R. Fog Computing: Principles, architectures, and applications. In *Internet of Things*; Elsevier BV: Amsterdam, The Netherlands, 2016; pp. 61–75.
11. Liu, L.; Chang, Z.; Guo, X.; Mao, S.; Ristaniemi, T. Multiobjective Optimization for Computation Offloading in Fog Computing. *IEEE Internet Things J.* **2018**, *5*, 283–294. [CrossRef]
12. Zhao, Z.; Bu, S.; Zhao, T.; Yin, Z.; Peng, M.; Ding, Z.; Quek, T.Q.S. On the Design of Computation Offloading in Fog Radio Access Networks. *IEEE Trans. Veh. Technol.* **2019**, *68*, 7136–7149. [CrossRef]
13. Zhu, Q.; Si, B.; Yang, F.; Ma, Y. Task offloading decision in fog computing system. *China Commun.* **2017**, *14*, 59–68. [CrossRef]
14. Chen, M.; Hao, Y. Task Offloading for Mobile Edge Computing in Software Defined Ultra-Dense Network. *IEEE J. Sel. Areas Commun.* **2018**, *36*, 587–597. [CrossRef]
15. Al-Khafajiy, M.; Baker, T.; Al-Libawy, H.; Maamar, Z.; Aloqaily, M.; Jararweh, Y. Improving fog computing performance via Fog-2-Fog collaboration. *Futur. Gener. Comput. Syst.* **2019**, *100*, 266–280. [CrossRef]
16. Yousefpour, A.; Ishigaki, G.; Jue, J.P. Fog Computing: Towards Minimizing Delay in the Internet of Things. In Proceedings of the 2017 IEEE International Conference on Edge Computing (EDGE), Honolulu, HI, USA, 25–30 June 2017; pp. 17–24.
17. Gao, X.; Huang, X.; Bian, S.; Shao, Z.; Yang, Y. PORA: Predictive Offloading and Resource Allocation in Dynamic Fog Computing Systems. *IEEE Internet Things J.* **2019**, *7*, 72–87. [CrossRef]
18. Tang, Q.; Xie, R.; Yu, F.R.; Huang, T.; Liu, Y. Decentralized Computation Offloading in IoT Fog Computing System With Energy Harvesting: A Dec-POMDP Approach. *IEEE Internet Things J.* **2020**, *7*, 4898–4911. [CrossRef]
19. Wang, Q.; Chen, S. Latency-minimum offloading decision and resource allocation for fog-enabled Internet of Things networks. *Trans. Emerg. Telecommun. Technol.* **2020**, *31*, 3880. [CrossRef]
20. Mukherjee, M.; Kumar, S.; Shojafar, M.; Zhang, Q.; Mavromoustakis, C.X. Joint Task Offloading and Resource Allocation for Delay-Sensitive Fog Networks. In Proceedings of the ICC 2019—2019 IEEE International Conference on Communications (ICC), Shanghai, China, 20–24 May 2019; pp. 1–7.
21. Mukherjee, M.; Kumar, V.; Kumar, S.; Matamy, R.; Mavromoustakis, C.X.; Zhang, Q.; Shojafar, M.; Mastorakis, G. Compu-tation offloading strategy in heterogeneous fog computing with energy and delay constraints. In Proceedings of the ICC 2020—2020 IEEE International Conference on Communications (ICC), Dublin, Ireland, 7–11 June 2020.
22. Sun, H.; Yu, H.; Fan, G.; Chen, L. Energy and time efficient task offloading and resource allocation on the generic IoT-fog-cloud architecture. *Peer Peer Netw. Appl.* **2020**, *13*, 548–563. [CrossRef]
23. Meng, X.; Wang, W.; Zhang, Z. Delay-Constrained Hybrid Computation Offloading With Cloud and Fog Computing. *IEEE Access* **2017**, *5*, 21355–21367. [CrossRef]
24. Yin, L.; Luo, J.; Luo, H. Tasks Scheduling and Resource Allocation in Fog Computing Based on Containers for Smart Manufacturing. *IEEE Trans. Ind. Inform.* **2018**, *14*, 4712–4721. [CrossRef]

25. Mukherjee, M.; Guo, M.; Lloret, J.; Iqbal, R.; Zhang, Q. Deadline-Aware Fair Scheduling for Offloaded Tasks in Fog Computing With Inter-Fog Dependency. *IEEE Commun. Lett.* **2019**, *24*, 307–311. [CrossRef]
26. Qayyum, T.; Malik, A.W.; Khattak, M.A.K.; Khalid, O.; Khan, S.U. FogNetSim++: A Toolkit for Modeling and Simulation of Distributed Fog Environment. *IEEE Access* **2018**, *6*, 63570–63583. [CrossRef]
27. Marsan, M.A.; Meo, M. Queueing systems to study the energy consumption of a campus WLAN. *Comput. Netw.* **2014**, *66*, 82–93. [CrossRef]
28. Li, F.; Wang, X.; Cao, J.; Wang, R.; Bi, Y. A State Transition-Aware Energy-Saving Mechanism for Dense WLANs in Buildings. *IEEE Access* **2017**, *5*, 25671–25681. [CrossRef]
29. Monil, M.A.H.; Qasim, R.; Rahman, R.M. Energy-Aware VM consolidation approach using combination of heuristics and migration control. In Proceedings of the Ninth International Conference on Digital Information Management (ICDIM 2014) 2014, Phitsanulok, Thailand, 29 September–1 October 2014; pp. 74–79.
30. Mosa, A.; Paton, N.W. Optimizing virtual machine placement for energy and SLA in clouds using utility functions. *J. Cloud Comput.* **2016**, *5*, 17. [CrossRef]
31. Mahadevamangalam, S. Energy-Aware Adaptation in Cloud Datacenters. Master's Thesis, Blekinge Institute of Technology, Karlskrona, Sweden, October 2018.
32. Monil, M.A.H.; Rahman, R.M. Implementation of modified overload detection technique with VM selection strategies based on heuristics and migration control. In Proceedings of the 2015 IEEE/ACIS 14th International Conference on Computer and Information Science (ICIS), Las Vegas, NV, USA, 28 June–1 July 2015; pp. 223–227.
33. Monil, M.A.H.; Rahman, R.M. VM consolidation approach based on heuristics fuzzy logic, and migration control. *J. Cloud Comput.* **2016**, *5*, 755. [CrossRef]
34. Abedin, S.F.; Alam, G.R.; Tran, N.H.; Hong, C.S. A Fog based system model for cooperative IoT node pairing using matching theory. In Proceedings of the 2015 17th Asia-Pacific Network Operations and Management Symposium (APNOMS), Busan, Korea, 19–21 August 2015; pp. 309–314.
35. Zohora, F.T.; Khan, M.R.R.; Bhuiyan, M.F.R.; Das, A.K. Enhancing the capabilities of IoT based fog and cloud infrastructures for time sensitive events. In Proceedings of the 2017 International Conference on Electrical Engineering and Computer Science (ICECOS), Palembang, Indonesia, 22–23 August 2017.
36. Lee, G.; Saad, W.; Bennis, M. An online secretary framework for fog network formation with minimal latency. In Proceedings of the 2017 IEEE International Conference on Communications (ICC), Paris, France, 21–25 May 2017; pp. 1–6.
37. El Kafhali, S.; Salah, K.; Alla, S.B. Performance Evaluation of IoT-Fog-Cloud Deployment for Healthcare Services. In Proceedings of the 2018 4th International Conference on Cloud Computing Technologies and Applications (Cloudtech), Brussels, Belgium, 26–28 November 2018.
38. Liu, L.; Chang, Z.; Guo, X. Socially Aware Dynamic Computation Offloading Scheme for Fog Computing System With Energy Harvesting Devices. *IEEE Internet Things J.* **2018**, *5*, 1869–1879. [CrossRef]
39. Rahbari, D.; Nickray, M. Scheduling of fog networks with optimized knapsack by symbiotic organisms search. In Proceedings of the 2017 21st Conference of Open Innovations Association (FRUCT), Helsinki, Finland, 6–10 November 2017; pp. 278–283.
40. Gupta, H.; Dastjerdi, A.V.; Ghosh, S.K.; Buyya, R. iFogSim: A toolkit for modeling and simulation of resource management techniques in the Internet of Things, Edge and Fog computing environments. *Softw. Pr. Exp.* **2017**, *47*, 1275–1296. [CrossRef]

sensors

MDPI

Article

Deep Reinforcement Learning-Based Task Scheduling in IoT Edge Computing

Shuran Sheng [1], Peng Chen [2,*], Zhimin Chen [3], Lenan Wu [1] and Yuxuan Yao [4]

1 School of Information Science and Engineering, Southeast University, Nanjing 210096, China; shengshuran@seu.edu.cn (S.S.); wuln@seu.edu.cn (L.W.)
2 State Key Laboratory of Millimeter Waves, Southeast University, Nanjing 210096, China
3 School of Electronic and Information, Shanghai Dianji University, Shanghai 201306, China; chenzm@sdju.edu.cn
4 Shannxi Key Laboratory of Integrated and Intelligent Navigation, Xi'an 710068, China; yyx_13022958068@163.com
* Correspondence: chenpengseu@seu.edu.cn

Abstract: Edge computing (EC) has recently emerged as a promising paradigm that supports resource-hungry Internet of Things (IoT) applications with low latency services at the network edge. However, the limited capacity of computing resources at the edge server poses great challenges for scheduling application tasks. In this paper, a task scheduling problem is studied in the EC scenario, and multiple tasks are scheduled to virtual machines (VMs) configured at the edge server by maximizing the long-term task satisfaction degree (LTSD). The problem is formulated as a Markov decision process (MDP) for which the state, action, state transition, and reward are designed. We leverage deep reinforcement learning (DRL) to solve both time scheduling (i.e., the task execution order) and resource allocation (i.e., which VM the task is assigned to), considering the diversity of the tasks and the heterogeneity of available resources. A policy-based REINFORCE algorithm is proposed for the task scheduling problem, and a fully-connected neural network (FCN) is utilized to extract the features. Simulation results show that the proposed DRL-based task scheduling algorithm outperforms the existing methods in the literature in terms of the average task satisfaction degree and success ratio.

Keywords: Internet of Things (IoT); edge computing; task scheduling; markov decision process (MDP); deep reinforcement learning (DRL)

check for updates

Citation: Sheng, S.; Chen, P.; Chen, Z.; Wu, L.; Yao, Y. Deep Reinforcement Learning-Based Task Scheduling in IoT Edge Computing. *Sensors* **2021**, *21*, 1666. https://doi.org/10.3390/s21051666

Academic Editor: Taehong Kim

Received: 31 January 2021
Accepted: 24 February 2021
Published: 28 February 2021

1. Introduction

Technology advancements in sensing, communications, and computing directly accelerate the recent development of the Internet of Things (IoT), leading to diverse IoT uses [1,2]. Most IoT-enabled applications are computationally-intensive, such as interactive gaming and augmented reality (AR) [3], and it is difficult for the devices themselves to fulfill such tasks due to the hardware constraints and power consideration. One feasible solution is to offload the tasks to the remote cloud for processing and return the results to the end devices. Although the cloud servers provide sufficient computation resources, a large amount of traffic delivered to the cloud would result in network congestion and unpredictable delay, which fails to meet the low latency requirement and degrades the quality of experience (QoE). The emerging edge computing technology overcomes the shortcomings of cloud computing [4,5].

Mobile edge computing enables various IoT applications and services performed at the network edge instead of being delivered to the remote cloud, which reduces the response time and alleviates the burden on the backhaul link. With edge computing, computationally-intensive IoT tasks are sent to the nearby VMs configured at the edge server to achieve low latency services [6–8]. However, the computation, storage, and network resources of the edge server are limited, and thus, task scheduling is vital to maximize the quality of experience (QoE) [9,10]. Task scheduling in edge computing is more challenging due to several aspects. First, the transmission delay is stochastic due to the dynamic wireless channel condition or network environment between the end devices and the edge node. Second, the available resources provided by the VMs are different in terms of their speed, ready time, and response time. Lastly, the task arrival rate, task size and delay requirement are diverse for various IoT applications, making task scheduling in edge computing more challenging.

Two special problems must be addressed for task scheduling in edge computing: time scheduling and resource allocation. Time scheduling determines the task execution order, and resource allocation is responsible for assigning tasks to suitable virtual machines (VMs) for execution. A number of task scheduling aspects in edge computing have been studied [11–16]. However, most existing works aim at resource allocation, while time scheduling has been seldomly studied. In [17], a general online scheduling model was proposed to minimize the task response time when tasks are offloaded to the edge servers. Based on Lyapunov optimization, a scheduling algorithm was proposed in [18] to minimize the communication delay and computing delay. In [19], a dual-scheduling framework in heterogeneous vehicular edge computing was proposed to adapt to the unstable capacity of servers and the task arrival rate. In [20], computationally-intensive data-parallel task offloading and scheduling were realized based on the first-come-first-serve (FCFS) mechanism to minimize the average completion time through a mixed integer non-linear programming (MINLP) algorithm. In [21], the shortest-job-first (SJF) scheduling method was applied in the task scheduling, where the task with the minimum delay is scheduled first. The authors in [22] investigated device-to-device (D2D) collaboration for task offloading by taking into account human mobility to optimize the task assignment and power allocation. In [23], the joint optimization problem of task allocation and the time scheduling problem were formulated as mixed-integer programming (MIP), and the logic based Benders decomposition (LBBD) approach was proposed to maximize the number of admitted tasks. A heuristic algorithm was proposed in [24] to address the energy-efficient and delay-sensitive task scheduling in IoT edge computing. In [25], the task scheduling and dispatching of networking and computing resources were investigated to maximize the number of completed tasks. These methods are based on an ideal mathematical model and optimized by a mixed-integer non-linear programming (MINLP) or heuristic algorithms. Although these model-oriented algorithms can achieve good results, they are not adapted to the dynamic environment where the task arriving rate and popularity are unknown in advance. Furthermore, the model-based task scheduling algorithms focus on the one step optimization rather than on the long-term objective. These algorithms assume the availability of resources is fixed during the scheduling period.

The Markov decision process (MDP) is an effective approach to model the sequential decision-making problem to achieve a long-term objective. Reinforcement learning (RL) has been developed as a promising approach to solve the MDP problems, where the agent makes sequential decisions by continually interacting with the environment [26,27]. The ultimate goal of the agent is to find an optimal policy to maximize the cumulative reward instead of the local optimal solution in real time. In RL, the mapping between the state and action is stored in a tabular form, which is not practical, especially for the large state space and continuous action space. Combined with the deep neural network (DNN), model-free deep reinforcement learning (DRL) is capable of making intelligent sequential decisions in sophisticated environments, and the table in RL is hence replaced by the function approximation of the DNN.

In recent years, DRL has been successfully applied to time scheduling and resource allocation in edge computing [28–30]. The computation resource allocation problem in edge computing is formulated as an MDP, and multiple replay memories were utilized for the deep Q-network (DQN) algorithm to minimize the total delay and resource utilization [31]. In [32], a DQN-based task scheduling was studied in cloud computing to maximize the number of successful tasks by considering the delay requirement. The authors in [33] investigated joint task offloading and resource allocation for computationally-intensive tasks in fog computing. The problem was formulated as a partially observable MDP, and the deep recurrent Q-network (DRQN) algorithm was applied to approximate the optimal value functions. In [34], a reinforcement learning algorithm was explored to address the delay-optimal task scheduling problem in cloud computing. In [35], a DRL-based approach was proposed to address the task scheduling and offloading problems in vehicular edge computing, while the latency demands were not considered. In [36], task scheduling with multiple resource allocation problems was tackled with DRL and imitation learning, where two objectives were defined.

In this paper, we design an intelligent task scheduling framework in edge computing. We focus on the heterogeneous VM resources for the task scheduling to maximize the long-term value of the QoE by considering the expected delay requirement. In achieving this goal, the DRL algorithm is applied, and the task satisfaction degree is determined as the reward. The action of the mechanism consists of two parts: one is determining the task execution order, and the other is assigning the task to the suitable VM. We formulate the task scheduling process in edge computing into an MDP, which is solved by a policy-based DRL algorithm. The main contribution of this article can be summarized as follows.

- Model-free DRL-based task scheduling is studied for task scheduling in edge computing, where the time scheduling and VM assignment are jointly optimized. The problem is formulated as an MDP problem, where the availability of VMs, task characteristics, and queue dynamics are taken into account.
- The action is represented as a VM-task pair, whose dimension may be extremely large. A new mechanism is designed in the MDP formulation, where the scheduling time step is decoupled from the real time step. By this mechanism, the action space stays linear with the product of the number of VMs and the queue size, and multiple tasks can be scheduled in one time step.
- Extensive simulation results demonstrate that the proposed DRL-based algorithm achieves a better task satisfaction degree in comparison with the baseline task scheduling algorithms.

The remainder of the paper is organized as follows. The system is presented in Section 2. In Section 3, the task scheduling in the edge computing problem is formulated as an MDP, and then, the DRL-based algorithm is applied. The simulation of the evaluation results is given in Section 4. Finally, the conclusions are given in Section 5.

2. System Model

In this section, the system architecture of the task scheduling in edge computing is introduced first, then the task model, task scheduling mechanism, and overall optimal objective are elaborated. Some notations are listed in Table 1.

Table 1. List of notations.

Symbol	Description
j_i	the task
a_i	the arriving time of j_i
z_i	the type of j_i
l_i	the size of j_i
d_i	the expected latency of j_i
v_j	the VM
M	the number of VMs
O	the maximum tasks in the waiting slot
$\mathbf{V}$	the state of the VM, with the shape of $M \times 2$
$\mathbf{Q}$	the state of the waiting tasks, with the shape of $4 \times O$
$\|b\|$	the number of tasks in the backlog queue
$tw_{i,j}$	the waiting time of j_i scheduled to v_j
$w_{i,j}$	the task satisfaction degree of j_i scheduled to v_j
$t_{i,j}$	the response time of j_i scheduled to v_j

2.1. System Architecture

We consider a task scheduling framework in an edge computing system, as illustrated in Figure 1. The computationally-intensive tasks generated by IoT applications, which are difficult to perform at local devices, are delivered to the server, which is deployed at the network edge close to the end devices. The edge server is configured with several VMs which vary significantly in their computational capacity and ready time to execute the next scheduled task. After arriving at the edge server, the tasks wait to be scheduled.

For simplicity, we only focus on the computational resource for task scheduling The scheduler monitors the status information of incoming tasks and the VMs that have an impact on the scheduling decision-making, including the task sizes, the expected completion time, the computing speed (in million instructions per second (MIPS)), and the waiting time. Based on the observation, the scheduler makes decisions on when to schedule (i.e., the scheduling order and the start time of each task) and where to schedule (i.e., which VM is allocated to each task). The tasks waiting to be scheduled are divided into two sets: one is the waiting set inside the circle in Figure 1, and the other is stored in the backlog queue. Each task of the waiting set occupies a waiting slot that can be fully observed, while only the number of tasks in the backlog queue can be observed by the scheduler. At each scheduling time step, the scheduler selects at most one task in the waiting slot to schedule In this article, we investigate task scheduling in edge computing for which only one edge server is deployed. The objective is to maximize the long-term task satisfaction of all tasks, which is:

$$max \sum_{t=1}^{T} \sum_{i \in J, j \in V} g_{i,j}, \tag{1}$$

where $g_{i,j}$ is the task satisfaction of the task i scheduled to VM j. To achieve the objective, we need to model from the following aspects.

2.2. Task Model

Computationally-intensive tasks arrive at the edge server dynamically and are classified into K types, $J = \{j_1, j_2, ..., j_K\}$. It is assumed that the tasks belonging to the same type have the same characteristics, including the task size (million instructions (MI)) and the delay requirement. The task types are ranked in ascending order by task size, and the popularity of the tasks is characterized by the Zipf distribution with the parameter popularity skewness β as $p_j = j^{-\beta} / \sum_{j=1}^{C} j^{-\beta}$. Therefore, a task i belonging to one of the K types can be denoted by a tuple as:

$$j_{i,k} = \langle a_i, z_i, l_i, d_i \rangle, \tag{2}$$

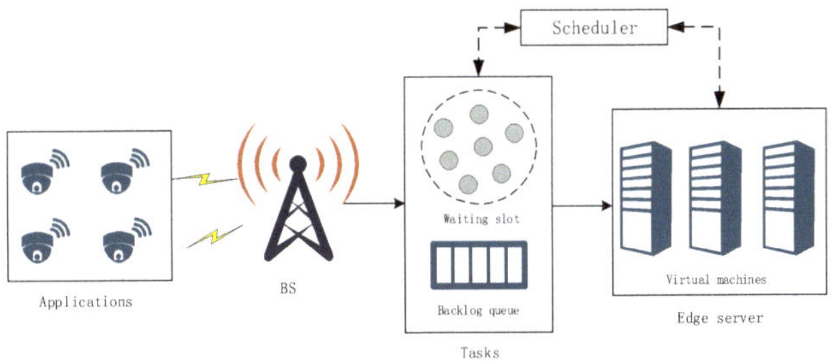

Figure 1. Illustration of the system model.

where a_i, l_i, and d_i are the arriving time, the size, and the expected delay of the task j_i, respectively.

2.3. Task Scheduling Mechanism

The edge server is configured with several VMs, denoted by $V = \{v_1, v_2, ...v_M\}$. These VMs are heterogeneous in terms of their computing capacity, denoted by $C = c_1, c_2, ..., c_M$. The task scheduler decides how to schedule tasks: determine the scheduling order and to which VM to assign. When a task is scheduled, it leaves the waiting slot, and the first task stored in the backlog queue is put into the waiting slot just vacated. It is assumed that each task is only processed on a single VM and that the computation resource of the VM will be fully utilized. The expected processing time in each VM is known before its execution. When a task is scheduled to a VM, its response time includes the waiting time in the waiting slot and the VM execution time. The execution time of task i in VM j can be computed as:

$$e_{i,j} = \frac{l_i}{v_j}. \tag{3}$$

If no tasks are executed by the VM, the start time of the current task is the arriving time; otherwise, the task begins being executed when the VM is available, i.e., all the earlier tasks have been finished. Let $s_{i,j}$ and $f_{i,j}$ denote the starting time and finishing time of task i on VM j. Therefore, the starting time depends on the finishing time of all the last tasks, which can be expressed as $s_{i,j} = max\left\{f_{l,j}, a_i\right\}$, and the task finishing time of task i can be calculated as:

$$f_{i,j} = s_{i,j} + e_{i,j}, \tag{4}$$

where $e_{i,j}$ is the time for VM j to process task i. The response time of task i on VM j is composed of two parts: the waiting time and the execution time:

$$t_{i,j} = w_{i,j} + e_{i,j}, \tag{5}$$

where $w_{i,j}$ is the waiting time of task i on VM j. If the task is processed immediately, there is no waiting time; otherwise, it is the time gap between the starting time and the arriving time and is given as:

$$w_{i,j} = s_{i,j} - a_i. \tag{6}$$

The response time is applied to evaluate the QoE of the tasks. For each task, the QoE is defined as the task satisfaction degree, which is the ratio of the expected latency and the response time. The task satisfaction degree of the task executed on VM j can be represented as:

$$g_{i,j} = d_i / t_{i,j}, \tag{7}$$

where d_i is the expected latency. It is obvious that the larger the ratio, the higher the task satisfaction degree is.

3. DRL Solution

The task scheduling problem is addressed by the model-free DRL according to the Markov decision process (MDP), which is an efficient mathematical model to model the sequent decision-making problem in a dynamic environment. This section gives the carefully designed MDP, and the policy-based DRL algorithm is applied to solve the task scheduling problem.

3.1. Preliminaries

DRL is an effective approach to deal with the Markov decision process (MDP) with a large-scale state space and action space. The ultimate goal of the DRL algorithm is to find an optimal policy π^* to maximize the expected return (long-term cumulative reward), which is considered as the state value function V. A sequence of decisions is made through the continuous interaction of the agent with the unknown environment. At the time of the scheduling time step n, the value function V under a policy π can be represented as [37]:

$$V_\pi = \mathbb{E}_\pi[G^n] = \mathbb{E}_\pi\left[r^n + \gamma r^{n+1} + \gamma^2 r^{n+1} + \cdots\right], \tag{8}$$

where $\gamma \in [0, 1]$ is a discounted factor, showing how important the future rewards are to the cumulative return, r is the instant reward obtained at each timestep, and $\mathbb{E}[\cdot]$ is the expectation operator.

In each interaction, the agent takes action based on the observed state s^n, then it receives a feedback reward r and a new state from the environment, as shown in Figure 2. The action-state value function of a state-action pair, namely the Q-function, is defined as:

$$\begin{aligned} Q_\pi(s, a; \boldsymbol{\theta}) &= \mathbb{E}_\pi\left[G^t | s^n = s, a^n = a\right] \\ &= \mathbb{E}_\pi\left[\sum_{k=n}^{\infty} \gamma^{k-n} r^k | s^n = s, a^n = a\right], \end{aligned} \tag{9}$$

in which $\boldsymbol{\theta}$ is the DNN paramter. Then, we have the optimal value:

$$V_*(s) = \max_\pi V_\pi(s), Q_*(s, a) = \max_\pi Q_\pi(s, a), \tag{10}$$

and the optimal policy:

$$\pi_* = \arg\max_\pi V_\pi(s), \pi_* = \arg\max_\pi Q_\pi(s, a). \tag{11}$$

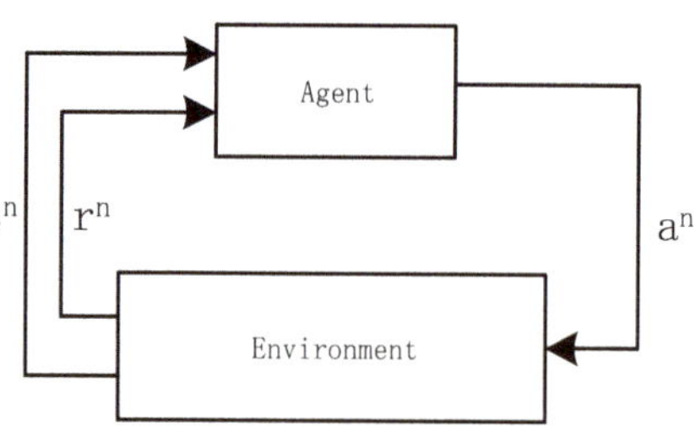

Figure 2. Interaction between the agent and the environment.

The goal of the DRL is to find an optimal behavior strategy for the agent to obtain optimal rewards. The optimal policy can be achieved by two methods: the value-based method and the policy-based method. The value-based methods aim to learn the Q function and then select an action with the maximum value, $\hat{a} = \arg\max_{a \in A} Q(s, a)$. The policy gradient methods instead target modeling and optimizing the policy $\pi_\theta(a|s)$ directly with a parameterized function with respect to θ [38]. In the policy gradient, the action is chosen following $\pi_\theta(a|s)$, which is a distribution of action probabilities with the softmax function:

$$\pi_\theta(s,a) = \frac{e^{\phi(s,a)^\top \theta}}{\sum_{k=1}^{K} e^{\phi(s,a_k)^\top \theta}}, \quad (12)$$

where $\phi(s,a)$ is the feature vector.

Compared with value-based methods, policy gradient methods directly predict the action and naturally explore it due to its stochastic policy representation. Moreover, it is more effective in high-dimensional or continuous action spaces. The objective of the policy gradient algorithm is:

$$\begin{aligned} J(\theta) &= \sum_{s \in S} d_{\pi_\theta}(s) V_{\pi_\theta}(s) \\ &= \sum_{s \in S} d_{\pi_\theta}(s) \sum_{a \in A} \pi_\theta Q_\pi(s,a), \end{aligned} \quad (13)$$

where $d_{\pi_\theta}(s)$ is the stationary distribution of the Markov chain for π_θ. The policy gradient is then given as:

$$\nabla_\theta J(\theta) = \mathbb{E}_{\pi_\theta}[\nabla_\theta log \pi_\theta(s,a) Q_\pi(s,a)], \quad (14)$$

in which ∇ is the gradient operator.

3.2. MDP Formulation

To apply DRL to solve the task scheduling in edge computing, we formulate the task scheduling process as an MDP, where the state space, action space, and state transition are carefully designed. The edge server is considered as the environment, and the scheduler plays the role of the agent, which interacts with the environment and makes decisions.

3.2.1. State Space

The state $s \in S$ describes the status information of the system, which is composed of three parts: the resource matrix, the task matrix, and the backlog queue length. Therefore, the state of the system can be given as:

$$S = \{s | s = (\mathbf{V}, \mathbf{Q}, |\mathbf{b}|)\}, \quad (15)$$

where $\mathbf{V}$ denotes the resource matrix, $\mathbf{Q}$ is the waiting matrix, and $\mathbf{b}$ indicates the backlog queue. The resource matrix represents the state of different VMs, including the processing capacity and the availability time of each VM for the next task. The waiting matrix can be observed by the scheduler, and at most one task is scheduled each scheduling time step. The tasks in the backlog queue cannot be scheduled at the current time step. As shown in Figure 3, each part of the state is elaborated as follows.

The resource matrix $\mathbf{V} \in \mathbb{R}^{n_{vm} \times 2}$. The first column represents the processing capacity (in MIPS) of the VMs, and the second column is the ready time for the next task that will be scheduled in the corresponding VM. For example, the VM v_1 is able to handle $c1$ MI per second, and $r1$ means that the task in the process will be completed in the future $r1$ time steps. The next task scheduled to individual VMs will start only if the value of $r1$ decreases to zero.

The tasks to be scheduled are divided into two parts: one is in the waiting slot, and the other is in the backlog queue. The tasks in the waiting slot are represented by a waiting matrix and can be scheduled at each scheduling time step. At most O tasks can be scheduled at each time step, and the tasks beyond O are stored in the backlog queue. In this case, the scheduler is able to observe the full status information of the waiting slot, while only the number of tasks at the backlog queue is visible. Therefore, the state of the waiting slot $\mathbf{Q}$ can be represented by a O-column matrix,

$$\mathbf{Q} = [\mathbf{q}_1, \mathbf{q}_2, ..., \mathbf{q}_o], \tag{16}$$

in which $\mathbf{q}_j$ is the size of the waiting slot, which is the length of the column of the waiting matrix, as shown in Figure 4. The row indicates the task characteristics of each task, including the task type, the task size (in MI), the task expected latency, and the waiting time before being scheduled, respectively. Thus, the waiting matrix $\mathbf{Q} \in \mathbb{R}^{4\times O}$. In particular, when the number of accepted tasks is less than the waiting slot size $(n_q) < O$, the empty position is padded with a fixed negative value to decrease the probability of being selected. In practice, the value can be set as $\mathbf{q}_j = [-1, -1, -1, -1, -1]^{\mathrm{T}}$, where T denotes the transposition of the vector.

The size of the backlog size indicates the maximum number of tasks that the edge server can accept. If a task is scheduled, it leaves the slot, and the first task in the backlog queue is inserted into the slot that was just vacated. When the tasks exceed the length of the backlog queue, the extra tasks are dropped out.

3.2.2. Action Space

The action space of the task scheduling includes two actions: one is to determine the execution order among O tasks, and the other is to assign one from M VMs for each task. Combining the two actions requires a large action space, resulting in the learning being too complicated. To keep the state space small, we decouple the scheduling time step n from the real time step t, and more than one task scheduling decision is made in each time step. At each time step, the time is frozen until an invalid action. The action is defined as $mq + n$, indicating that the task j_n in the waiting slot is scheduled to VM v_m. Furthermore, an "invalid" action means the scheduler selects a void task, then time step t proceeds to the next time step $t+1$. Therefore, the size of the action space decreases to MO. Therefore, the scheduling process performs according to the two time steps: the scheduling time step and the real time step. At the start of each real time step, the scheduler fetches new tasks if one arrives, while the scheduling time step is the scheduler's decision sequences. By decoupling the two time steps, the action space stays linear in MO. At each scheduling time step n, the action is represented as a (v_m, j_n) pair, which is:

$$A = \left\{ A_e | A_e = (v_m, j_n)|_{n \in \{-1,1,2,...,O\}}^{m \in \{-1,1,2,..,M\}} \right\}, \tag{17}$$

where $(-1,-1)$ is the invalid action A_ψ, indicating a void task is scheduled, and the others are the valid action A_φ, indicating that task j_n is scheduled to VM v_m.

At the beginning of the real time step t, new tasks are put in the waiting slot if there is an empty position; otherwise, they are put in the backlog queue. For each scheduling time step n, the scheduler makes a decision by observing the system state. If a valid action A_φ is selected, the scheduled task is removed from the waiting slot, and the first task in the backlog queue is placed in the waiting slot that was just vacated. If an invalid action A_ψ is selected, the time step proceeds to the next time step.

3.2.3. State Transition

The state transits to the next based on the state and action (s, a). As shown in Figure 3, the cases of state transition are explained as follows.

(a) The scheduler selects a valid action, and the backlog queue is not empty. For example, in Figure 3(a), $a = 19$, that is $A_e = \{(v_3, j_3)\}$, where the subscript indicates the index of the VM and task in the waiting slot. Then, the task j_3 is scheduled to the VM v_4 and will be executed after $r4$ time steps. The value of ready time $r4$ for the next scheduled tasks changes by pulsing the execution time to process j_3. Furthermore, the first tasks b_1 in the backlog queue are put into the position that just stores j_3 and the number of tasks of the backlog queue minus one simultaneously. It is noted that the waiting time of all the tasks stays unchanged within the same time step.

(b) An invalid action is chosen, meaning no task is scheduled and the backlog queue is empty at the current time step, as shown in Figure 3(b). In this case, the time step proceeds to the next time step to accept new tasks. New tasks move to the waiting slot firstly, and the extra tasks are put into the backlog queue. The tasks are dropped if the number of new tasks is larger than the backlog queue size. Meanwhile, the waiting time in both the waiting matrix and the backlog plus one and the ready time of VM v_m are set to a value of $max\{r_m - 1, 0\}$.

(c) The scheduler selects a valid action. After that, both the waiting slot and the backlog queue are empty. The time step goes to the next time step and fetches new tasks. In this case, only the ready time of all the VMs changes.

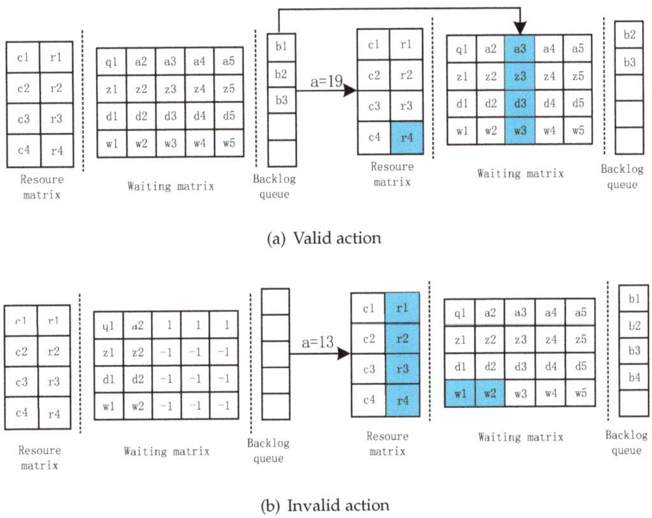

Figure 3. Illustration of the state transition with two examples of a valid action and an invalid action. (**a**) Valid action; the time step is frozen; (**b**) invalid action; the time step proceeds to the next time step.

3.2.4. Reward

As mentioned above, the objective is to maximize the LTSD, as presented in Equation (1). The reward is designed to guide the scheduler toward the goal of the optimal policy $\pi = p(a|s)$. For a valid action, the reward is the ratio of the response time and the expected latency requirement. We give zero rewards if the invalid action is selected; thus, the reward function is designed as:

$$r = \begin{cases} w_j, a \in A_\varphi \\ 0, a \in A_\psi \end{cases}. \tag{18}$$

3.3. REINFORCE Implementation

REINFORCEis a Monte Carlo policy gradient algorithm that updates the policy parameter θ based on the expected return over trajectories $\tau = (s^0, a^0, r^1, s^1, a^1, r^2, \cdots a^T, r^{T+1}, s^{T+1})$. The policy gradient at each time step t in the trajectory of each episode is converted to:

$$\nabla_\theta J(\theta) = \mathbb{E}_{\pi_\theta}[\nabla_\theta log\pi_\theta(s^n, a^n)G^n] \tag{19}$$

The parameter θ is updated according to the gradient ascent, which is:

$$\theta^{n+1} = \theta^n + \alpha G^n \nabla_\theta log\pi_\theta(s^n, a^n), \tag{20}$$

in which α is the learning rate. Equation (19) indicates that if G^n is positive, we want to increase the log probability of selecting action a^n in state s^n; otherwise, we decrease the log probability. The agent is thus guided to find the optimal policy.

Based on the MDP formulation in Section 3.2, at each scheduling time step t, the policy network accepts the system state $\mathbf{s^n} = (\mathbf{V}, \mathbf{Q}, |\mathbf{b}|)$ and generates the probability of selecting an action as the output. The action is selected as $\pi_\theta(a^n|s^n)$ and then represented as a VM-task pair $A_e = (v_m, j_k)$. If $A_e = (-1, -1)$, this means that no task is scheduled at the current scheduling time step, and no reward is obtained. In this case, the real time step moves forward, and the next state is generated based on Case (b) in Section 3.2.3. In the case of a valid action (i.e., $A_e \in A_\varphi$), the real time step also proceeds to fetch new tasks if both the waiting slots $\mathbf{q}$ and the backlog queue are empty. The next state changes as described in Case (c) in Section 3.2.3. Additionally, if the backlog queue is not empty after a valid action is selected, the first task in the backlog queue is put into the waiting slots, and the scheduling time step adds one. The next state is obtained by Case (a) in Section 3.2.3. The states, actions, and rewards constitute an episodic trajectory to compute the cumulative reward for further training. When updating the parameter θ, the cross-entropy is applied to calculate the difference between the predicted action distribution $\pi_\theta(a|s)$ and the target (label) action. The outcome of the cross-entropy multiplied by the expected discounted cumulative reward is used as the loss function to optimize the policy network parameter θ. The cross-entropy is calculated as:

$$L_{ce}^n = -y_a \log(\pi_\theta(a^n|s^n)), \tag{21}$$

where y_a^n is the label action in scheduling time step n, and the final loss function of the policy network is given as:

$$L_\theta = \frac{1}{T}\sum_{n=1}^{T} G^n L_{ce}^n, \tag{22}$$

where G_n is the discounted cumulative reward at time step n during the episode. The proposed algorithm is illustrated in Algorithm 1.

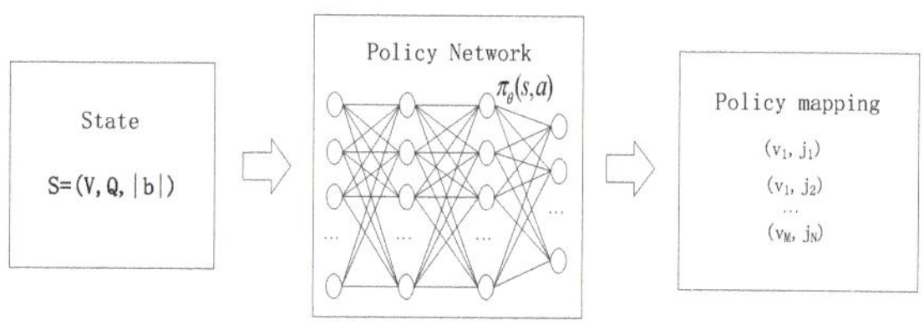

Figure 4. Illustration of the proposed REINFORCE network. The policy is mapped to a VM-task pair.

Algorithm 1: Task scheduling and allocation with the REINFORCE algorithm.

Input: episode times E, learning rate α, discount factor γ
Initialize: policy network parameter $\boldsymbol{\theta}$ and $\pi_\theta(s,a)$;
for *k=1 to E* **do**
    Obtain initial state $s^0 = (\mathbf{V}, \mathbf{Q}, |\mathbf{b}|)$;
    Initialize real time step $t = 0$ and scheduling time step $n = 0$;
    Initialize the backlog queue $\mathbf{b} = 0$;
    Initialization done = *False*;
    while *not done* **do**
        Put the state s^n into the policy network, and select action a^n following $\pi_\theta(a|s)$;
        Map a^n to the VM-task pair of $A_e = (v_m, j_k)$;
        Calculate reward r^1 with Equation (18);
        if $A_e \in A_\psi$ **then**
            time step $t+1$, scheduling time step $n+1$, $s^{n+1} \leftarrow$ Case (b) in 3.2.3
        end
        if $A_e \in A_\varphi$ **then**
            if $\mathbf{q} = 0$, *and* $\mathbf{b} = 0$ **then**
                time step $n+1$, $s^{n+1} \leftarrow$ Case (c) in 3.2.3
            else
                scheduling time step $n+1$,$s^{n+1} \leftarrow$ Case (1) in 3.2.3
            end
        end
        if *done* **then**
            break
        end
    end
    Collect a trajectory $T = (s^0, a^0, r^1, s^1, a^1, r^2, \cdots a^T, r^{T+1}, s^{T+1})$;
    Calculate G^n for each time step in trajectory T;
    Calculate the cross-entropy by Equation (21);
    Update the policy network parameter $\boldsymbol{\theta}$ according to Equation (22);
    Output: Policy network $\pi_\theta(a|s)$.
endfor

4. Simulation Results

In this section, numerical results are presented to evaluate the performance of the proposed task scheduling and allocation algorithm. All simulation results were obtained using Python 3 running with Pytorch. We further compared the proposed algorithm with two baselines.

4.1. Simulation Setting

A four-layer DNN structure was applied to realize the task scheduling and allocation policy. Both hidden layers had 64 neurons, and the rectified linear unit (ReLU) was applied as the activation function. The dimension of the output layer had $(M+1) \times O$ neurons. The discount factor γ was set as 0.99 during the training, indicating that the future steps influence the current action. The learning rate was set as 10^{-4}, and the Adam optimizer was used for gradient descent. The hyperparameters were kept fixed throughout the simulation. The detailed hyperparameter setting is shown in the table. The convergence of the proposed algorithm for different discounted factors γ during the training period is shown in Figure 5. The reward increases with the growth of γ because a higher value of γ indicates higher weights of the future rewards. In this paper, only the computational resource is considered for the task scheduling in the edge system, where the transmission delay is used to calculate the residual computational delay, so the environment parameters only include the task characteristics and the VM resources, which are described as follows

- The tasks generated by the IoT device are sent to the BS suffer from the communication transmission and arrive at the edge server at a certain rate. We assumed that the expected latency ranges from 5 s to 10 s and the transmission delay ranges from 1s to 5 s. The range of the task size was set as $[500, 4000]$ MI. In general, any arriving rate is applicable, because it is unknown in advance and is not included in the input state feature; for convenience, the tasks arrive at the edge server according to a Poisson distribution, and the average arrival rate varies from three request/s to seven requests/s, so the task arriving interval follows an exponential distribution with $[0.14, 0.33]$.
- The processing capacity of the VMs was set in the range $[1000, 2000]$ MIPS.
- The size of the waiting slot was set as $O = 5$, and the length of the backlog queue was set as $|\mathbf{b}| = 5$.
- There were five types of tasks in the simulation, and the task characteristics, including the size and the expected delay, are shown in Table 2.

Table 2. Task characteristics. MI, million instructions.

Type	Size (MI)	Expect Delay (s)
1	500	5
2	1375	6
3	2250	7
4	3125	8
5	4000	10

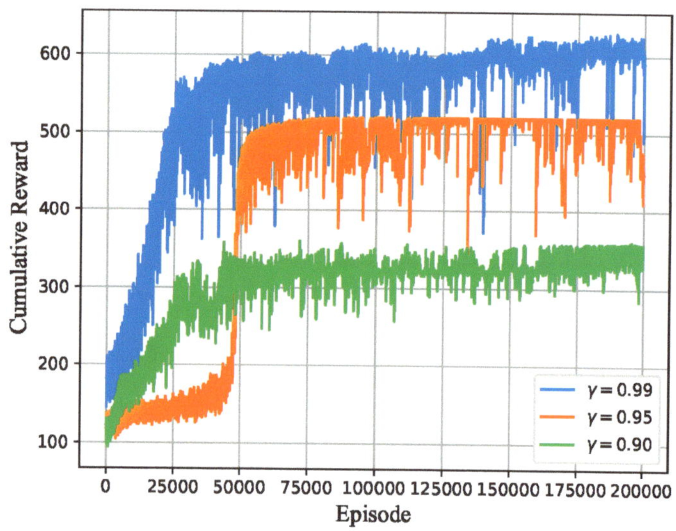

Figure 5. Cumulative reward per episode.

4.2. Performance Evaluation

Some factors, including the task arriving rate, the number of VMs, and the task popularity skewness on the task satisfaction and the success ratio, were studied. Simulation results are shown in Figures 6 and 7.

In Figure 6, we evaluate the influence of the task arriving rate λ and the number of VMs on the cumulative task satisfaction degree. The tasking arriving rate λ ranged from three to seven, and the VM number increased from three to five, while the popularity skewness was set as 0.3.

From Figure 6, it can be seen that the cumulative task satisfaction degree of the proposed DRL-based task scheduling and allocation algorithm decreases with the increment of the task arriving rate. The reason is that the higher arriving rate indicates more tasks wait to be scheduled in the edge system within the same time step, which increases the waiting time of the tasks. In terms of the number of VMs, it is apparent that the average task satisfaction degree increases when the number of VMs increases. This is because the tasks can be scheduled to more VMs, leading to a reduced waiting time.

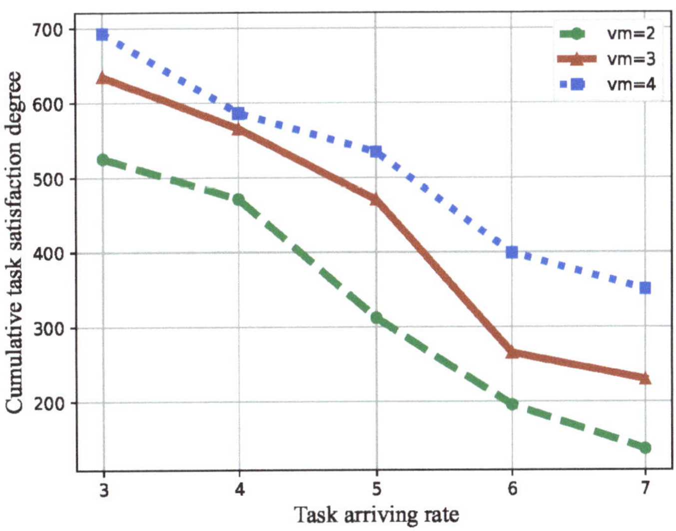

Figure 6. Cumulative task satisfaction degree versus task arriving rate and the number of VM.

In Figure 7, the effect of the task popularity skewness β on the task satisfaction degree is represented. The value of β increased from 0.1 to 0.9, while the number of VMs was set to three.

As shown in Figure 7, increasing β enlarges the cumulative task satisfaction degree. The popularity skewness indicates different popularities of each type of task. As β increases, the proportion of small-sized tasks increases, while the popularity of large-sized tasks decreases, which reduces the overall waiting time for tasks.

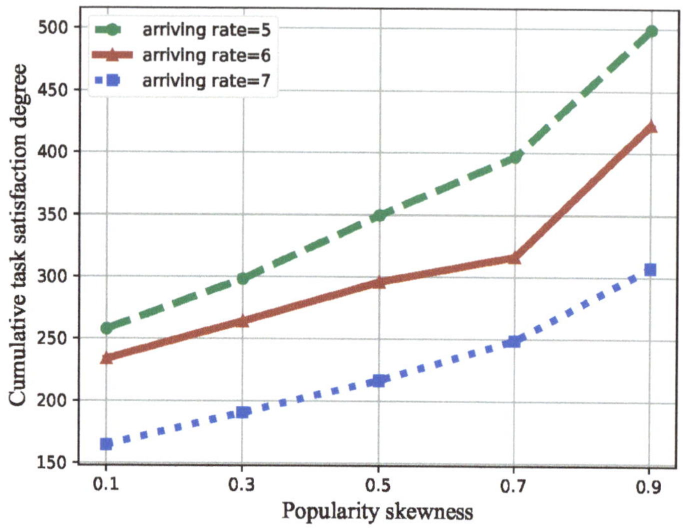

Figure 7. Cumulative task satisfaction degree versus the popularity skewness.

4.3. Performance Comparison with the Benchmark Methods

To better evaluate the performance of the proposed DRL-based task scheduling algorithm, the FCFS [20] algorithm and the SJF [21] algorithm were selected as the two benchmark methods. In both FCFS and SJF, the scheduled task is assigned to the VM with the maximum task instant reward. This means that the scheduled tasks are assigned to the VM greedily. Therefore, the two benchmarks can be expressed as greedy-FCFS and greedy-SFJ.

We compared our proposed algorithm with greedy-FCFS and greedy-SJF concerning the average task satisfaction degree and the task success ratio. The average task satisfaction degree reflects the overall quality of the algorithm, which considers the effect of a single value of each task satisfaction degree on the total task satisfaction degree. If the task's response time is less than its expected delay requirement, that is $w_{i,j} >= 1$, we say that the task is completed perfectly. The task success ratio is defined by the ratio of the number of perfectly completed tasks to the total number of tasks, which is:

$$\epsilon_s = \frac{N_s}{\sum_{j \in J} N_T}, \tag{23}$$

where N_s is the number of perfectly completed tasks.

Figures 8 and 9 give the performance comparison under different task arriving rates λ. When the task arriving rate increases, both the average task satisfaction degree and task success ratio present a declining trend for all algorithms. Compared to the greedy-FCFS scheduling algorithm and the greedy-SJF scheduling algorithm, our proposed algorithm has a significant improvement. Specifically, the proposed algorithm can improve by around 50% and 25% the average task satisfaction degree compared to greedy-FCFS and greedy-SJF, respectively. The reason is that, in FCFS, the earlier arriving tasks are scheduled first, which will cause a long waiting time for the subsequent tasks if the earlier arriving tasks require too much CPU resource of the VMs. In greedy-SFJ, tasks with the shortest execution time have higher scheduling priority even though they arrive later, which is not good for long tasks. Neither greedy-FCFS nor greedy-SJF take into account the expected delay demand.

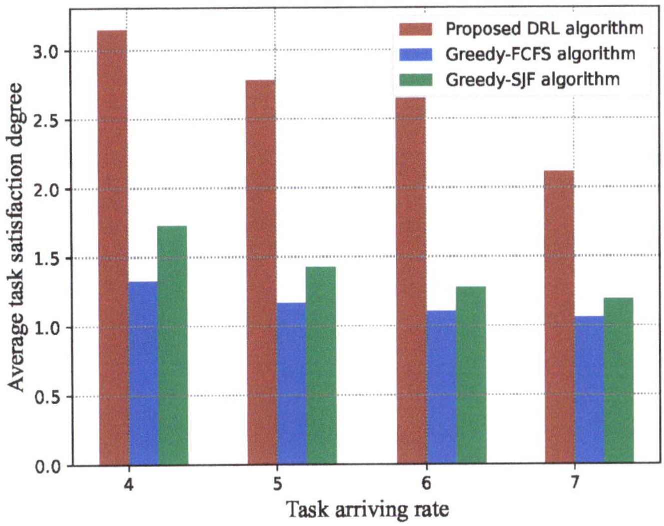

Figure 8. Average task satisfaction degree versus task arriving rate. FCFS, first-come-first-serve; SJF, shortest-job-first.

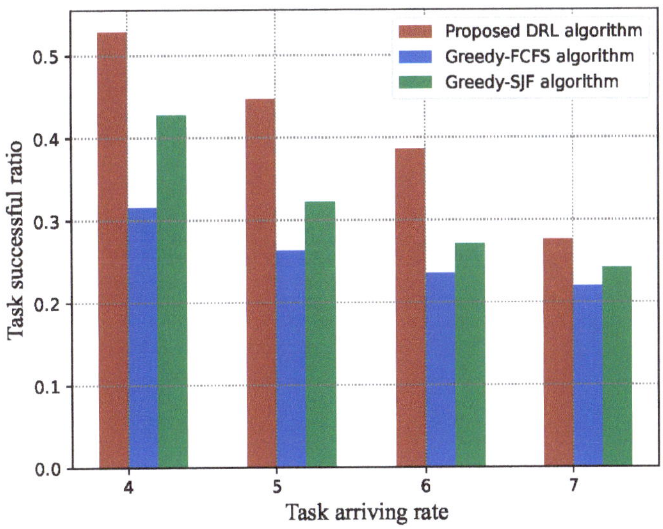

Figure 9. Success ratio versus task arriving rate.

The performance comparison of the proposed DRL-based task scheduling algorithm and the baselines toward the task popularity is illustrated in Figures 10 and 11. The cumulative task satisfaction degree and success ratio increase with the increment of task popularity skewness β. Additionally, from Figure 10, we can see that our proposed algorithm can significantly improve the task satisfaction. The gap enlarges with the increasing of the popularity factor value β compared to the greedy-FCFS algorithm and the greedy-SJF algorithm. This is because the small-sized tasks account for a larger proportion as β increases, resulting in a longer waiting time, leading to the performance degradation of the greedy-SJF algorithm.

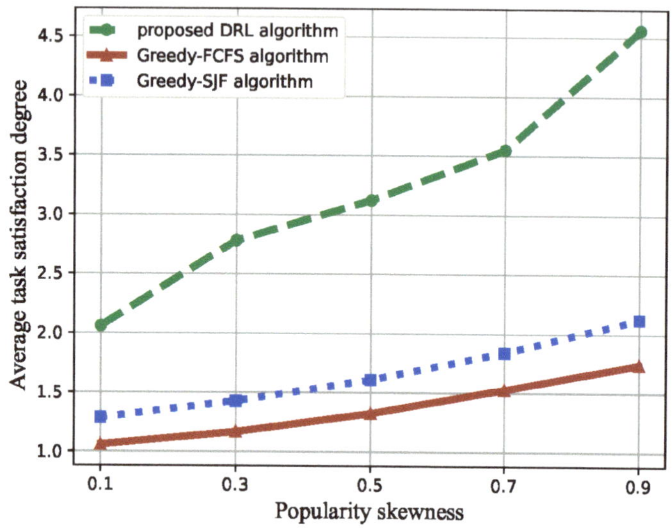

Figure 10. Average task satisfaction degree versus the popularity skewness.

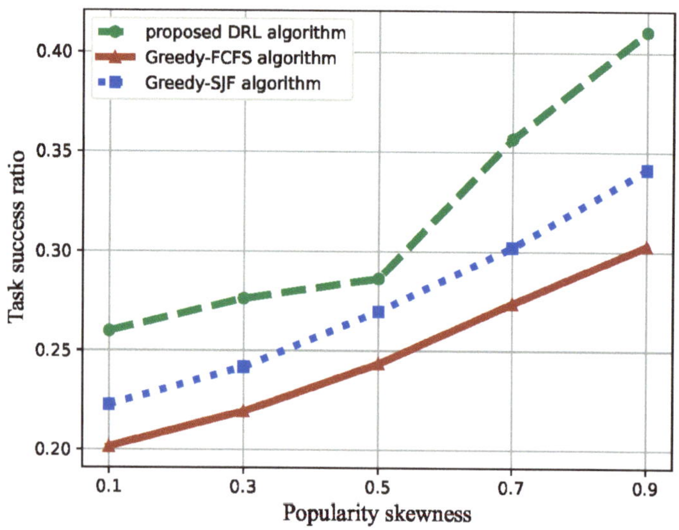

Figure 11. Success ratio versus the popularity skewness.

The performance of the proposed algorithm, the average task satisfaction degree, and the success ratio with different VMs are illustrated in Tables 3 and 4, where the number of VMs was set to 2, 3, and 4, respectively. The proposed algorithm achieves better performance than the greedy-FCFS and greedy-SJF methods on the task satisfaction degree and success ratio. This is because the scheduler always selects the VM that minimizes the response time for the current scheduled task without taking the future tasks into account.

Table 3. Comparison of the average task satisfaction degree.

Number of VMs	DRL	Greedy-FCFS	Greedy-SJF
VM = 2	1.5175	0.9072	0.9752
VM = 3	2.5502	1.4163	1.5091
VM = 4	4.7276	1.9956	2.2305

Table 4. Comparison of the task success ratio.

Number of VMs	DRL	Greedy-FCFS	Greedy-SJF
VM = 2	0.3362	0.1717	0.1740
VM = 3	0.3741	0.2924	0.3204
VM = 4	0.5688	0.4602	0.5613

5. Conclusions

This paper proposes the computationally-intensive task scheduling problem in the IoT edge system, where the task execution order and task allocation are jointly optimized. We formulate the optimization problems as an MDP model, where the state, action, reward, and state transition are carefully designed. To reduce the dimension of the action space, the scheduling time step is decoupled from the real time step. A policy-based deep reinforcement learning algorithm is applied to solve the MDP. It demonstrates that our proposed algorithm has good convergence performance. Moreover, extensive simulations are conducted to evaluate the cumulative task satisfaction degree and success ratio. The results show that the proposed algorithm outperforms other benchmark methods. Future work will focus on collaborative cloud computing and edge computing, where the communication delay will be taken into account.

Author Contributions: Conceptualization, S.S., P.C., and L.W.; methodology, S.S.; software, S.S.; validation, S.S.; formal analysis, S.S. and P.C.; investigation, S.S.; writing—original draft preparation, S.S.; writing—review and editing, P.C., Z.C., L.W., and Y.Y. All authors read and agreed to the published version of the manuscript.

Funding: This research was supported in part by the foundation of Shaanxi Key Laboratory of Integrated and Intelligent Navigation (Grant No. SKLIIN-20190204), the Open Program of State Key Laboratory of Millimeter Waves at Southeast University (Grant No. K202029), the National Natural Science Foundation of China (Grant No. 61801112), and the Natural Science Foundation of Jiangsu Province (Grant No. BK20180357).

Institutional Review Board Statement: Not applicable.

Informed Consent Statement: Not applicable.

Data Availability Statement: Not applicable.

Conflicts of Interest: The authors declare no conflict of interest.

References

1. Lin, J.; Yu, W.; Zhang, N.; Yang, X.; Zhang, H.; Zhao, W. A Survey on Internet of Things: Architecture, Enabling Technologies, Security and Privacy, and Applications. *IEEE Internet Things J.* **2017**, 4, 1125–1142. doi:10.1109/JIOT.2017.2683200.
2. Weyrich, M.; Ebert, C. Reference Architectures for the Internet of Things. *IEEE Softw.* **2016**, 33, 112–116. doi:10.1109/MS.2016.20.

3. Deng, S.; Huang, L.; Taheri, J.; Zomaya, A.Y. Computation Offloading for Service Workflow in Mobile Cloud Computing. *IEEE Trans. Parallel Distrib. Syst.* **2015**, *26*, 3317–3329. doi:10.1109/TPDS.2014.2381640.
4. Abbas, N.; Zhang, Y.; Taherkordi, A.; Skeie, T. Mobile Edge Computing: A Survey. *IEEE Internet Things J.* **2018**, *5*, 450–465 doi:10.1109/JIOT.2017.2750180.
5. Shi, W.; Cao, J.; Zhang, Q.; Li, Y.; Xu, L. Edge Computing: Vision and Challenges. *IEEE Internet Things J.* **2016**, *3*, 637–646 doi:10.1109/JIOT.2016.2579198.
6. Yan, J.; Bi, S.; Zhang, Y.J.A. Offloading and Resource Allocation With General Task Graph in Mobile Edge Computing: A Deep Reinforcement Learning Approach. *IEEE Trans. Wirel. Commun.* **2020**, *19*, 5404–5419. doi:10.1109/TWC.2020.2993071.
7. Ren, J.; Yu, G.; He, Y.; Li, G.Y. Collaborative Cloud and Edge Computing for Latency Minimization. *IEEE Trans. Veh. Technol.* **2019**, *68*, 5031–5044. doi:10.1109/TVT.2019.2904244.
8. Fan, Q.; Ansari, N. Application Aware Workload Allocation for Edge Computing-Based IoT. *IEEE Internet Things J.* **2018**, *5*, 2146–2153. doi:10.1109/JIOT.2018.2826006.
9. Samie, F.; Bauer, L.; Henkel, J. From Cloud Down to Things: An Overview of Machine Learning in Internet of Things. *IEEE Internet Things J.* **2019**, *6*, 4921–4934. doi:10.1109/JIOT.2019.2893866.
10. Lin, L.; Liao, X.; Jin, H.; Li, P. Computation Offloading Toward Edge Computing. *Proc. IEEE* **2019**, *107*, 1584–1607 doi:10.1109/JPROC.2019.2922285.
11. Li, Z.; Yang, Z.; Xie, S.; Chen, W.; Liu, K. Credit-Based Payments for Fast Computing Resource Trading in Edge-Assisted Internet of Things. *IEEE Internet Things J.* **2019**, *6*, 6606–6617. doi:10.1109/JIOT.2019.2908861.
12. Wang, P.; Yao, C.; Zheng, Z.; Sun, G.; Song, L. Joint Task Assignment, Transmission, and Computing Resource Allocation in Multilayer Mobile Edge Computing Systems. *IEEE Internet Things J.* **2019**, *6*, 2872–2884. doi:10.1109/JIOT.2018.2876198.
13. Zhang, J.; Hu, X.; Ning, Z.; Ngai, E.C..; Zhou, L.; Wei, J.; Cheng, J.; Hu, B.; Leung, V.C.M. Joint Resource Allocation for Latency-Sensitive Services Over Mobile Edge Computing Networks With Caching. *IEEE Internet Things J.* **2019**, *6*, 4283–4294. doi:10.1109/JIOT.2018.2875917.
14. Zhang, X.; Zhong, Y.; Liu, P.; Zhou, F.; Wang, Y. Resource Allocation for a UAV-Enabled Mobile-Edge Computing System: Computation Efficiency Maximization. *IEEE Access* **2019**, *7*, 113345–113354. doi:10.1109/ACCESS.2019.2935217.
15. AlQerm, I.; Pan, J. Enhanced Online Q-Learning Scheme for Resource Allocation with Maximum Utility and Fairness in Edge-IoT Networks. *IEEE Trans. Netw. Sci. Eng.* **2020**, *7*, 3074–3086. doi:10.1109/TNSE.2020.3015689.
16. Tran, T.X.; Pompili, D. Joint Task Offloading and Resource Allocation for Multi-Server Mobile-Edge Computing Networks. *IEEE Trans. Veh. Technol.* **2019**, *68*, 856–868. doi:10.1109/TVT.2018.2881191.
17. Tan, H.; Han, Z.; Li, X.; Lau, F.C.M. Online job dispatching and scheduling in edge-clouds. In Proceedings of the IEEE INFOCOM 2017 - IEEE Conference on Computer Communications, Atlanta, GA, USA, 1–4 May 2017; pp. 1–9. doi:10.1109/INFOCOM.2017.8057116.
18. Zhang, Y.; Du, P.; Wang, J.; Ba, T.; Ding, R.; Xin, N. Resource Scheduling for Delay Minimization in Multi-Server Cellular Edge Computing Systems. *IEEE Access* **2019**, *7*, 86265–86273.
19. Chen, X.; Thomas, N.; Zhan, T.; Ding, J. A Hybrid Task Scheduling Scheme for Heterogeneous Vehicular Edge Systems. *IEEE Access* **2019**, *7*, 117088–117099. doi:10.1109/ACCESS.2019.2934890.
20. Chiang, Y.; Zhang, T.; Ji, Y. Joint Cotask-Aware Offloading and Scheduling in Mobile Edge Computing Systems. *IEEE Access* **2019**, *7*, 105008–105018. doi:10.1109/ACCESS.2019.2931336.
21. Li, C.; Tang, J.; Tang, H.; Luo, Y. Collaborative cache allocation and task scheduling for data-intensive applications in edge computing environment. *Future Gener. Comput. Syst.* **2019**, *95*, 249–264.
22. Saleem, U.; Liu, Y.; Jangsher, S.; Li, Y.; Jiang, T. Mobility-Aware Joint Task Scheduling and Resource Allocation for Cooperative Mobile Edge Computing. *IEEE Trans. Wirel. Commun.* **2021**, *20*, 360–374. doi:10.1109/TWC.2020.3024538.
23. Alameddine, H.A.; Sharafeddine, S.; Sebbah, S.; Ayoubi, S.; Assi, C. Dynamic Task Offloading and Scheduling for Low-Latency IoT Services in Multi-Access Edge Computing. *IEEE J. Sel. Areas Commun.* **2019**, *37*, 668–682. doi:10.1109/JSAC.2019.2894306.
24. Abdel-Basset, M.; Mohamed, R.; Elhoseny, M.; Bashir, A.K.; Jolfaei, A.; Kumar, N. Energy-Aware Marine Predators Algorithm for Task Scheduling in IoT-based Fog Computing Applications. *IEEE Trans. Ind. Informatics* **2020**, *early access*. doi:10.1109/TII.2020.3001067.
25. Meng, J.; Tan, H.; Li, X.; Han, Z.; Li, B. Online Deadline-Aware Task Dispatching and Scheduling in Edge Computing. *IEEE Trans. Parallel Distrib. Syst.* **2020**, *31*, 1270–1286. doi:10.1109/TPDS.2019.2961905.
26. Hou, L.; Lei, L.; Zheng, K.; Wang, X. A *Q* -Learning-Based Proactive Caching Strategy for Non-Safety Related Services in Vehicular Networks. *IEEE Internet Things J.* **2019**, *6*, 4512–4520. doi:10.1109/JIOT.2018.2883762.
27. Xiang, H.; Peng, M.; Sun, Y.; Yan, S. Mode Selection and Resource Allocation in Sliced Fog Radio Access Networks: A Reinforcement Learning Approach. *IEEE Trans. Veh. Technol.* **2020**, *69*, 4271–4284. doi:10.1109/TVT.2020.2972999.
28. Meng, F.; Chen, P.; Wu, L.; Cheng, J. Power Allocation in Multi-User Cellular Networks: Deep Reinforcement Learning Approaches. *IEEE Trans. Wirel. Commun.* **2020**, *19*, 6255–6267. doi:10.1109/TWC.2020.3001736.
29. Chen, J.; Chen, S.; Wang, Q.; Cao, B.; Feng, G.; Hu, J. iRAF: A Deep Reinforcement Learning Approach for Collaborative Mobile Edge Computing IoT Networks. *IEEE Internet Things J.* **2019**, *6*, 7011–7024. doi:10.1109/JIOT.2019.2913162.
30. Xu, Z.; Tang, J.; Yin, C.; Wang, Y.; Xue, G.; Wang, J.; Gursoy, M.C. ReCARL: Resource Allocation in Cloud RANs with Deep Reinforcement Learning. *IEEE Trans. Mob. Comput.* **2020**, *early access*. doi:10.1109/TMC.2020.3044282.

31. Xiong, X.; Zheng, K.; Lei, L.; Hou, L. Resource Allocation Based on Deep Reinforcement Learning in IoT Edge Computing. *IEEE J. Sel. Areas Commun.* **2020**, *38*, 1133–1146. doi:10.1109/JSAC.2020.2986615.
32. Wei, Y.; Pan, L.; Liu, S.; Wu, L.; Meng, X. DRL-Scheduling: An Intelligent QoS-Aware Job Scheduling Framework for Applications in Clouds. *IEEE Access* **2018**, *6*, 55112–55125. doi:10.1109/ACCESS.2018.2872674.
33. Baek, J.; Kaddoum, G. Heterogeneous Task Offloading and Resource Allocations via Deep Recurrent Reinforcement Learning in Partial Observable Multifog Networks. *IEEE Internet Things J.* **2021**, *8*, 1041–1056. doi:10.1109/JIOT.2020.3009540.
34. Guo, M.; Guan, Q.; Ke, W. Optimal Scheduling of VMs in Queueing Cloud Computing Systems With a Heterogeneous Workload. *IEEE Access* **2018**, *6*, 15178–15191. doi:10.1109/ACCESS.2018.2801319.
35. Zhan, W.; Luo, C.; Wang, J.; Wang, C.; Min, G.; Duan, H.; Zhu, Q. Deep-Reinforcement-Learning-Based Offloading Scheduling for Vehicular Edge Computing. *IEEE Internet Things J.* **2020**, *7*, 5449–5465. doi:10.1109/JIOT.2020.2978830.
36. Guo, W.; Tian, W.; Ye, Y.; Xu, L.; Wu, K. Cloud Resource Scheduling with Deep Reinforcement Learning and Imitation Learning. *IEEE Internet Things J.* **2020**, *8*, 3576–3586. doi:10.1109/JIOT.2020.3025015.
37. Sutton, R.S.; Barto, A.G. *Reinforcement Learning: An Introduction*; MIT Press: Cambridge, MA, USA, 2018; pp. 54–55.
38. Sutton, R.S.; McAllester, D.A.; Singh, S.P.; Mansour, Y. Policy gradient methods for reinforcement learning with function approximation. *Adv. Neural Inf. Process. Syst.* **1999**, *99*, 1057–1063. doi:10.1109/JIOT.2020.3025015.

sensors

MDPI

Article

Efficient Implementation of NIST LWC ESTATE Algorithm Using OpenCL and Web Assembly for Secure Communication in Edge Computing Environment

BoSun Park and Seog Chung Seo *

Department of Financial Information Security, Kookmin University, Seoul 02707, Korea; 20175206@kookmin.ac.kr
* Correspondence: scseo@kookmin.ac.kr; Tel.: +82-02-910-4742

Abstract: In edge computing service, edge devices collect data from a number of embedded devices, like sensors, CCTVs (Closed-circuit Television), and so on, and communicate with application servers. Since a large portion of communication in edge computing services are conducted in wireless, the transmitted data needs to be properly encrypted. Furthermore, the application servers (resp. edge devices) are responsible for encrypting or decrypting a large amount of data from edge devices (resp. terminal devices), the cryptographic operation needs to be optimized on both server side and edge device side. Actually, the confidentiality and integrity of data are essential for secure communication. In this paper, we present two versions of security software which can be used on edge device side and server side for secure communication between them in edge computing environment. Our softwares are basically web-based application because of its universality where the softwares can be executed on any web browsers. Our softwares make use of ESTATE (Energy efficient and Single-state Tweakable block cipher based MAC-Then-Encrypt)algorithm, which is a promising candidate of NIST LWC (National Institute of Standards and Technology LightWeight Cryptography) competition and it provides not only data confidentiality but also data authentication. It also implements the ESTATE algorithm using Web Assembly for efficient use on edge devices, and optimizes the performance of the algorithm using the properties of the underlying block cipher. Several methods are applied to efficiently operate the ESTATE algorithm. We use conditional statements to XOR the extended tweak values during the operation of the ESTATE algorithm. To eliminate this unnecessary process, we use a method of expanding and storing the tweak value through pre-computation. The measured results of the ESTATE algorithm implemented with Web Assembly and the reference C/C++ ESTATE algorithm are compared. ESTATE implemented as Web Assembly is measured in web browsers Chrome, FireFox, and Microsoft Edge. For efficiency on server side, we make use of OpenCL which is parallel computing framework in order to process a number of data simultaneously. In addition, when implementing with OpenCL, using conditional statements causes performance degradation. We eliminated the conditional statement using the loop unrolling method to eliminate the performance degradation. In addition, OpenCL operates by moving the data to be encrypted to the local memory because the local memory has a high operation speed. TweAES-128 and TweAES-128-6, which have the same structure as AES algorithm, can apply the previously existing studied T-table method. In addition, the input value 16-byte is processed in parallel and calculated. In addition, since it may be vulnerable to cache-timing attack, it is safely operated by applying the previously existing studied T-table shuffling method. Our softwares cover the necessary security service from edge devices to servers in edge computing services and they can be easily used for various types of edge computing devices because they are all web-based applications.

Keywords: web; Web Assembly; OpenCL; LWC; fast implementation

Citation: Park, B.; Chung, S.S. Efficient Implementation of NIST LWC ESTATE Algorithm Using OpenCL and Web Assembly for Secure Communication in Edge Computing Environment. *Sensors* **2021**, *21*, 1987. https://doi.org/doi:10.3390/s21061987

Academic Editor: Taehong Kim

Received: 31 January 2021
Accepted: 6 March 2021
Published: 11 March 2021

1. Introduction

Existing cloud computing methods provide overall services, such as data processing and transmission in servers and data centers. However, with the increase of users using cloud computing services, the amount of data that has to be processed by the server has increased. So, there is a system load in the process of data processing and transmission To solve this problem, an edge computing method was created. Edge computing method is a method that processes data in devices, such as smartphones, unlike the method in which servers and data centers handle all services. Edge computing method reduces the system load because it only processes data generated by the device. It is also relatively efficient compared to cloud computing because it collects and processes data on its own. In the case of cloud computing, if a server is paralyzed, it is a fatal blow, but, because edge computing performs its own computing, it can effectively respond to failures. Therefore, we propose a web-based application edge computing method using Web Assembly. The existing edge computing method provides services by processing data sent from a server using a method optimized for hardware, such as ARM, RISC-V, and AVR. However, when edge computing is used using a variety of hardware, there is a disadvantage of having to implement a service and cryptographic algorithm according to each hardware. However, this method can be used generally in PCs (Personal Computer), smartphones, and IoT (Internet of Things) devices that can use web-based applications, such as web browsers and web apps. In addition, there is an advantage that can be used in common in various web-based applications without additional modification on implementation.

In addition, in edge computing method, communication between server and edge computing, communication between edge computing and users, and communication between edge computing will be achieved. For secure data communication, it is necessary to encrypt data and verify that the transmitted data is transmitted without change. So, encryption algorithm and authentication algorithm must be used separately. However, we use the LWC ESTATE (LightWeight Cryptography Energy efficient and Single-stateTweakable block cipher based MAC-Then-Encrypt) algorithm, which can do this process at once. In addition, it provides edge computing service by implementing encryption and authentication service of ESTATE algorithm with Web Assembly, which has better performance than JavaScript for communication using web-based applications.

We propose an efficient implementation of the ESTATE algorithm that uses OpenCL parallel processing to efficiently transfer data through the ESTATE algorithm as a web-based application that provides edge computing services on the server. Even if the server system load is reduced due to the edge computing method, the final processed data is stored on the server. It is a web-based application that provides edge computing services on the server and needs to transfer data using the ESTATE algorithm. Therefore, there is a need for a way to efficiently operate the ESTATE algorithm on the server. Therefore, we applied several additional methods to ensure that the ESTATE algorithm works efficiently for each environment.

Contribution

The contribution of this paper is as follows:

1. Web-based application edge computing method using Web Assembly
   As the number of users using cloud computing services increases, so does the amount of data that must be processed. So, there is a system load in the process of providing the service. So, the edge computing approach was created. The edge computing method transmits and processes data to hardware, such as ARM, RISC-V, and AVR, to reduce system load. However, this method has the disadvantage of having to implement the service differently using each hardware environment and programming language. So, we propose a web-based application edge computing method using Web Assembly. Web Assembly was created to show similar performance to a low-level language. The web-based application edge computing method has the advantage that it can be used in common without additional modification in PCs, smartphones, and IoT devices that can use web-based applications. In addition, the edge computing method communicates data between server and web-base application, web-base application and user, and web-base application. So, the ESTATE

algorithm that can generate the encryption process and tag for authentication at once is implemented using Web Assembly to provide edge computing services. Check how far Web Assembly has caught up with the low-level language in terms of performance. Web Assembly was run on Chrome, FireFox, and Microsoft Edge. At Chrome, FireFox, and Microsoft Edge, Web Assembly is approximately 11%, 10%, 9% slower than Reference C code, TweAES-128-6 is about 5%, 2%, 6% slower, and TweGIFT-128 is about 22%, 54%, and 17% slower than Reference C code.

2. LWC ESTATE parallel processing using OpenCL
   ESTATE (Energy efficient and Single-state Tweakable block cipher based MAC-Then-Encrypt) algorithm is designed to be used in a limited environment, but the data are finally stored on the server. Therefore, ESTATE algorithm optimization is also required in the server. ESTATE algorithm divides AD (Associated Data) and messages into 128-bit blocks, encrypts them one block at a time, and affects the next process, so it cannot process a large amount of data through parallel processing at once. Servers have to send data to multiple platforms, so if they are processed sequentially, the communication speed becomes slow. So, we propose a method of simultaneously generating multiple ciphertexts and tags to be sent to multiple web-based applications for edge computing using OpenCL parallel processing. As a result, the implemented TweAES-128, TweAES-128-6, and TweGIFT-128 implemented in OpenCL showed performance improvement of 6.69×, 7.31×, and 1.47×, respectively, compared to the reference C code.

3. Optimization method for safe and efficient operation of ESTATE algorithm
   ESTATE algorithm uses TweAES-128, TweAES-128-6, and TweGIFT-128. We propose several methods for safe and efficient operation, and apply the previously existing studied methods. In the operation process of TweAES-128, TweAES-128-6, and TweGIFT-128, there is a process of XOR operation by expanding the 4-bit tweak value. TweAES-128 and TweAES-128-6 expand to 8-bit, and TweGIFT-128 expand to 32-bit. However, only 0∼7, 15 are used as 4-bit tweak values. Therefore, we propose a way to store and use 8-bit, 32-bit extended tweak values for 94-bit tweak values through pre-computation. In the OpenCL implementations of TweAES-128, TweAES-128-6, and TweGIFT-128, to remove the performance load, we use a loop unwind method to remove the load and implement it using local memory with a relatively fast working speed. The operation process of TweAES-128 and TweAES-128-6 is the same as AES algorithm. Therefore, the T-table method, which was previously existing studied, was applied. In addition, AES algorithm is vulnerable to cache-timing attack, and TweAES-128 and TweAES-128-6 with the same structure will be vulnerable. Therefore, TweAES-128 and TweAES-128-6 are safely operated by applying the T-table shuffling method, which is the method previously existing studied. TweAES-128 and TweAES-128-6, which applied the table shuffling method previously existing studied, show about 7% and 51% performance overhead, respectively. It simply shuffles the T-table, so it shows little performance overhead.

The remainder of this paper is organized as follows. Section 2 provides a basic overview of the web environment, Web Assembly, OpenCL, Edge computing, and LWC ESTATE. Section 3 describes the relate work of OpenCL and Web Assembly. Section 4 describes the method proposed in the paper. Section 5 describes the performance measurement results. Finally, Section 6 concludes the paper.

2. Background

2.1. Edge Computing

Several companies have used cloud computing methods [1] to provide computing services, such as servers, storage, software, and analytics, over the Internet. Cloud computing is a method focused on centralizing services to several large data centers. However, such cloud computing also begins to have problems. As the number of people using cloud services increases, the amount of data processed by servers and data centers increases, and data delays occur in the process of analyzing and transmitting collected data. That is why edge computing [2–4] was created to solve problems, such as data processing speed, capacity, and security. Edge computing is performing computing at or near the physical location of a user or data source. In the case of cloud computing, data is processed in the data center, whereas edge computing is a method of processing data in devices, such as smartphones. Edge computing method has several advantages. When using cloud computing, the larger the amount of data to be processed, the higher the system load, but in the case of edge computing, data load can be reduced because only data generated by the device is processed. In addition, cloud computing has to strengthen security from the process of data transmission and delivery with a central server architecture, but edge computing is relatively more secure than cloud computing because it collects and processes data on its own. In addition, when the server is paralyzed when using cloud computing, the overall damage is seriously affected, but when using edge computing, it can effectively respond to failures because it performs its own computing. Figure 1 is the structure of the edge computing method. It shows a structure that does not process data in a server or data center but sends data to peripheral devices that will perform edge computing services and sends data to the user's device after processing.

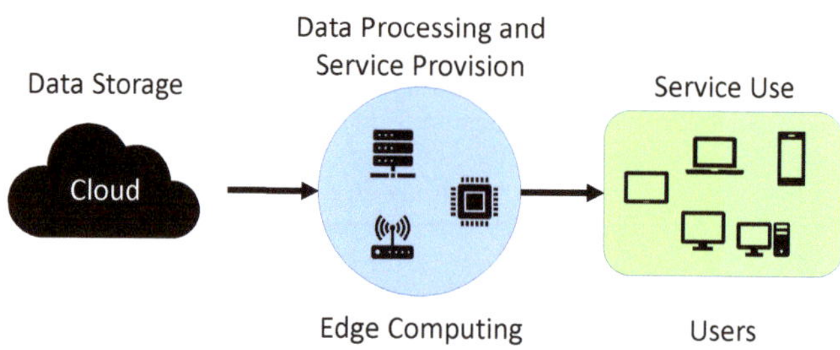

Figure 1. Edge computing structure.

2.2. *Web Environment*

Due to the continuous development of the web environment, various functions are being performed in the web environment. Due to the advancement of web technology, information on the web is displayed the same on different platforms to which networks are connected, such as PCs or smart devices. Web-based applications run within a web browser without communicating with the operating system. Due to the development of internet technology, and hardware performance improve, web technologies and libraries are continuously being created so that more complex and heavy calculations and functions can be made in a web environment. There are various web browsers in which these functions can be used, and various web browsers, such as Chrome, FireFox, and Microsoft Edge, exist. Each web browser has a JavaScript engine that renders JavaScript code and a rendering engine that provides visual services to users through web screens. Chrome uses V8 and Blink, FireFox uses SpiderMonkey and Gecko, and Microsoft Edge uses Chakra and EdgeHTML as JavaScript engines and rendering engines. There is Node.js [5], a software platform used for network application development. Node.js includes a built-in http server library, so it can be operated without additional software on the web server, and through this, more control over the operation of the web server is possible. In addition, web socket communication is possible using Node.js. Figure 2 shows the process of communication between the user who uses the web and the web server, and the process of storing data generated while using the web in the database.

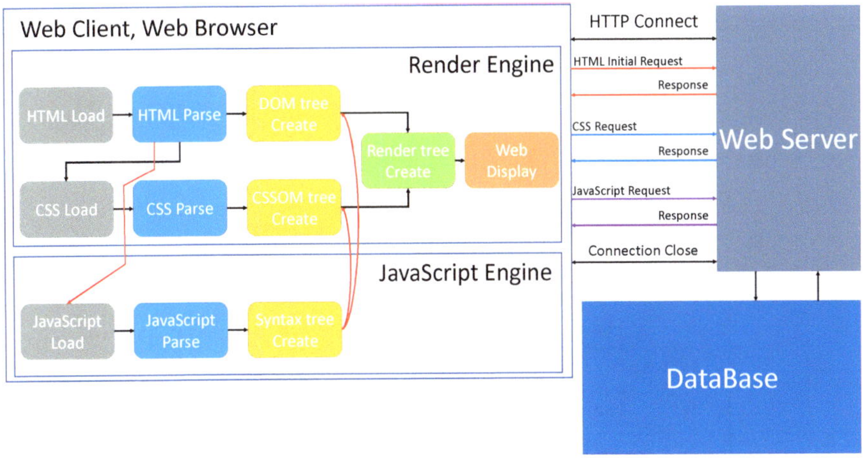

Figure 2. Web operation process [6].

2.3. Web Assembly

With the development of web-based applications, various organizations, companies, and individuals develop web-based applications to provide various services, and various web technologies are being developed. In addition, many users access web-based applications to use the various services and functions provided. These web-based applications are mainly developed in JavaScript, which is a cross-platform language, to display the same information to users on multiple platforms. However, as the number of web users increases, the amount of data that needs to be collected and processed increases, so it is important to increase the speed of processing data to reduce the load. In order to compute faster even in the web environment, Web Assembly [7,8] was created and it is constantly evolving. In addition, Web Assembly can be used by converting to languages with data types, such as C [9], C++ [10], and Typescript [11]. Therefore, it is possible to use previously implemented codes without additional modification. Due to the existence of data types, unlike JavaScript, mathematical operations allow the desired value to be computed without additional computation. Figure 3 shows the process of converting to Web Assembly using programming languages that have data types, such as C, C++, and Rust [12].

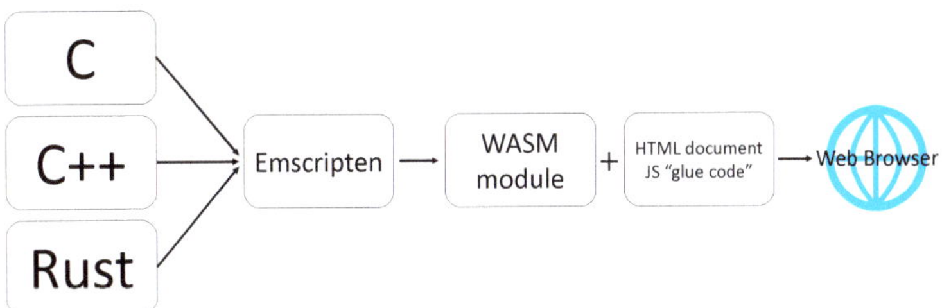

Figure 3. Web Assembly conversion process [6].

2.4. OpenCL

OpenCL is an open general purpose parallel computing framework for creating programs that run on heterogeneous platforms, such as CPUs and GPUs. OpenCL provides task-based, data-based parallel computing. OpenCL can be used in AMD, Intel CPU, Intel integrated graphics, and NVIDIA graphics card products. Table 1 and Figure 4 show the four types of memory used in OpenCL and their respective functions.

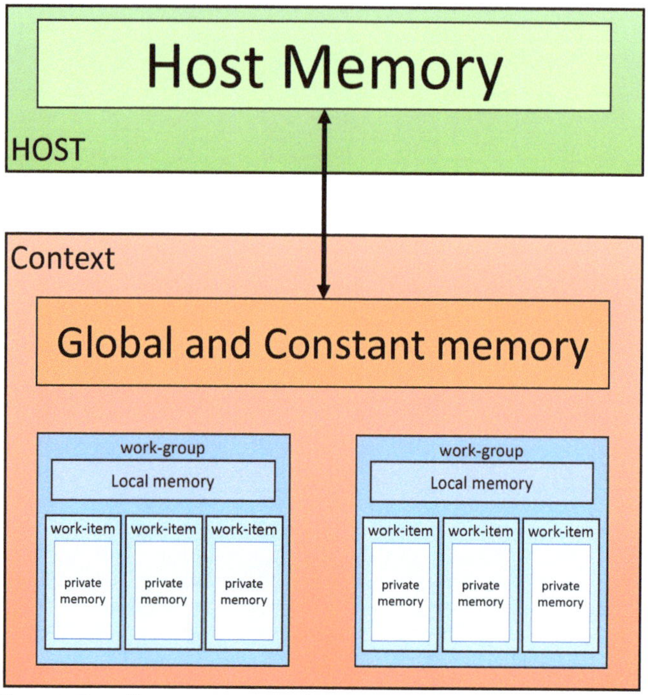

Figure 4. OpenCL memory structure [13].

Figure 5 shows the OpenCL platform model consisting of one host and one or more devices. The OpenCL platform always contains only one host. Each device has one or more compute units, and each computational unit has one or more processing element (PE). The device is where the kernel runs. Devices are provided by CPU, GPU, DSP, and hardware manufacturers. And the actual calculation for the device is done in PE.

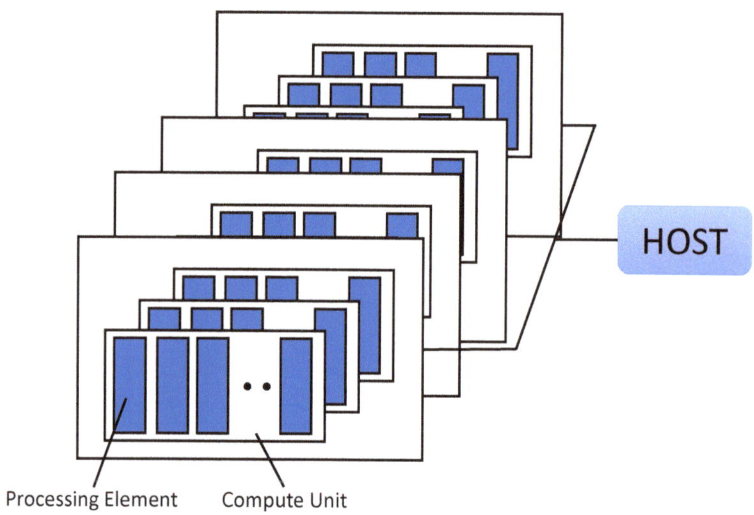

Figure 5. OpenCL platform model [13].

Table 1. OpenCL memory characteristics [14,15].

Memory	Characteristics
Global Memory	(1) Read and write from all work items (2) Placed in device's main memory
Constant Memory	(1) Read only from all work items (2) Placed in device's main memory
Local Memory	(1) Can be shared and used among work items in a work group (2) In many cases, a shared memory disposed on each operation unit is used.
Private Memory	(1) Dedicated memory area for work items (2) Often times you use registers used by processing elements.

2.5. Lightweight Cryptography (Lwc) Estate

Table 2 is a table of notations, operations, and algorithms used in the ESTATE algorithm.

Energy efficient and Single-state Tweakable block cipher-based MAC-Then-Encrypt (ESTATE) [16], one of the second round candidates for lightweight encryption algorithms, adopts FCBC-like [17] authentication and is a tunable block cipher-based authentication encryption system using OFB [18] encryption. ESTATE is based on the MAC-then-Encrypt paradigm [19]. ESTATE does not require field multiplications and has single-state, inverse-free, and RUP secure construction features [16]. In addition, ESTATE is divided into ESTATE mode and sESTATE mode. In ESTATE mode, TweAES-128 and TweGIFT-128 are proposed and used as core algorithms. TweAES-128 and TweGIFT-128 are modified versions of AES-128 [20] and GIFT-128 [21], respectively.

$$\forall X \in \bigcup_{m=1}^{n} 0,1^m, X \mapsto \begin{cases} 0^{n-|X|-1}\|1\|X & if|X| < n, \\ X & otherwise, \end{cases} \quad (1)$$

$$(E_1;E_2)?a:b:c:d := \begin{cases} a & if E_1 \wedge E_2 \\ b & if E_1 \wedge \neg E_2 \\ c & if \neg E_1 \wedge E_2 \\ d & if \neg E_1 \wedge \neg E_2 \end{cases}. \quad (2)$$

If the last block of the message and AD is smaller than 128-bit, padding is performed using Equation (1). Equation (2) is to determine the tweak value used in the ESTATE encryption process.

Table 2. Energy efficient and Single-state Tweakable block cipher-based MAC-Then-Encrypt (ESTATE) notation [16].

Notation	Denote	Notation	Denote
$\|A\|$	length(bit) of A	K	$K \in \{0,1\}^k$
$(X_{k-1},\ldots,X_0) \xleftarrow{8} x$	n-bit block parsing of X	T	$T \in \{0,1\}^t$
i	$0 \leq i \leq k-2$	M	$M \in \{0,1\}^m$
$\|X_i\|$	$\|X_i\| = n$	$\tilde{E}$	TweAES-128 or TweGIFT-128
$\|X_{k-1}\|$	$1 \leq \|X_{k-1}\| \leq n$	$\tilde{F}$	TweAES-128-6
$A \oplus B$	the bitwise XOR of A and B	$X \lll i$	i-bit left
$A\|\|B$	the concatenation of A and B	$X \ggg i$	i-bit right
n	128-bit block size	k	128-bit key size
t	128-bit tag size	τ	4-bit tweak size

Algorithm 1 is the functions used in the overall encryption and decryption operation in ESTATE mode. MAC function [16] is a function that creates a tag. The FCBC* [16] function is a function that determines the tweak value according to AD, message length, and encryption process conditions. OFB function [16] is a function that creates a cipher text using the created Tag value. Figures 6–8 show ESTATE mode operation process when both

AD and Message are used as input values, when only Message is used, and when only AD is used.

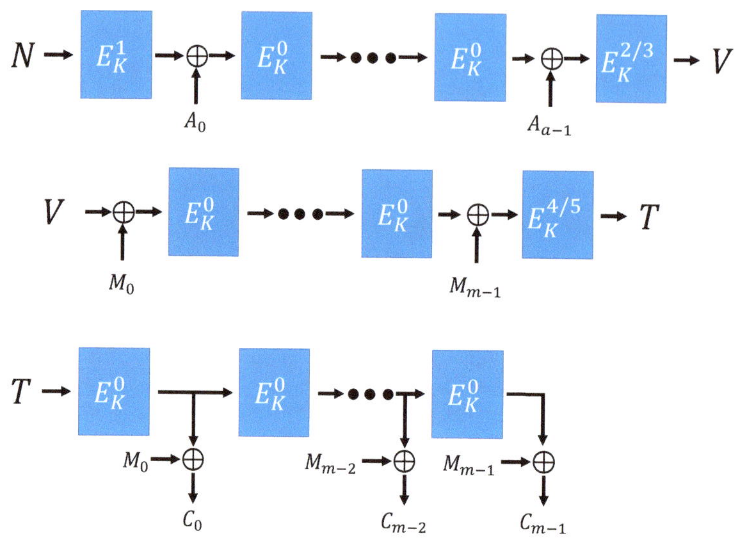

Figure 6. ESTATE mode (Using AD and Message) [16].

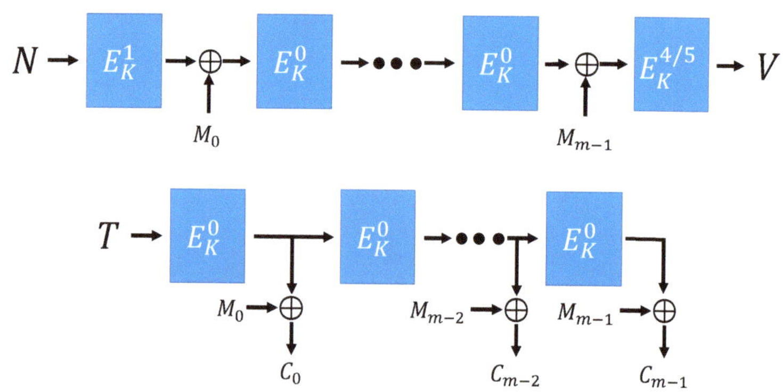

Figure 7. ESTATE mode (no AD, using Message) [16].

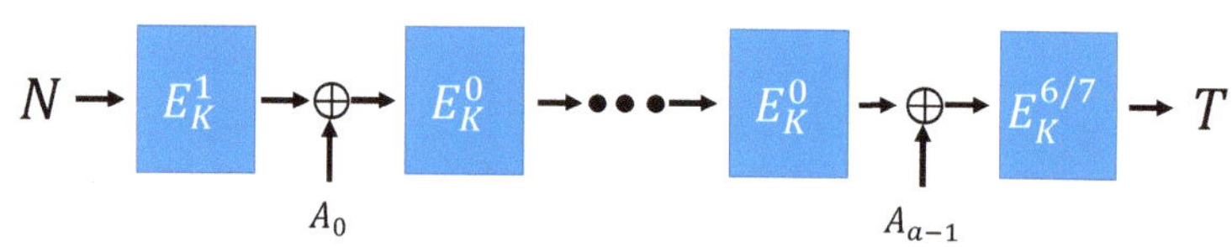

Figure 8. ESTATE mode (using AD, no Message) [16].

Algorithm 2 is the overall operation process of sESTATE mode. sESTATE mode has the same operation process as ESTATE mode. Figures 9–11 show the process of sESTATE mode operation.

Algorithm 1 ESTATE Encryption, Tag Creation, Authentication, and Decryption Algorithm [16].

function ESTATE.ENC$[\tilde{E}](K, N, A, M)$
  $T \leftarrow MAC[\tilde{E}](K, N, A, M)$
  $C \leftarrow OFB[\tilde{E}](K, T, M)$
  retrun (C, T)

function MAC$[\tilde{E}](K, N, A, M)$
  if $|A| = 0$ and $|M| = 0$ then
    return $T \leftarrow \tilde{E}_K^8(N)$
  $T \leftarrow \tilde{E}_K^1(N)$
  if $|A| > 0$ then
    $A_{a-1}\|\cdots\|A_0 \leftarrow A$
    $t \leftarrow (|M| > 0\,; |A_{a-1}| = n)\ ?\ 2 : 3 : 6 : 7$
    $T \leftarrow$ FCBC$^*[\tilde{E}](K, T, M, t)$
  if $|M| > 0$ then
    $M_{m-1}\|\cdots\|M_0 \leftarrow M$
    $t \leftarrow (|M_{m-1}| = n)\ ?\ 4 : 5$
    $T \leftarrow$ FCBC$^*[\tilde{E}](K, T, M, t)$
  return T

function ESTATE.DEC$[\tilde{E}](K, N, A, C, T)$
  $M \leftarrow OFB[\tilde{E}](K, T, C)$
  $T' \leftarrow MAC[\tilde{E}](K, N, A, M)$
  return $(T' = T)\ ?\ M : \perp$

function FCBC$^*[\tilde{E}](K, T, D, t)$
  $D_{d-1}\|\cdots\|D_0 \leftarrow D$
  for $i = 0$ to $d-2$ do
    $T \leftarrow \tilde{E}_K^0(T \oplus D_i)$
  $T \leftarrow \tilde{E}_K^t(T \oplus \text{ozp}(Dd-1))$
  return T

function OFB$[\tilde{E}](K, T, M)$
  $M_{m-1}\|\cdots\|M_0 \leftarrow M$
  for $i = 0$ to $m-1$ do
    $T \leftarrow \tilde{E}_K^0(T)$
    $C_i \leftarrow \text{chop}(T, |M_i|) \oplus M_i$
  return $(C_{m-1}\|\cdots\|C_0)$

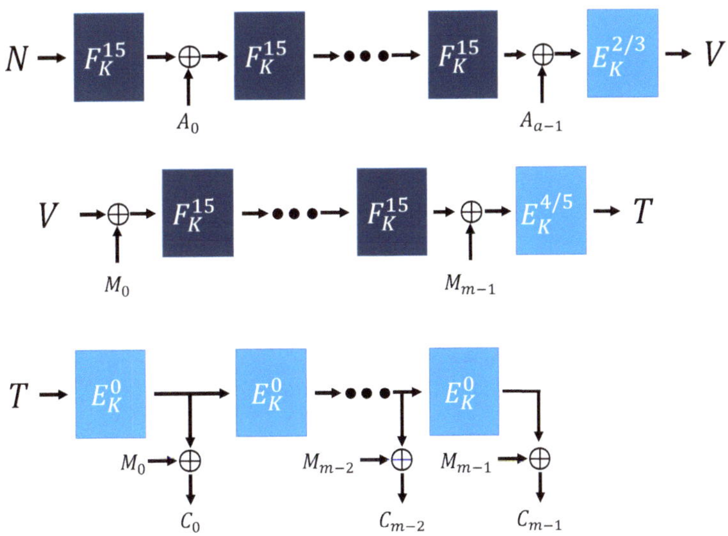

Figure 9. sESTATE mode (using AD and Message) [16].

Algorithm 2 sESTATE Encryption, Tag Creation, Authentication, and Decryption Algorithm [16].

function ESTATE.ENC$[\tilde{E},\tilde{F}](K,N,A,M)$
  $T \leftarrow MAC[\tilde{E},\tilde{F}](K,N,A,M)$
  $C \leftarrow OFB[\tilde{E}](K,T,M)$
  retrun (C,T)

function MAC$[\tilde{E},\tilde{F}](K,N,A,M)$
  if $|A|=0$ and $|M|=0$ then
    return $T \leftarrow \tilde{E}^8_K(N)$
  $T \leftarrow \tilde{F}^{15}_K(N)$
  if $|A|>0$ then
    $A_{a-1}\|\cdots\|A_0 \leftarrow A$
    $t \leftarrow (|M|>0\,;|A_{a-1}|=n)\ ?\ 2:3:6:7$
    $T \leftarrow$ FCBC$^*[\tilde{E},\tilde{F}](K,T,M,t)$
  if $|M|>0$ then
    $M_{m-1}\|\cdots\|M_0 \leftarrow M$
    $t \leftarrow (|M_{m-1}|=n)\ ?\ 4:5$
    $T \leftarrow$ FCBC$^*[\tilde{E},\tilde{F}](K,T,M,t)$
  return T

function ESTATE.DEC$[\tilde{E},\tilde{F}](K,N,A,C,T)$
  $M \leftarrow OFB[\tilde{E}](K,T,C)$
  $T' \leftarrow MAC[\tilde{E},\tilde{F}](K,N,A,M)$
  return $(T'=T)\ ?\ M:\perp$

function FCBC$^*[\tilde{E},\tilde{F}](K,T,D,t)$
  $D_{d-1}\|\cdots\|D_0 \leftarrow D$
  for $i=0$ to $d-2$ do
    $T \leftarrow \tilde{F}^{15}_K(T \oplus D_i)$
  $T \leftarrow \tilde{E}^t_K(T \oplus \text{ozp}(Dd-1))$
  return T

function OFB$[\tilde{E}](K,T,M)$
  $M_{m-1}\|\cdots\|M_0 \leftarrow M$
  for $i=0$ to $m-1$ do
    $T \leftarrow \tilde{E}^0_K(T)$
    $C_i \leftarrow \text{chop}(T,|M_i|) \oplus M_i$
  return $(C_{m-1}\|\cdots\|C_0)$

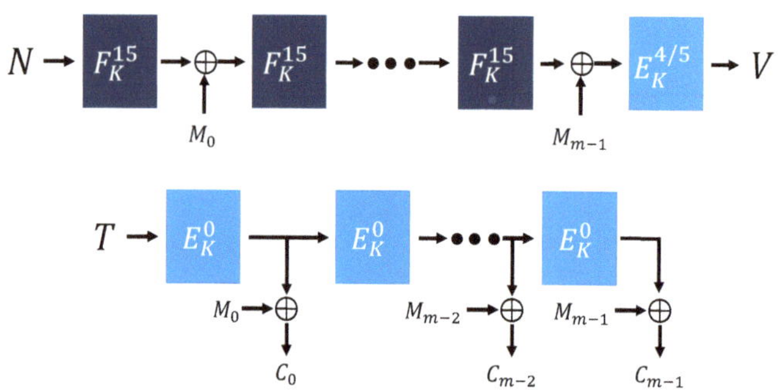

Figure 10. sESTATE mode (no AD, using Message) [16].

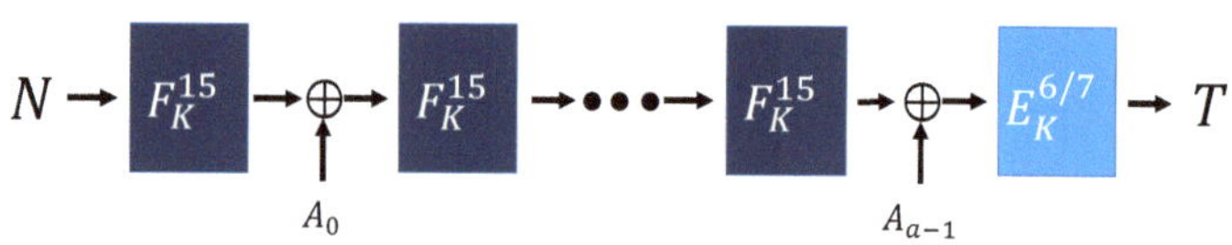

Figure 11. sESTATE mode (using AD, no Message) [16].

2.5.1. TweAES-128, TweAES-128-6

Algorithm 3 are functions of TweAES-128, a cryptographic algorithm used in ESTATE mode. The overall process is the same as AES-128, and a process of XOR operation is added by expanding the 4-bit tweak value to an 8-bit tweak value for every even round except the last round. TweAES-128-6 is proposed and used as a cryptographic algorithm to be used while designing the sESTATE mode in the ESTATE algorithm. TweAES-128-6 has the same operation process as TweAES-128. The difference is that the TweAES-128 runs 10 rounds, while the TweAES-128-6 only runs 6 rounds.

Algorithm 3 TweAES-128 Algorithm [16].

```
function TweAES(K, T, M)
    (W_43, ..., W_0) ← KeyGen(K)
    X ← X ⊕ (W_3, W_2, W_1, W_0)
    for i = 1 to 9 do
        X ← SubBytes(X)
        X ← ShiftRows(X)
        X ← MixColumns(X)
        X ← X ⊕ (W_{4i+3}, W_{4i+2}, W_{4i+1}, W_{4i})
        if i%2 = 0 then
            X ← AddTweak(X, T)
    X ← SubBytes(X)
    X ← ShiftRows(X)
    X ← X ⊕ (W_43, W_42, W_41, W_40)
    return X

function TweAES-6(K, T, X)
    (W_43, ..., W_0) ← KeyGen(K, X)
    X ← X ⊕ (W_3, W_2, W_1, W_0)
    for i = 1 to 6 do
        X ← SubBytes(X)
        X ← ShiftRows(X)
        X ← MixColumns(X)
        X ← X ⊕ (W_{4i+3}, W_{4i+2}, W_{4i+1}, W_{4i})
        if i%2 = 0 and i < 6 then
            X ← AddTweak(X, T)
    return X

function AddTweak(X, T)
    (X_127, ..., X_0) ←(1) X
    (T_3, ..., T_0) ←(1) T
    T_⊕ ← T_0 ⊕ T_1 ⊕ T_2 ⊕ T_3
    for i = 0 to 3 do
        T_{i+4} ← T_i ⊕ T_⊕
    for i = 0 to 7 do
        X_{8i} ← X_{8i} ⊕ T_i
    return X
```

2.5.2. TweGIFT-128

Algorithm 4 is the overall process of TweGIFT-128 used in ESTATE mode. TweGIFT-128 has the same structure as GIFT-128, and XOR operation is added by expanding the 4-bit tweak value to 32-bit tweak value every $(round + 1)\%5 == 0$th rounds. TweGIFT-128's tweak expansion process is the same as TweAES-128. First, expand it to 8-bit in the same way, and then use the expanded 8-bit value to store the same value in the remaining 24-bits and expand it to a total of 32-bits.

Algorithm 4 TweGIFT-128 Algorithm [16].

```
function TweGIFT(K, T, X)
    C ← 000000
    for i = 0 to 39 do
        X ← SubCells(X)
        X ← PermBits(X)
        (K, X) ← AddRoundKey(K, X)
        (C, X) ← AddRoundConstant(C, X)
        if (i + 1)%5 = 0 and i < 39 then
            X ← AddTweak(X, T)
    return X

function AddTweak(X, T)
    (X_127, ..., X_0) ←(1) X
    (T_3, ..., T_0) ←(1) T
    T_⊕ ← T_0 ⊕ T_1 ⊕ T_2 ⊕ T_3
    for i = 0 to 3 do
        T_{i+4} ← T_i ⊕ T_⊕
    T_{15..8} ← T_{7..0}
    T_{23..16} ← T_{7..0}
    T_{31..24} ← T_{7..0}
    for i = 0 to 31 do
        X_{4i} ← X_{4i} ⊕ T_i
    return X
```

3. Related Work

3.1. Existing Crypto Implementation Using OpenCL

Due to the development of multi-core processes, parallel processing technology is being used in various fields. In addition, the use of OpenCL for parallel processing is increasing, and it is efficient for processing large amounts of data. Therefore, studies are being conducted to rapidly encrypt a large amount of data using an cryptographic algorithm using OpenCL.

In Reference [22], we use OpenCL to improve encryption speed using the AES encryption algorithm. They used the NVIDIA GeForce GTX 1060 to measure performance. Table 3 is a table comparing the results measured in Reference [22] with previous studies As a result, their research results show that the XTS (XEX-based tweaked-codebook mode with ciphertext stealing) mode is 12.86% and the CTR (Counter) mode is 14.71%, compared to the previous studies.

Table 3. AES (Advanced Encryption Standard) fast implementation study results comparison.

Paper	GPU	Language	Mode	Throughput (Gbps)
Yuan et al. [23]	ATI HD 7670M	OpenCL	CTR	5.04 Gbps
Wang et al. [24]	NVIDIA GTX 285	OpenCL	XTS	8.59 Gbps
Wang et al. [24]	NVIDIA GTX 285	CUDA	XTS	9.74 Gbps
Conti et al. [25]	NVIDIA GT 555M	OpenCL	CTR	10.00 Gbps
Biagio et al. [26]	NVIDIA GT 8800	CUDA	CTR	12.50 Gbps
Sanida et al. [22]	NVIDIA GTX 1060	OpenCL	XTS	12.53 Gbps
Sanida et al. [22]	NVIDIA GTX 1060	OpenCL	CTR	14.71 Gbps

In Reference [27], various cryptographic algorithms are implemented in OpenCL and used for image encryption. Table 4 is an information table that implements AES (Advanced Encryption Standard), DES (Data Encryption Standard), BlowFish, and RSA (Ron Rivest, Adi Shamir, Leonard Adleman) using OpenCL in Reference [27]. Table 5 is the result of measurement by CPU and GPU for each cryptographic algorithm implemented using OpenCL. As a result, AES, DES, BlowFish, and RSA show performance improvements of 8 times, 2.5 times, 11.13 times, and 5 times, respectively.

Table 4. Memory size, line of code for cryptographic algorithm implementation using OpenCL [27].

Cryptographic Algorithm	Key Size	Constant Space	Compilation Time
AES [20]	128-bit	844 KB	2.7 ms
DES [28]	192-bit	1294 KB	5.3 ms
BlowFish [29]	256-bit	252 B	3.5 ms
RSA [30]	128-bit	6 KB	1031 ms

Table 5. Measuring cryptographic algorithm results using OpenCL [27].

Device	AES	DES	BlowFish	RSA
AMD FX 6100 3.0 GHz (CPU 6 Cores)	240 Mbps	144 Mbps	736 Mbps	4 Mbps
NVIDIA GTX 550 (GPU)	1920 Mbps	368 Mbps	8192 Mbps	20 Mbps
Ratio of Performance Improvement	8 times	2.5 times	11.13 times	5 times

In Reference [31], AES-256 encryption and decryption implementation using OpenCL parallel processing is compared with AES-256 implemented using sequential processing. As a result, when 10,240,000 work items are used, the implementation using OpenCL parallel processing shows performance improvement of about 240 times for encryption and 481 times for decryption. In addition, as measured by AMD Radeon HD 8850M and AMD

Radeon HD 8570, AMD Radeon HD 8570 shows performance improvement of 3.8 times and 3.3 times in encryption and decryption, respectively.

3.2. Web Assembly

Web Assembly shows better performance than JavaScript in web-based applications, and due to continuous development, it will continue to be close to the performance of low-level languages, such as C language. In addition, research on Web Assembly is actively underway.

In Reference [6], the revised CHAM, *P*-256-wNAF (window Non-Adjacent Form), SHA-256 (Secure Hash Algorithm), and HMAC (Hash-based Message Authentication Code) algorithms are compared after implementation using Web Assembly and JavaScript for more efficient encryption, key exchange, and authentication in the web environment. Table 6 shows the performance measurement results for cryptographic algorithms, and it can be seen that it is more efficient when Web Assembly implements cryptographic algorithms than JavaScript. In addition, in the case of wNAF used for key exchange, the Atomic block method was applied to be safe from side-channel attack (SCA) [32]. Web Assembly shows that it can operate efficiently and safely because its performance overhead ratio is lower than that of JavaScript.

Table 6. Web Assembly and JavaScript performance measurement and comparison through cryptographic algorithm implementation (cpb: Cycle Per Byte) [6].

	Chrome		**FireFox**		**Microsoft Edge**	
	Web Assembly	**JavaScript**	**Web Assembly**	**JavaScript**	**Web Assembly**	**JavaScript**
revised CHAM-64/128 [33]	120 cpb (2.1 times)	260 cpb	120 cpb (2.1 times)	260 cpb	120 cpb (2 times)	240 cpb
revised CHAM-128/128 [33]	60 cpb (3 times)	180 cpb	60 cpb (1.6 times)	100 cpb	70 cpb (1.8 times)	130 cpb
revised CHAM-128/256 [33]	70 cpb (3 times)	210 cpb	70 cpb (2.1 times)	150 cpb	70 cpb (2.8 times)	200 cpb
wNAF	27 cpb (11 times)	300 cpb	30 cpb (12 times)	365 cpb	27 cpb (11 times)	322 cpb
wNAF [34] (Atomic block [35])	42 cpb (10 times)	447 cpb	37 cpb (10 times)	405 cpb	37 cpb (14 times)	522 cpb
wNAF (Improved Atomic block [6])	32 cpb (11 times)	365 cpb	32 cpb (12 times)	387 cpb	30 cpb (14 times)	437 cpb
SHA-256 [36]	27 cpb (7.5 times)	203 cpb	20 cpb (10.8 times)	216 cpb	20 cpb (11 times)	221 cpb
HMAC [37]	92 cpb (7.5 times)	697 cpb	93 cpb (24.8 times)	2315 cpb	97 cpb (7.1 times)	693 cpb

Reference [38] converts the Picnic algorithm [39] to Web Assembly, measures the performance in Chrome, FireFox, and Microsoft Edge, and compares it with the C/C++ implementation. As a result, the Picnic algorithm implemented by Web Assembly is about 2∼3 times slower than the C/C++ implementation.

3.3. Cache Timing Attack

There are various attack methods, such as differential attack and side-channel attack, to find out important information about encryption algorithm. In addition, there is an attack method that finds out the key value, which is important information of the cryptographic algorithm through the cache access time of the CPU, and research on this is being actively studied as interest in it increases. Ref. [40] proved the vulnerability through an attack to find the last round key against the T-table AES algorithm of OpenSSL 1.1.0f [41]. So, in

Reference [40], they study and apply the T-table shuffling method to be safe against Flush + Reload, a kind of cache-timing attack [42]. In Reference [40], they randomly shuffle the array containing values from 0 to 255 using the Fisher-Yates function [43]. Then, the values stored in 4 256-byte T-tables are shuffled and used by using the shuffled array values. In Reference [40], the T-table was shuffled using the Fisher-Yates function in the AES T-table, and the test shows that it is safe against Flush + Reload cache-timing attacks.

4. Proposed Implementation for Secure Communication in Edge Computing Services

4.1. Overall Architecture of Proposed Software

The existing edge computing method processes data received from a server or user or data to be sent, and communicates through encryption and authentication. Therefore, in hardware, such as ARM, AVR, and RISC-V used in edge computing for encryption and authentication, secure communication is implemented by implementing encryption algorithms and authentication algorithms using programming languages suitable for each environment. However, since each environment uses different performance, different functions, and different programming languages, even the same algorithm needs to be implemented in each hardware. So, we use Web Assembly to implement encryption and authentication so that it can be used generally on each device. In addition, it uses the LWC ESTATE algorithm, which has both an encryption function and an authentication process. Web Assembly is designed for performance similar to a low-level language in a web environment. The ESTATE algorithm implemented by Web Assembly can be used in general without additional modification in PCs, smartphones, and IoT devices where web apps and web browsers can be used. Therefore, once created, it can be used in multiple devices for secure communication. In addition, the finally processed data is stored on the main server. Therefore, we propose additional optimization methods to use the ESTATE algorithm efficiently in the server. The operation process of the ESTATE algorithm has a characteristic that affects the next process using the previous value. Therefore, it is difficult to process a large amount of data at the same time. However, if the main server processes data sequentially, even if the edge computing method is used, the communication process eventually shows slow performance. So, we propose a method of using OpenCL parallel processing so that multiple ciphertexts and tags to be sent to multiple web-based applications can be created at the same time. In addition, to safely and efficiently operate the ESTATE algorithm, an additional method is proposed, and the previously existing studied methods are applied. During operation of TweAES-128, TweAES-128-6, and TweGIFT-128 used in the ESTATE algorithm, the 4-bit tweak value is checked for each specific round through conditional statements, and then expanded to perform XOR (exclusive OR) operation on the encrypted data. Therefore, we propose a method of storing and using the extended tweak values for 9 4-bit tweak values through pre-computation. So, tweak values are extended to 8-bit and 32-bit, respectively, through pre-computation. In the implementation of OpenCL, if there is a conditional statement, there is a load in the operation process. The ESTATE algorithm uses conditional statements due to the type of input value, tweak value check for each specific round, and tweak value XOR operation for each specific round. So, when we implement TweAES-128, TweAES-128-6, and TweGIFT-128 using OpenCL, we implement it using the loop unrolling method to eliminate performance degradation. In addition, it operates using local memory, which has a high operation speed. TweAES-128 and TweAES-128-6 are similar in operation to the AES algorithm. Therefore, it operates faster by applying the existing T-table method. In addition, there are studies that the AES algorithm is vulnerable to cache-timing attacks. Since TweAES-128 and TweAES-128-6, which have the same structure as the AES algorithm, can be vulnerable, they are safely operated by applying the T-table shuffling method, which is an the existing cache-timing attack response algorithm.

4.2. Edge Computing and Estate Implementation Using Web Assembly

We propose a web-based application edge computing method using Web Assembly. Web Assembly was created to show performance similar to low-level language in web environment. The existing edge computing method provides services by optimizing each environment and functions in hardware, such as ARM, AVR, and RISC-V. However, this method is difficult to use in general because it uses programming languages and functions used in each environment, such as ARM, AVR, and RISC-V, and additional cost is consumed because additional implementation is required for each device. The web-based application edge computing method proposed by us can be used in PCs, smartphones, IoT devices, etc. that can basically use web-based applications. In addition, even if the platform is different, it is efficient because it can be used generally without additional modification in terms of implementation. In addition, in order to implement the algorithm with Web Assembly, the existing code implemented in a programming language with a data type can be converted and used, so there is no additional cost. In addition, if you use a library, such as Node.js, so that web socket communication is possible without adapting the communication process to each hardware, communication becomes easy. Web-based application In the edge computing method, communication between server and web-based application, communication between web-based application and user, and communication between web-based application are made. Encryption and authentication functions are required to safely send data in various communication processes. So, we use the ESTATE algorithm, which has encryption and authentication functions. Therefore, as shown in Figure 12, in a web-based application using Web Assembly, a ciphertext and a tag for authentication are created using the ESTATE algorithm, and data is safely delivered to the user.

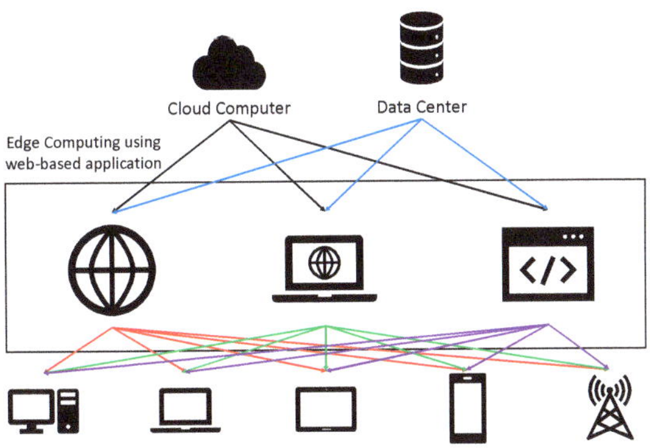

Figure 12. Edge computing structure using Web Assembly.

4.3. Parallel Implementation of Estate Using OpenCL

The ESTATE algorithm uses TweAES-128 and TweGIFT-128 to encrypt each block of 128-bit size. Then, the next step is performed using the previously encrypted result value. Therefore, it is impossible to use a method of processing a large amount of data at once through parallel processing. It is designed for use in a limited environment, but the finally communicated data is stored on the server. Therefore, it is necessary to implement ESTATE according to the server environment so that the server can use ESTATE efficiently. We use OpenCL to simultaneously calculate and transmit ciphertext and tag generation to be sent to multiple web-based applications.

Instead of sequentially processing multiple data using the ESTATE algorithm, it uses a method of simultaneously processing using OpenCL parallel processing as shown in Figure 13. When implemented using OpenCL parallel processing, performance degradation occurs when conditional statements exist. TweAES-128, TweAES-128-6, and TweGIFT-128 use conditional statements to check the type of input value, check whether or not padding, check the tweak value, and perform the extended tweak value XOR operation for each round. We use the loop unrolling method to remove the conditional statement in order to remove the performance load in the OpenCL implementation. In addition, the local memory has the fastest operation speed among OpenCL memories. For this reason, data is moved to local memory and encrypted to improve performance. Algorithm 5 is an OpenCL code algorithm that reduces the performance load by eliminating conditional statements using a loop unrolling method.

Algorithm 5 TweAES-128, TweAES-128-6, TweGIFT-128 proposed by applying loop unrolling method.

```
function loop unrolling TweAES-128(K, T, X)
  (W_43, ..., W_0) ← KeyGen(K, X)
  X ← X ⊕ (W_3, W_2, W_1, W_0)
  for i = 1 to 4 do
    X ← SubBytes(X)
    X ← ShiftRows(X)
    X ← MixColumns(X)
    X ← X ⊕ (W_{4i+3}, W_{4i+2}, W_{4i+1}, W_{4i})

    X ← SubBytes(X)
    X ← ShiftRows(X)
    X ← MixColumns(X)
    X ← X ⊕ (W_{8i+3}, W_{8i+2}, W_{8i+1}, W_{8i})
    AddTweak(X, T)

  X ← SubBytes(X)
  X ← ShiftRows(X)
  X ← MixColumns(X)
  X ← X ⊕ (W_39, W_38, W_37, W_36)
  X ← SubBytes(X)
  X ← ShiftRows(X)
  X ← X ⊕ (W_43, W_42, W_41, W_40)

function loop unrolling TweAES-6(K, T, X)
  (W_43, ..., W_0) ← KeyGen(K, X)
  X ← X ⊕ (W_3, W_2, W_1, W_0)
  for i = 1 to 2 do
    X ← SubBytes(X)
    X ← ShiftRows(X)
    X ← MixColumns(X)
    X ← X ⊕ (W_{4i+3}, W_{4i+2}, W_{4i+1}, W_{4i})

    X ← SubBytes(X)
    X ← ShiftRows(X)
    X ← MixColumns(X)
    X ← X ⊕ (W_{8i+3}, W_{8i+2}, W_{8i+1}, W_{8i})
    AddTweak(X, T)

  X ← SubBytes(X)
  X ← ShiftRows(X)
  X ← MixColumns(X)
  X ← X ⊕ (W_23, W_22, W_21, W_20)
  X ← SubBytes(X)
  X ← ShiftRows(X)
  X ← X ⊕ (W_43, W_42, W_41, W_40)

function loop unrolling TweGIFT-128(K, T, X)
  C ← 000000
  for i = 0 to 7 do
    for j = 0 to 3 do
      X ← SubCells(X)
      X ← PermBits(X)
      (K, X) ← AddRoundKey(K, X)
      (C, X) ← AddRoundConstant(C, X)

    X ← SubCells(X)
    X ← PermBits(X)
    (K, X) ← AddRoundKey(K, X)
    (C, X) ← AddRoundConstant(C, X)
    AddTweak(X, T)

  for i = 35 to 39 do
    X ← SubCells(X)
    X ← PermBits(X)
    (K, X) ← AddRoundKey(K, X)
    (C, X) ← AddRoundConstant(C, X)
```

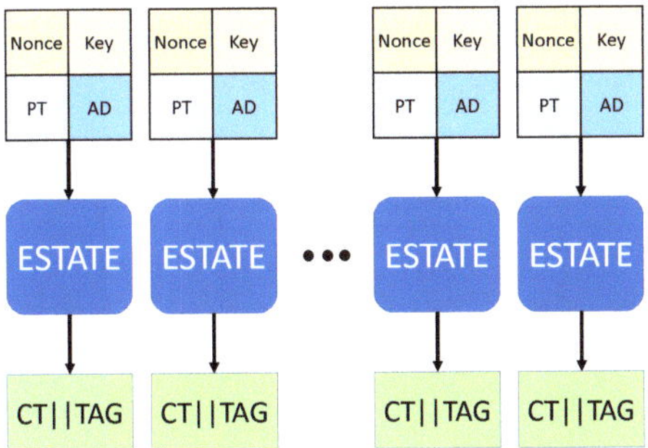

Figure 13. Structure of ESTATE algorithm operation using parallel processing.

4.4. Safe and Efficient Implementation of TweAES-128, TweAES-128-6, TweGIFT-128 of Estate Algorithm

TweAES-128 and TweGIFT-128 are used in ESTATE mode, and TweAES-128-6 is used in sESTATE mode. TweAES-128, TweGIFT-128, and TweAES-128-6 have the same operation process as AES-128 and GIFT-128, but additionally, the process of XOR operation by expanding the 4-bit tweak value is added. However, in TweAES-128, TweAES-128-6, and TweGIFT-128, only 0∼7, 15 are used as tweak values. Therefore, we propose a method to extend the 4-bit tweak value to 8-bit and 32-bit in advance to fit each algorithm and use it after storage. This method eliminates the unnecessary process of repeatedly checking and expanding tweak value. In addition, TweAES-128 and TweAES-128-6 have the same structure as the AES algorithm, so the existing studied T-table method to quickly compute AES can be applied. In addition, it is possible to perform faster operation by processing the 16-byte input value used in both algorithms in parallel.

As shown in Figure 14, the operation process of TweAES-128 and TweAES-128-6 used in the ESTATE algorithm uses an efficient method of simultaneously calculating 16-byte input values through OpenCL parallel processing. In addition, T-table shuffling method, which is the method studied in Reference [40], is applied to the T-table used in ESTATE TweAES-128 and TweAES-128-6 to safely operate against cache-timing attack.

Using method in Reference [40], mix the index value of 0∼255 to shuffle the T-table. Then, the T-table is shuffled using the mixed index value. Algorithm 6 is a process that will be used every round of ESTATE TweAES-128 and TweAES-128-6.

Algorithm 6 ESTATE TweAES-128, TweAES-128-6 Proposal Method Applying T-table Shuffling

Te0-sf : Te0[shuffle-array]
Te1-sf : Te1[shuffle-array]
Te2-sf : Te2[shuffle-array]
Te3-sf : Te3[shuffle-array]

function 1-round(S0∼S3, RK)
  S0 = Te0-sf[S0 $\gg$ 24] $\oplus$ Te1-sf[S1 $\gg$ 16 & 0xff] $\oplus$ Te2-sf[S2 $\gg$ 8 & 0xff] $\oplus$ Te3-sf[S3 & 0xff] $\oplus$ RK
  S1 = Te0-sf[S1 $\gg$ 24] $\oplus$ Te1-sf[S2 $\gg$ 16 & 0xff] $\oplus$ Te2-sf[S3 $\gg$ 8 & 0xff] $\oplus$ Te3-sf[S0 & 0xff] $\oplus$ RK
  S0 = Te0-sf[S2 $\gg$ 24] $\oplus$ Te1-sf[S3 $\gg$ 16 & 0xff] $\oplus$ Te2-sf[S0 $\gg$ 8 & 0xff] $\oplus$ Te3-sf[S1 & 0xff] $\oplus$ RK
  S0 = Te0-sf[S3 $\gg$ 24] $\oplus$ Te1-sf[S0 $\gg$ 16 & 0xff] $\oplus$ Te2-sf[S1 $\gg$ 8 & 0xff] $\oplus$ Te3-sf[S2 & 0xff] $\oplus$ RK

function 1-round with AddTweak(S0∼S3, RK, tweak)
  S0 = Te0-sf[S0 $\gg$ 24] $\oplus$ Te1-sf[S1 $\gg$ 16 & 0xff] $\oplus$ Te2-sf[S2 $\gg$ 8 & 0xff] $\oplus$ Te3-sf[S3 & 0xff] $\oplus$ RK
  S1 = Te0-sf[S1 $\gg$ 24] $\oplus$ Te1-sf[S2 $\gg$ 16 & 0xff] $\oplus$ Te2-sf[S3 $\gg$ 8 & 0xff] $\oplus$ Te3-sf[S0 & 0xff] $\oplus$ RK
  S0 = Te0-sf[S2 $\gg$ 24] $\oplus$ Te1-sf[S3 $\gg$ 16 & 0xff] $\oplus$ Te2-sf[S0 $\gg$ 8 & 0xff] $\oplus$ Te3-sf[S1 & 0xff] $\oplus$ RK
  S0 = Te0-sf[S3 $\gg$ 24] $\oplus$ Te1-sf[S0 $\gg$ 16 & 0xff] $\oplus$ Te2-sf[S1 $\gg$ 8 & 0xff] $\oplus$ Te3-sf[S2 & 0xff] $\oplus$ RK
  AddTweak(S0∼S3, tweak)

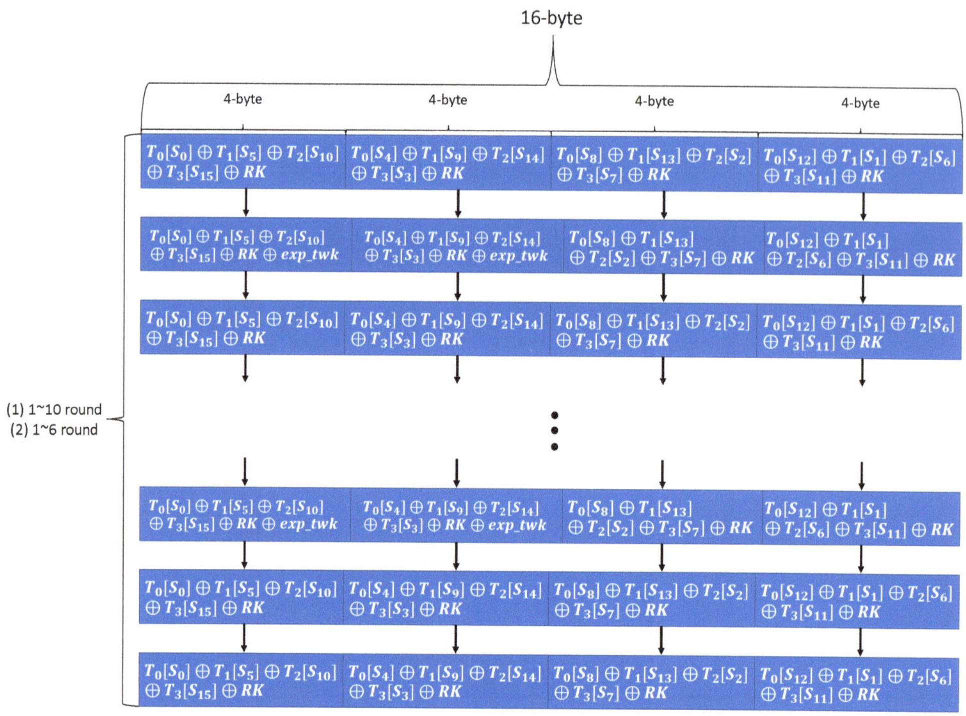

Figure 14. Parallel operation process of TweAES-128 or TweAES-128-6 using T-table method.

5. Results

Table 7 is an environment in which the results were measured by applying the methods proposed by us to the ESTATE algorithm using OpenCL parallel processing, the ESTATE algorithm implemented with Web Assembly, and the reference C ESTATE algorithm.

Table 8 is a comparison result of OpenCL parallel processing, AES T-table, extended tweak pre-computation, and ESTATE algorithm applying loop unrolling methods and the reference C code ESTATE algorithm for sequential processing. We measured the process of creating a total of 6,400 ciphertexts and tags, respectively. As a result, in ESTATE TweAES-128, TweAES-128-6 and TweGIFT-128, OpenCL was 6.69 times, 7.31 times, and 1.47 times faster than the reference C/C++ code, respectively.

Table 7. Performance measurement environment.

Operationg System	CPU	RAM	SW	Languages and API	Used Input Value	ESTATE Operation Count
Window 10 Education	Intel i5-8250U 1.6 GHz	8 GB	(1) Chrome (2) FireFox (3) Microsoft Edge	(1) C/C++ (2) Web Assembly (3) OpenCL	Nonce: 25,600-byte AD: 51,200-byte Message: 512,000-byte	6400

Table 8. Performance comparison of OpenCL implementation and C/C++ reference code applying the proposed method (ns: nanosecond).

Algorithm	OpenCL	Reference C/C++	Performance Improvement
ESTATE TweAES-128	19,088,500 ns	127,842,877 ns	6.69 times
ESTATE TweAES-128-6	15,966,333 ns	116,813,270 ns	7.31 times
ESTATE TweGIFT-128	1,958,343,000 ms	2,897,251,400 ns	1.47 times

Table 9 shows the result of comparing the algorithm to which the T-table shuffling method was applied and the algorithm not applied. This is a measurement result of the process of shuffling and calculating the 1024-byte T-table. Due to shuffling, performance overhead occurs because memory must be accessed twice, unlike the method not applied. As a result, ESTATE TweAES-128 and TweAES-128-6 show performance overhead of 7% and 51%, respectively.

Table 9. Performance overhead measurement result through application of T-table shuffling method (ns: nanosecond).

Algorithm	Applied T-Table Shuffling Method	Normal Method	Performance Overhead
ESTATE TweAES-128	20,589,394 ns	19,088,500 ns	7%
ESTATE TweAES-128-6	24,192,899 ns	15,966,333 ns	51%

Table 10 shows how much performance overhead occurs compared to C language by implementing the ESTATE algorithm in Web Assembly to use the edge computing method using Web Assembly. Measurements were made for C and Web Assembly using the same input values. Web Assembly was measured on Chrome, FireFox, and Microsoft Edge. As a result, TweAES-128, TweAES-128-6, and TweGIFT-128 implemented as Web Assembly have 11%, 5%, 22% performance overhead in Chrome, 10%, 2%, 54 in FireFox. It shows % performance overhead, and 9%, 6%, and 17% performance overhead in Microsoft Edge. The reason the performance overhead ratio is different for each web browser is that the rendering engine and JavaScript engine used for each web browser are different. However, in the case of TweAES-128 and TweAES-128-6, the performance overhead is not large, so it can be seen that it is efficient to perform edge computing through a web-based application using Web Assembly.

Table 10. Performance overhead measurement result of ESTATE algorithm using Web Assembly (ns: nanosecond).

Algorithm	Reference C/C++ Code	Web Assembly		
		Chrome (Performance Overhead)	FireFox (Performance Overhead)	Microsoft Edge (Performance Overhead)
ESTATE TweAES-128	127,842,877 ns	142,775,000 ns (11%)	141,000,000 ns (10%)	140,374,999 ns (9%)
ESTATE TweAES-128-6	116,813,270 ns	123,155,001 ns (5%)	120,000,000 ns (2%)	124,045,001 ns (6%)
ESTATE TweGIFT-128	2,897,251,400 ns	3,560,440,001 ns (22%)	4,490,000,000 ns (54%)	3,401,205,000 ns (17%)

6. Conclusions

The existing edge computing method takes over the role of cloud computing services in hardware, such as ARM, AVR, and RISC-V. Therefore, there is a disadvantage of having to implement separately using a function and programming language suitable for each environment used in ARM, AVR, and RISC-V. In this paper, we propose a web-based application edge computing method using Web Assembly in order to use an efficient edge computing method.

1. Implementation of ESTATE algorithm using Web Assembly
Web Assembly was created to show similar performance to low-level language in a web environment. Cryptographic algorithms using web-based applications can use web-based applications, and can be used without additional modification in PCs, smart phones, and IoT devices used as edge devices. Therefore, even if the platforms used are different, it is also cost-effective because it can be used generally without additional modification in terms of implementation. In addition, web-based application edge computing communicates with various platforms, so, to send data securely, we implement and use the ESTATE algorithm, which has both encryption and authentication processes, in Web Assembly. We can see how Web Assembly has caught up with the performance of low-level languages. ESTATE Web Assembly implementation compares performance with reference C/C++ code. Web Assembly implementation is measured in web browsers Chrome, FireFox, and Microsoft Edge. As a result, TweAES-128, TweAES-128-6, and TweGIFT-128 implemented as Web Assembly have 11%, 5%, 22% performance overhead in Chrome, 10%, 2%, 54 in FireFox. It shows % performance overhead, and 9%, 6%, and 17% performance overhead in Microsoft Edge. As a result, it is slower than C/C++, which is a low-level language, but it can be used efficiently because it can be used without special modifications on devices that can use web-based applications.
2. ESTATE algorithm using OpenCL parallel processing
Data processed by the web-based application edge computing method are eventually stored on the main server. Therefore, in order to use the ESTATE algorithm efficiently, it is necessary to implement it according to the server environment. So, we propose a method of simultaneously processing ciphertext and tag generation to be sent to multiple platforms using OpenCL parallel processing. Through OpenCL parallel processing, each byte value is processed simultaneously instead of sequentially for the 16-byte input value used for one encryption process. OpenCL has a load when using conditional statements. In the ESTATE algorithm, a conditional statement is used to XOR the extended tweak value every specific round. Therefore, the loop unrolling method was used to remove the performance load by removing the process of using conditional statements. In addition, data is stored in a local memory with a fast operation speed and encrypted to perform efficient operation. For performance comparison, we compare the OpenCL parallel processing implementation and the reference C/C++ sequential processing implementation. As a result, the OpenCL implementation shows about 6.69 times, 7.31 times, and 1.47 times performance improvement in ESTATE TweAES-128, TweAES-128-6, and TweGIFT-128 than the reference C/C++ implementation.
3. Method for efficient and safe operation of ESTATE algorithm
Additional methods are applied to safely and efficiently operate the ESTATE algorithm itself. The ESTATE algorithm uses conditional statements to check the type of input value to be encrypted, check whether it is the last block, check the tweak value, and calculate the extended tweak value for each specific round. The 8-bit and 32-bit extended tweak values used in TweAES-128, TweAES-128-6, and TweGIFT-128 are stored and used in advance through pre-calculation. This method reduces the performance load by removing unnecessary conditional statements. In addition, TweAES-128 and TweAES-128-6 have the same operation process as the AES algorithm, so they may be vulnerable to cache-timing attacks. So, we apply the T-table shuffling method, which is a previously studied method, to operate safely. We reduced the performance load by applying the proposed methods to minimize the performance load even when the T-table shuffling method is applied. As a result of applying the T-table shuffling method, TweAES-128 and TweAES-128-6 show about

7% and 51% performance overhead, respectively, than those without applying the T-table shuffling method.

4. Future Work
Web-based application using Web Assembly can be used in various devices without additional modification, so it can reduce the system load of the server and is effective in responding to failures. Web Assembly is currently continuously developing, and, since various devices, such as PCs, smart phones, and smart devices, are developing more and more, web technology is also developing accordingly. Currently, technologies using high-end hardware, such as Web Assembly's SIMD technology and WebGPU, are being developed. In addition, it is being developed so that Web Assembly and WebGPU can be used together. When these technologies become stable in the future, many web developers will develop web services using various technologies, such as SIMD and WebGPU. Therefore, it can be used in various ways in terms of crypto security, and various studies will be conducted using web technologies developed in the field of crypto security. Therefore, the web-based application edge computing method can also be developed, and performance will be improved. Currently, there are various NIST LWC (National Institute of Standards and Technology LightWeight Cryptography) Round 2 candidate algorithms. However, the OpenCL parallel processing method we used is a method applicable to other candidate algorithms. Even if the LWC algorithm other than ESTATE is used to send data to multiple devices, the service can be provided more efficiently by using the method of simultaneously processing multiple ciphertexts and tags through the OpenCL parallel processing method.

Author Contributions: Writing—original draft, B.P.; Writing—review and editing, S.C.S. All authors have read and agreed to the published version of the manuscript.

Funding: This work was supported by the National Research Foundation of Korea (NRF) grant funded by the Korea government (MSIT) (No. 2019R1F1A1058494).

Institutional Review Board Statement: Not applicable.

Informed Consent Statement: Not applicable.

Data Availability Statement: Not applicable.

Conflicts of Interest: The authors declare no conflict of interest.

References

1. Hayes, B. *Cloud Computing;* Communications of the ACM: New York, NY, USA, July 2008.
2. Yu, W.; Liang, F.; He, X.; Hatcher, W.G.; Lu, C.; Lin, J.; Yang, X. A Survey on the Edge Computing for the Internet of Things. *IEEE Access* **2018**, *6*, 6900–6919. [CrossRef]
3. Ai, Y.; Peng, M.; Zhang, K. Edge computing technologies for Internet of Things: A primer. *Digit. Commun. Netw.* **2018**, *4*, 77–86. [CrossRef]
4. Wang, X.; Han, Y.; Leung, V.C.M.; Niyato, D.; Yan, X.; Chen, X. Convergence of Edge Computing and Deep Learning: A Comprehensive Survey. *IEEE Commun. Surv. Tutor.* **2020**, *22*, 869–904. [CrossRef]
5. Tilkov, S.; Vinoski, S. Node.js: Using JavaScript to build high-performance network programs. *IEEE Internet Comput.* **2010**, *14*, 80–83. [CrossRef]
6. Park, B.; Song, J.; Seo, S.C. Efficient Implementation of a Crypto Library Using Web Assembly. *Electronics* **2020**, *9*, 1839. [CrossRef]
7. Rossberg, A.; Titzer, B.L.; Haas, A.; Schuff, D.L.; Gohman, D.; Wagner, L.; Zakai, A.; Bastien, J.F.; Holman, M. Bringing the web up to speed with WebAssembly. *Commun. ACM* **2018**, *61*, 107–115. [CrossRef]
8. Rossberg, A. WebAssembly Specification Release 1.1, 2020. Available online: https://webassembly.github.io/spec/core/ (accessed on 25 February 2021).
9. Ritchie, D.M. The development of the C language. *ACM Sigplan Not.* **1993**, *28*, 201–208. [CrossRef]
10. Smith, E. The C++ Language. In *Introduction to the Tools of Scientific Computing;* Springer: Berlin, Germany, 2020; pp. 133–148.
11. Doglio, F. An Introduction to TypeScript. In *Introducing Deno;* Springer: Berlin, Germany, 2020; pp. 27–62.
12. Bhattacharjee, J. Basics of Rust. In *Practical Machine Learning with Rust;* Springer: Berlin, Germany, 2020; pp. 1–30.
13. Sjölander, Erik. Krypteringsalgoritmer i OpenCL: AES-256 och ECC ElGamal, 2012. Available online: https://www.diva-portal.org/smash/get/diva2:555565/FULLTEXT01.pdf (accessed on 25 February 2021).

14. Munshi, A.; Gaster, B.; Mattson, T.G.; Ginsburg, D. *OpenCL Programming Guide*; Pearson Education: London, UK, 2011.
15. Munshi, A. The opencl specification. In Proceedings of the 2009 IEEE Hot Chips 21 Symposium (HCS), Stanford, CA, USA 26 May 2009; pp. 1–314.
16. Chakraborti, A.; Datta, N.; Jha, A.; Mancillas-López, C.; Nandi, M.; Sasaki, Y. ESTATE: A Lightweight and Low Energy Authenticated Encryption Mode. *IACR Trans. Symmetric Cryptol.* **2020**, *2020*, 350–389. [CrossRef]
17. Black, J.; Rogaway, P. *CBC MACs for Arbitrary-Length Messages: The Three-Key Constructions*; Springer: Berlin, Germany, 2000 pp. 197–215.
18. Dworkin, M. *Recommendation for Block Cipher Modes of Operation. Methods and Techniques*; Technical Report; National Inst of Standards and Technology: Gaithersburg, MD, USA, 2001.
19. Yao, J.; Zimmer, V. Cryptography. In *Building Secure Firmware*; Springer: Berlin, Germany, 2020; pp. 767–823.
20. Heron, S. Advanced encryption standard (AES). *Netw. Secur.* **2009**, *2009*, 8–12. [CrossRef]
21. Banik, S.; Pandey, S.K.; Peyrin, T.; Sasaki, Y.; Sim, S.M.; Todo, Y. GIFT: A Small Present—Towards Reaching the Limit of Lightweight Encryption. In Proceedings of the Cryptographic Hardware and Embedded Systems—CHES 2017—19th International Conference, Taipei, Taiwan, 25–28 September 2017; pp. 321–345. [CrossRef]
22. Sanida, T.; Sideris, A.; Dasygenis, M. Accelerating the AES Algorithm using OpenCL. In Proceedings of the 2020 9th International Conference on Modern Circuits and Systems Technologies (MOCAST), Bremen, Germany, 18 September 2020; pp. 1–4.
23. Yuan, Y.; He, Z.; Gong, Z.; Qiu, W. Acceleration of AES encryption with OpenCL. In Proceedings of the 2014 Ninth Asia Joint Conference on Information Security, Wuhan, China, 29 January 2015; pp. 64–70.
24. Wang, X.; Li, X.; Zou, M.; Zhou, J. AES finalists implementation for GPU and multi-core CPU based on OpenCL. In Proceedings of the 2011 IEEE International Conference on Anti-Counterfeiting, Security and Identification, Xiamen, China, 29 July 2011; pp. 38–42.
25. Conti, V.; Vitabile, S. Design exploration of aes accelerators on fpgas and gpus. *J. Telecommun. Inf. Technol.* **2017**, *1*, 28–38.
26. Di Biagio, A.; Barenghi, A.; Agosta, G.; Pelosi, G. Design of a parallel AES for graphics hardware using the CUDA framework. In Proceedings of the 2009 IEEE International Symposium on Parallel & Distributed Processing, Rome, Italy, 10 July 2009; pp. 1–8.
27. D Amato, J.P.; Vénere, M.J. Encrypting video and image streams using OpenCL code on-demand. *CLEI Electron. J.* **2014**, *17*, 6.
28. Matsui, M. Linear cryptanalysis method for DES cipher. In *Workshop on the Theory and Application of of Cryptographic Techniques*; Springer: Berlin, Germany, 1993; pp. 386–397.
29. Schneier, B. *Description of a New Variable-Length Key, 64-Bit Block Cipher (Blowfish)*; Springer: Berlin, Germany, 1993; pp. 191–204.
30. Rivest, R.L.; Shamir, A.; Adleman, L. A method for obtaining digital signatures and public-key cryptosystems. *Commun. ACM* **1978**, *21*, 120–126. [CrossRef]
31. Inampudi, G.R.; Shyamala, K.; Ramachandram, S. Parallel implementation of cryptographic algorithm: AES using OpenCL on GPUs. In Proceedings of the 2018 2nd International Conference on Inventive Systems and Control (ICISC), Coimbatore, India, 28 June 2018; pp. 984–988.
32. Standaert, F.X. Introduction to side-channel attacks. In *Secure Integrated Circuits and Systems*; Springer: Berlin, Germany, 2010; pp. 27–42.
33. Roh, D.; Koo, B.; Jung, Y.; Jeong, I.; Lee, D.; Kwon, D.; Kim, W. Revised Version of Block Cipher CHAM. In Proceedings of the Information Security and Cryptology—ICISC 2019—22nd International Conference, Seoul, Korea, 4–6 December 2019; pp. 1–19. [CrossRef]
34. King, B. *wNAF*, an Efficient Left-To-Right Signed Digit Recoding Algorithm*; Springer: Berlin, Germany, 2008; pp. 429–445.
35. Chevallier-Mames, B.; Ciet, M.; Joye, M. Low-cost solutions for preventing simple side-channel analysis: Side-channel atomicity *IEEE Trans. Comput.* **2004**, *53*, 760–768. [CrossRef]
36. Fips Pub. *Secure Hash Standard (SHS)*; Fips Pub: Edmonds, WA, USA, March 2012; Volume 180.
37. Turner, J.M. The keyed-hash message authentication code (hmac). *Fed. Inf. Process. Stand. Publ.* **2008**, *198*, 1.
38. Rösch, J. Efficient Implementation of Picnic. 23 September 2020. Available online: https://is.muni.cz/th/pbn05/Efficient_implementation_of_Picnic.pdf (accessed on 25 February 2021.)
39. Chase, M.; Derler, D.; Goldfeder, S.; Orlandi, C.; Ramacher, S.; Rechberger, C.; Slamanig, D.; Zaverucha, G. *The Picnic Signature Scheme Design Document (Version 1.0)*; NIST Post-Quantum Cryptogr. Stand. Round; NIST: Gaithersburg, MD, USA, 2017; Volume 3.
40. Daehyeon Bae, J.H.; Ha, J. Implementation of AES Resistant to Cache Side-Channel Attack Using T-Table Shuffling Method. *Conf. Inf. Secur. Cryptogr. Winter* **2020**, *30*, 579–583.
41. Young, E.A.; Hudson, T.J.; Engelschall, R.S. OpenSSL. 9 November 2001. Available online: http://www.openssl.org/ (accessed on 25 February 2021)
42. Bernstein, D.J. Cache-Timing Attacks on AES. 14 April 2005. Available online: https://cr.yp.to/antiforgery/cachetiming-20050414.pdf (accessed on 25 February 2021).
43. Fisher, R.A.; Yates, F. *Statistical Tables: For Biological, Agricultural and Medical Research*; Oliver and Boyd: Edinburgh, UK, 1938.

sensors

MDPI

Article

Identification of IoT Actors

Suada Hadzovic [1,*], Sasa Mrdovic [1] and Milutin Radonjic [2]

1 Faculty of Electrical Engineering, University of Sarajevo, 71000 Sarajevo, Bosnia and Herzegovina; sasa.mrdovic@etf.unsa.ba
2 Faculty of Electrical Engineering, University of Montenegro, 81000 Podgorica, Montenegro; mico@ucg.ac.me
* Correspondence: shadzovic@rak.ba

Abstract: The Internet of Things (IoT) is a leading trend with numerous opportunities accompanied by advantages as well as disadvantages. Parallel with IoT development, significant privacy and personal data protection challenges are also growing. In this regard, the General Data Protection Regulation (GDPR) is often considered the world's strongest set of data protection rules and has proven to be a catalyst for many countries around the world. The concepts and interaction of the data controller, the joint controllers, and the data processor play a key role in the implementation of the GDPR. Therefore, clarifying the blurred IoT actors' relationships to determine corresponding responsibilities is necessary. Given the IoT transformation reflected in shifting computing power from cloud to the edge, in this research we have considered how these computing paradigms are affecting IoT actors. In this regard, we have introduced identification of IoT actors according to a new five-computing layer IoT model based on the cloud, fog, edge, mist, and dew computing. Our conclusion is that identifying IoT actors in the light of the corresponding IoT data manager roles could be useful in determining the responsibilities of IoT actors for their compliance with data protection and privacy rules.

Keywords: Internet of things; IoT actor; data manager; GDPR; computing

Citation: Hadzovic, S.; Mrdovic, S.; Radonjic, M. Identification of IoT Actors. *Sensors* **2021**, *21*, 2093. https://doi.org/10.3390/s21062093

Academic Editor: Taehong Kim

Received: 31 January 2021
Accepted: 9 March 2021
Published: 17 March 2021

1. Introduction

The Internet of Things (IoT) already occupies a significant area, and its perspective is practically unlimited. According to Cisco, there will be 29.3 billion networked devices by 2023 [1].

The IoT vision is still evolving as an enabling technology because the IoT keeps developing and new IoT applications are being proposed. Consequently, there is no common IoT definition [2]. In this regard, there is an open call for the contribution of knowledge and perception of the ever-changing definition of the IoT [3].

Because standardization is a process that accompanies the production of new IoT platforms, sensors, and actuators, it has been carried out from the very beginning of the application of this technology. The high complexity of the IoT ecosystem encompasses a wide spectrum of solutions and standards, which is clear from the fact that, in 2016, there were more than 900 IoT-related standards [4].

At the beginning of concept development, typical examples of IoT applications were mostly related to common day to day objects and processes. However, intensive growth of IoT applications moved the focus of implementation towards industrial automation, smart cities, public safety, medical and healthcare systems, and many others. In such circumstances, proper regulation in the IoT domain becomes very important.

To enable synergies for new business models and to reduce barriers, IoT stakeholders must work together and address issues such as interoperability, privacy, and security, and many others, while policymakers need to understand these complex relationships and clearly identify IoT actors and their responsibilities.

In this regard, as a starting point, we considered some of the most relevant recommendations for the IoT, developed by the International Telecommunication Union (ITU).

Accordingly, we analysed the IoT ecosystem and business models given in the ITU Recommendations ITU T Y.4000 [5], ITU T Y.4100 [6], and ITU T Y.4114 [7].

To get a complete picture, we have conducted the mappings between the IoT actors and business roles that were given in the selected ITU recommendations and from various other perspectives (regulatory, product development and technology consultancy companies, industry organizations, and other).

The scope of the IoT involves different sectors and, consequently, various authorities are present, such as electronic communications regulatory authorities; data authorities; regulators and ministries for energy, health, air security, traffic, transport; and others. Consequently, more open, collaborative, and cross-sectoral regulation is needed. IoT policies and regulations are still under development and it is important to encourage a coordinated regulatory approach that includes all sectors. This concept of "collaborative regulation" and "fifth-generation regulation", originally developed by the ITU, is the only viable solution in this Data Age, where IoT development enables the generation of a huge amount of data by various data sources [8].

Given the wide range of regulatory challenges, we focused our work on regulatory challenges with an emphasis on the data protection and privacy aspects. In this regard, in the IoT legislative landscape of EU, the General Data Protection Regulation (GDPR) [9] and draft ePrivacy Regulation [10] are standing as trust drivers, and correct identification of roles such as data controller, joint controllers, or data processor and allocation of corresponding responsibilities would be extremely demanding. The situation in the case of IoT is even more difficult if we keep in mind that IoT is characterized by joint controllers who have complex and different shares of corresponding responsibilities.

A review of existing IoT architectures in Alshohoumi et al. [11] identified sixteen different IoT architectures that were developed during the period from 2008 to 2018, emphasizing the gradual evolution of IoT architecture across the years. It is shown that, from the layered architecture perspective, IoT extends from an early three-layer architecture model to the eight-layer architecture model. Based on analysis of seven IoT architectures, authors in Lynn et al. [12] summarized that key features in IoT reference architectures include data management, security and privacy, analytics, data visualization and user interface, supported computing paradigms, scalability, and interoperability.

Different standardization groups work continuously on reference models for IoT architectures. Acknowledging the existence of many IoT architectures, we attempt to find a correlation between the basic model of the network for the IoT presented in Recommendation ITU-T Y.4113 [13], Cisco IoT simplified architecture [14], the conceptual model of fog and mist computing for IoT given in the publication by National Institute of Standards and Technology (NIST) [15], the architectural reference models of devices for IoT applications in Recommendation ITU-T Y.4460 [16], and the IoT value chain as a useful tool for regulatory authorities.

There is existing related study into the identification of various stakeholders in the IoT value chain model on Smart Cities, where authors proposed taxonomy that categorizes and lists the relevant technology and regulatory characteristics of Smart City services [17]. Regarding computing paradigms, the authors in Yousefpour et al. [18] provide an extensive tutorial on fog computing and related computing paradigms and identify relevant operators and computing hardware locations. The authors in Ray [19] and Šojat and Skala [20] give an introduction to dew computing and identify dew computing locations.

Different from these studies, the contribution of this paper is three-fold: (1) We provide a detailed mapping between the IoT actors identified from various perspectives. (2) As part of the identification of IoT components and IoT actors, we identify new data manager roles related to dew, mist, edge, fog, and cloud computing. (3) We have compiled it into a new five-computing layer IoT model based on the cloud, fog, edge, mist, and dew computing, including identified IoT actors and additional roles according to the different computing paradigms and the GDPR. This IoT model can serve as a valuable support in clarifying IoT components, IoT actors, and corresponding GDPR roles.

The paper is organized as follows: Section 2 gives a short overview of GDPR and draft ePrivacy Regulation, with a focus on identification of GDPR actors. Section 3 presents a detailed description of the IoT components. Section 4 is focused on computing paradigms in IoT and the most common fog, edge, mist, and dew computing hardware locations. Next, in Section 5, mappings are provided between the IoT actors identified in selected ITU recommendations and those given from various perspectives (regulatory, product development and technology consultancy companies, industry organizations, and others). Section 6 introduces the proposed IoT model with the identified IoT actors, relevant GDPR actors, and the new data manager roles related to dew, mist, edge, fog, and cloud computing. Section 7 gives a short overview of current regulatory status and the need for a collaborative regulatory approach. Section 8 gives a short overview of data brokers, gatekeepers, and other actors in light of recent legislatives. In Section 9, conclusions are drawn.

2. The Data Protection and Privacy Legislative Landscape in the EU

The IoT is an important part of the current Data Age reality where, parallel to IoT development, significant privacy and personal data protection challenges are also growing. The GDPR and draft ePrivacy Regulation in the EU are dealing with these issues.

2.1. GDPR and Implementation Guidelines

Core activities regarding data are creation, collection, storage, aggregation and organization, processing and analysis, marketing and distribution, and use. Support activities regarding data are data laws, regulations, and policies; data security and privacy-related service; ICT (Information and Communication Technology) connectivity and infrastructure services; and data skills enhancement services. Data laws, regulations, and policies can address data and data rights ownership, data classification and metadata, data protection and security, data privacy, data transparency and consent, and data commercialization [21].

GDPR entered into force in 2016, and all organizations who target or collect data related to people in the EU were required to be compliant with GDPR as of May 25, 2018 [9]. Data protection by design and by default is mandated; consequently, data protection should be considered both at the stage of the determination of the means of the processing as well as at the time of the actual processing (Article 25). The concept and interaction of data controller and data processor are central.

Given the existing complexity in defining corresponding roles, relevant guidelines have been published to clarify the roles of the controller, joint controller, and processor, and the distribution of responsibilities among them.

The European Data Body Supervisor issued guidelines to the EU institutions regarding their role in the processing of personal data on 7 November 2019. Although the guideline is limited to EU institutions, it can be very useful for all business in determining their role as controller, joint controller, or processor under the GDPR. It is clarified that an entity does not need to have access to personal data to be a controller as long as it has an influence on processing, determines the purposes and means of processing, or receives the anonymous statistics based on personal data collected and processed by another entity. The duties of controllers and processors are explained in Annexes 2 and 3, while the flowchart is given in Annex 1 for a situation in which the distribution of roles of processors and controllers has not been determined by a legal act [22].

The European Data Protection Board has issued guidelines on controller and processor concepts in the GDPR for open public consultations from 2 September 2020 to 19 October 2020. The guidelines clarify that joint control can be based on a joint decision of two or more multiple entities or through a convergent decision of two or more entities [23].

Proposed guidelines for meeting the GDPR principles is also given in European Telecommunications Standards Institute (ETSI) Technical Report ETSI TR 103 591 (2019-10) [24].

Under the GDPR, supervisory authorities and the EU Commission are allowed to issue standard clauses to be included in the contract between processors and controllers providing a way to ensure that the contract complies with the GDPR [25].

The first European Commission evaluation and review of the GDPR, published on 24 June 2020, emphasizes the importance of clarifying how to apply proven principles to specific technologies such as IoT, artificial intelligence, blockchain, and facial recognition The implementation of the GDPR, as opposed to large digital companies and integrated companies, has been recognized as an essential element for the protection of individuals. The right to data portability, which enables individuals to switch between different service providers, is considered one of the Commission's priorities, particularly with the increasing use of the IoT [26].

A short overview of GDPR actors is given in Table 1.

Table 1. General Data Protection Regulation (GDPR) actors.

GDPR Actor	Description by the GDPR [9]
Controller	Article 4 point (7) "controller means the natural or legal person, public authority, agency or other body which, alone or jointly with others, determines the purposes and means of the processing of personal data". Article 35 paragraph 1. "Where a type of processing in particular using new technologies, and taking into account the nature, scope, context and purposes of the processing, is likely to result in a high risk to the rights and freedoms of natural persons, the controller shall, prior to the processing, carry out an assessment of the impact of the envisaged processing operations on the protection of personal data."
Joint Controller	Article paragraph 26 "Where two or more controllers jointly determine the purposes and means of processing, they shall be joint controllers".
Processor	Article 4 point (8) "processor means a natural or legal person, public authority, agency, or other body which processes personal data on behalf of the controller".
Third Party	Article 4 point (10) "third party means a natural or legal person, public authority, agency or body other than the data subject, controller, processor and persons who, under the direct authority of the controller or processor, are authorised to process personal data".
Data Protection Officer (DPO)	Article 37 paragraph 1. "The controller and the processor shall designate a data protection officer in any case where:" ... b) "the core activities of the controller or the processor consist of processing operations which, by virtue of their nature, their scope and/or their purposes, require regular and systematic monitoring of data subjects on a large scale;" Article 37 6. "The data protection officer may be a staff member of the controller or processor, or fulfil the tasks on the basis of a service contract"
Supervisory Authority	Article 51 paragraph 1 "Each Member State shall provide for one or more independent public authorities to be responsible for monitoring the application of this Regulation, ... "
Lead Supervisory Authority	Article 56 paragraph 1 " ... the supervisory authority of the main establishment or of the single establishment of the controller or processor shall be competent to act as lead supervisory authority for the cross-border processing carried out by that controller or processor ... " Article 56 paragraph 6 "The lead supervisory authority shall be the sole interlocutor of the controller or processor for the cross-border processing carried out by that controller or processor. "

2.2. Draft ePrivacy Regulation

In January 2017, the European Commission published a Proposal of Regulation on Privacy in Electronic Communications (draft ePrivacy). On January 5, 2021, the Council of the European Union released the 14th draft version of the ePrivacy Regulation. [10]. It is broader than the GDPR because it applies not only to the processing of personal data but also to the processing of any electronic communications data and other data collected from the end user's device. The goal is to safeguard the integrity of end user devices and the privacy and confidentiality of their communications.

In the proposed text, Recital 12 states that, to ensure full protection of the rights to privacy and confidentiality of communications and the promotion of a reliable and secure IoT, the proposed regulation should apply to machine-to-machine communications transmission.

Recital 17 states that electronic communication networks and service providers should be permitted to process electronic communications metadata after obtaining the consent of the end user or, where necessary, to provide an electronic communications service under an end user contract, and these where necessary to protect an interest that is essential for the lives of end-users.

Proposed taxonomy for personal data in the context of the telecommunication sector in [27] depicts what kind of data is accessible by different actors. For example, Fixed Network Operators, Mobile Network Operators, and Mobile Virtual Network Operators have access to content data in clear text but cannot use it. Over to the top (OTT) service providers can access the data for the service they provide. Device manufacturers/Operating system providers can access data before it leaves the device.

3. IoT Components

The IoT consists of various components or building blocks, and there are a variety of approaches to identify IoT components such as those included in the basic model of the network for the IoT, identified in Recommendation ITU-T Y.4113, which consists of Device, IoT area network, Gateway, Access Network, Core network, IoT Platform, and IoT application server [13]. The IoT infrastructure identified in the paper [28] consists of IoT Devices, IoT Platform, Fog nodes, Cloud nodes, and IoT Applications.

To develop the effective legislation further, regulatory authorities need to have a better understanding of IoT building components, who the IoT actors are, what are their relationships, and how IoT building components add value to the IoT solution and for the end-user. In this regard, the value chain model could be an example of a useful analytical tool for regulators. At the same time, the IoT value chain presents a challenge as IoT is evolving, and it involves various IoT actors and building components with dynamic and unclear relationships between them. Some of the components of the IoT value chain are: Device, Connectivity, as identified in Mackenzie and Rebbeck [29,30] and Paradis [31], Applications, identified in Mackenzie and Rebbeck [29,30], IoT platform enablement, identified in [30], System Integration, identified in Mackenzie and Rebbeck [29,30], Service enablement, identified in Paradis [31], Service Provision, identified in [30], and Customer, identified in Paradis [31].

From the above examples of identification of IoT components, it can be concluded that the identified IoT components sometimes have a certain degree of overlap. In the following, we would like to point out some IoT components that we believe would be useful for regulators to understand the relevant processes and relationships among IoT actors, noting that the list of possible IoT components is not exhaustive.

3.1. *Thing*

As for the IoT, the ITU has recognized the Thing as an object of the physical world or the information world, capable of being identified and integrated into communication networks [5]. Sometimes, Thing is integrated into a smart device itself, or Thing stands alone and a separate product is connected, making it a smart device. Although the ITU basic network [13] does not present Thing alone, our point of view is that Thing needs to be visible as a building block.

3.2. *Device*

The IoT is full of new terms, such as the Mote, which stands for Remote, where Motes make up a significant portion of the IoT [32]. The ITU has defined Mote as a miniature computing device equipped with sensors and signal transceivers operating in a specific radio band and used to transmit sensed data [33].

While acknowledging the existence of various terms, our focus is on Device as an elementary IoT building block. Device is identified by the ITU as a piece of equipment with the mandatory capabilities of communication and optional capabilities of sensing, actuation, data capture, data storage, and data processing [5].

Regarding actuation and sensing capabilities, an actuator performs physical actions caused by an input signal and a sensor senses chemical compounds or monitors environmental conditions and sends an electronic signal proportional to the sensed value. The Global System for Mobile Communications Association (GSMA) considered generalized IoT Device Architecture as a combination of IoT Device Host and IoT Device where IoT Device consists of the IoT Device Application and Communications Module (consisting of Communications Module Firmware, Radio Baseband Chipset, and Universal Integrated Circuit Card) [34]. The IoT Device is a combination of software and hardware. IoT device hardware typically consists of thing and modules for data acquisition, data processing, and communication, while IoT device software consists of operating systems and device applications [35].

Our focus is on the ITU classification of devices regarding processing capabilities [16], as follows:

- Devices with no processing capabilities (a low-cost device with no microcontrollers and without processing capabilities);
- Devices with low processing capabilities (a low-cost device with very limited microcontrollers and processing capabilities, used only for reading or writing data from/to sensors/actuators and sending or receiving those data);
- Devices with high processing capabilities (devices with processing capabilities for making decisions, running algorithms, and directly coordinating with other devices).

If a correlation of the processing and communications is pursued, bearing in mind that the combination of high processing and low connectivity is not usual, it is possible to list three types of device [16]:

- Device with low processing and low connectivity (LPLC);
- Device with low processing and high connectivity (LPHC);
- Device with high processing and high connectivity (HPHC).

3.3. Gateway

A Gateway [36] interconnects devices with communication networks and performs the necessary translation between the protocols used in the communication networks and those used by devices. Acknowledging that some IoT solutions do not require a gateway, the gateway must be identified as a basic IoT component.

3.4. Connectivity Network

Many connectivity technologies can be used in IoT. They range from wired to wireless technologies as a trade-off between bandwidth, range, and power consumption. Consequently, one of the classifications [37] may be as follows:

- High range, high power consumption, and high bandwidth (Cellular, Satellite);
- High range, low power consumption, and low bandwidth (LPWANs);
- Low range, low power consumption, and high bandwidth (Ethernet, Bluetooth, Wi-Fi).

Besides traditional connectivity networks, other networks also appear. For example, sensor control networks are used increasingly for a variety of applications. The ITU has defined a sensor control network as a sensor network consisting of Motes intended to control one or more actuators [33]. The IoT area network, defined by the ITU, is a network of IoT devices and gateways interconnected through local connections [13].

3.5. IoT Platforms

The ITU has defined the IoT platform as a technical infrastructure that provides the integration of generic and specific capabilities. These capabilities, in conjunction with the capabilities of the core network, may be exposed to IoT application servers. The core network enables communication functionalities for supporting the data transfer to devices and gateways via the access network. Some of those functionalities can be used by service providers [5].

There is no standard IoT platform configuration, and there are a variety of IoT platforms. Currently, stated in IoT Analytics IoT Platforms Company Landscape 2020, there are officially 620 IoT Platform companies on the open market [38]. The possible classification of IoT platforms [39] fall into five main types:

- Connectivity platforms;
- Device management platforms;
- Cloud platforms;
- Application enablement platforms;
- Advanced analytics platforms.

The IoT Platform market is concentrated around a few well-known key providers because the market share of the top 10 platforms is 58%. The focus area is primarily on Manufacturing/Industrial use, Energy, Mobility, Smart Cities, Health, Supply Chain, etc. There are multiple benefits for IoT Platform vendors to create open ecosystems and cross-vertical and cross-value chain collaborations through the IoT platform itself or by creating formal alliances and partnerships with vendors at multiple levels of the value chain. Some of the benefits are cost reduction, enhancing security, shortening time to get into the market, the speed of innovation increases, etc. [40].

In ITU Recommendation ITU-T Y.4208 has identified a new IoT component, the edge platform, which is usually a kind of cloud platform. It is about transferring some IoT capabilities from IoT application server and IoT platform to the edge platform, aiming to support edge computing. The edge platform is situated between the access network and the core network [41].

3.6. IoT Application Server

According to ITU definitions, the IoT application server runs applications and communicates with devices, gateways, and the IoT platform via the core network (or directly, in the case of communicating with the IoT platform) directly to deliver application services [13].

Application is a structured set of capabilities that provide value-added functionality supported by one or more services, while Service is a structured set of capabilities for applications support [30].

3.7. IoT Application

Application is a structured set of capabilities that provide value-added functionality supported by one or more services [42]. IoT application can be referred to as application provided by an IoT application provider.

3.8. IoT Service

Service is a structure set of capabilities for applications support [42]. However, there is still no clear definition of the IoT service given that IoT services are constantly evolving and taking different forms [43]. The IoT service can be referred to as a service provided by an IoT service provider.

3.9. IoT User

The "IoT user" actor is an IoT actor that uses all possible services related with things, such as monitoring, location tracking, and service discovery, defined by ITU Recommendation ITU-T Y.4100 [6]. IoT user is defined by the Body of the European Regulators of Electronic Communications (BEREC) as the purchaser of an IoT service who incorporates the IoT service as one component in his own products and/or services [43].

3.10. End User

End user is defined by the ITU as the actual user of the products or services offered by the enterprise. The end user consumes the product or service [44].

3.11. IoT Data Protection and Privacy

Data protection and privacy must be ensured in the IoT. Designation of a data protection officer is needed because the IoT fulfils the requirements of Article 37, paragraph 1 GDPR, in such a way that processing operations , by virtue of their nature, scope, and/or their purposes, require regular and systematic monitoring of data subjects on a large scale [9].

3.12. IoT Security

Security must be ensured for data in use (device level), idle data (stored data), and data in motion (data transported across a network). Some of the possible consequences of inadequate IoT security could be loss of privacy, danger to health and safety, theft of data from the system or theft of material items, danger of reputation, loss of productivity, and noncompliance with laws or regulations, etc. Therefore, IoT security is a central issue and must be implemented along with the entire IoT system, which means at the device level, in the network, cloud, etc.

IoT devices are going to be more vulnerable (for example, low-cost nodes with low budget for security, low compute power for encryption) and easily accessible to attackers (for example, smart light bulbs, smart thermostats) than traditional IT systems. Additionally, the exponential growth of IoT connected devices means a larger area for attackers. Consequently, IoT security is more challenging than cybersecurity. It starts with cybersecurity, and further security measures are needed [45].

It is recommended that device manufacturers perform certain cybersecurity activities to provide the necessary cybersecurity functionality of IoT devices and to provide related information to customers. In this regard, the comprehensive guidelines for IoT device manufacturers, issued by the National Institute of Standards and Technology (NIST) classified those specific activities that have primarily a pre-market impact and activities with primarily a post-market impact [46] while providing, as a starting point, a set of IoT device cybersecurity capabilities for manufacturers [47].

There are many available IoT security certification schemes; one example is the Eurosmart IoT Security Certification Scheme for the IoT Device, defined by the Cybersecurity Act, with a focus on the Substantial security assurance level. The goal is ensuring that certified IoT devices comply with specified requirements throughout their life cycle [48].

The ETSI released ETSI EN 303 645 in June 2020, which is a consumer IoT Security standard specifying 13 provisions for the security of Internet-connected consumer devices and their associated services [49]

The Cyber Security Act established the EU wide cybersecurity framework for ICT products, services, and processes on 27 June 2019. The European Commission will be required to conduct periodic assessment if specific cybersecurity requirements become mandatory for certain ICT processes, services, and products. In this regard, from a consumer perspective, the ENISA (The European Union Agency for Cybersecurity) Advisory Group's working group on cybersecurity calls for mandatory certification schemes for certain ICT products, services, and processes instead of the current EU-wide voluntary certification scheme. Responsibility for implementation and supervision of the schemes is assigned to National cybersecurity certification authorities [50].

There are different certification requirements for IoT devices, which can be classified as follows [34]:

- Regulatory certification (FCC, EC);
- Industry Certification;
- Telecoms: the two main Telecoms Industry certification schemes are the Global Certification Forum (GCF) and the PCS Type Certification Review Board (PTCRB);
- Operator Certification (Deutsche Telecom, Verizon, AT&T).

4. Computing in the IoT Ecosystem

General issues of the network for the IoT are identified in ITU Recommendation ITU Y. 4113 [13], as follows:

- Packet loss and higher latency;
- Unreliability of short-range radio communications in the IoT area network;
- Network overload due to large amount of traffic to be processed.

The largest amount of computing capability is tied to cloud computing. However, cloud computing poses a substantial problem in supporting time-critical and location-aware IoT applications because it relies on remote and centralized resource provision. As more devices are expected to be connected to the Internet, problems with high latency, poor security, poor reliability, high network bandwidth, storage costs, communication power, and more are expected to grow, as discussed in Silva et al. [51] and Jiang et al. [52].

The authors in Li and Wang [53] point out the shortcomings of cloud computing in solving possible problems encountered with IoT and explain the possibility for solving these problems and introduce fog computing. Recognizing that cloud and fog roles are complementary, the authors in Bonomi et al. [54] claim that there is a fruitful interplay between the cloud and the fog, especially when it comes to analytics and data management.

4.1. Computing Models

The term fog computing, introduced by Cisco in 2014, is tied to the decentralization of computing infrastructure. Cisco's simplified IoT architecture consists of basic building blocks with security across the entire architecture and data management aligned with each layer of the core functional stack. The three IoT data management layers are [14]:

- The edge layer, where data management takes place within the sensors themselves;
- The fog layer, where data management takes place within the gateways and transit network;
- The cloud layer, where data management takes place within the cloud data centre.

On March 2018, NIST released a publication presenting a fog and mist computing conceptual model, together with their relationships to cloud computing models. In the model, the fog node is presented as a physical component, such as gateway, server, router, switch etc., or a virtual component, such as a virtual machine, etc. Mist nodes are located at the edge of the network, directly within the network fabric where they use microcontrollers and microcomputers. It is underlined that edge computing is considered as a network layer that envelops end devices and their users [15].

On the other hand, Cisco considers fog computing, micro data centres, multi-access edge computing, cloudlets, and emergency response units as five types of edge computing [55].

As an extension of the existing client-server architecture, a new four-tier architecture has been proposed in Ray [19]. This architecture consists of a cloud, fog, edge, and dew layer and makes it easier for the user to access web data from any sources (edge, fog, or cloud) through minimal or no Internet access.

4.2. Fog, Edge, Mist, and Dew Hardware Computing Locations

The authors in [18] provide an extensive tutorial on fog computing and related computing paradigms. The observed computing paradigms included cloud computing, cloudlet computing, mobile computing, edge computing, mist computing, and other similar computing paradigms. Although there are many computing paradigms, it is emphasized that some paradigms are a subset of others; for example, mobile computing is a sub-set of mist computing and edge computing is a sub-set of fog computing, etc. Though not identified in [18], there is another type of computing, namely, dew computing, which aims to enable content when there is no Internet connection.

Many researchers are trying to figure out where all these computing nodes are located. Dew computing uses mostly on-premises computers, while fog mainly includes routers and sensors in the IoT [56]. Dew is identified by the authors in Ray [19] as a server inside the user's PC. According to Šojat and Skala [20], Dew computing happens in information processing devices located, for example, in refrigerators, car motors, traffic-controls, lights, theatres, and industries. In addition, it is emphasized that the benefits of integrating dew devices into the cloud-fog-dew hierarchy are very significant.

Mist computing puts computing power at the far edge of the network, and usually consists of microchips or microcontrollers built into the device [57]. Mist computing uses microcomputers and microcontrollers for sending data to fog nodes and the cloud if needed. In the mist layer, sensor data pre-processing is performed, and only the essential data is sent to the gateway, server, or router, which saves bandwidth and battery power [58].

Edge computing can be used to process data in near real-time by processing data closer to the edge, directly on devices that have attached sensors or gateway devices that are close to the sensors. Edge computing is less scalable compared to fog computing and supports low interoperability, making some IoT devices incompatible with some operating systems and cloud service.

Authors in the paper [59] discussed fog computing, mobile edge computing, and cloudlet computing in detail, together with comparisons of their features. In this regard, fog computing node devices are identified as routers, access points, switches, and gateways while, for mobile edge computing, node devices are identified as servers running in base stations and cloudlets running a virtual machine. A lack of standardization and different interpretations by different consumers was also emphasized.

Fog computing is characterized by placing computing capability in a connection between device sensors and a cloud server, usually in a device that acts as a gateway, connecting the sensors and managing the connection to the cloud. Computing decentralization is achieved by processing data in a fog node, and it can be any device capable of computing, data storage, and network connectivity. Fog computing can reduce latency and process larger amounts of data compared to edge computing due to its ability to process requests in real-time [60]. Fog computing and edge computing differ in intelligence location identification and power computation. In the case of edge computing, processing power and intelligence are placed in devices such as built-in automation controllers while, in the case of fog computing, intelligence is placed in the local area network while edge devices and the gateways along local area networks are used for power processing [61]. The fog layer supports local data storage, data filtering, compression, merging, and intermediate analytics to save backbone bandwidth, reduce the cloud load, and improve system performance [58]. The fog node is identified as a mini cloud, located at the edge of the network, where the most common fog locations are in high-performance devices, such as smart gateways or routers [60].

The authors in [18] identified hardware locations for cloud computing as being large data centres; hardware locations for fog computing are devices with virtualization capacity as servers, routers, switches, access points; hardware locations for edge computing are edge devices with computing capability; and hardware locations for mist computing are IoT devices (e.g., sensors, cell phones, home appliance devices). Additionally, cloud service providers are identified as operators for cloud computing, users and cloud service providers as operators for fog computing, network infrastructure providers or local business as operators for edge computing, and self-organized or local business as operators for mist computing.

In the situation of there being no consensus on the distances among more computing paradigms, it can be concluded from the above computing location identifications examples that sometimes computing hardware locations have some degree of overlap. Based on the analysis, our approach is summarized in Table 2.

Table 2. Computing hardware locations and corresponding Internet of Things (IoT) actors.

	Cloud Computing	Fog Computing	Edge Computing	Mist Computing	Dew Computing
Computing hardware location	Large Cloud, Data centres	Mini Cloud smart gateways or routers.	The first hop from the IoT device Wi-Fi access points, switches, or gateways	The far edge of the IoT network It usually has microchips or microcontrollers built into the device	Server located inside the user's PC and Information processing devices
IoT actor	Cloud Data Manager Cloud Service Provider	Fog Data Manager Network equipment provider Network providers or other business	Edge Data Manager Network equipment provider Network providers or other business	Mist Data Manager Device provider Network providers or other business	Dew Data Manager Device provider Users Self-organized, local, or other business

As all these computing technologies have some advantages and disadvantages, the use of all these types of computing will be key to ensuring the ability of applications and systems to scale alongside a growing network of devices [62].

According to the vision expressed in Roberts [63], the biggest IoT transformation will be in shifting power in the network from the centre to the edge. Therefore, the IoT will allow devices to directly communicate with each other rather than communicate through cloud-based management servers or central hubs.

5. IoT Actors

The IoT ecosystem consist of multiple coexisting and competing platforms and products, along with a variety of business players interacting with each other. In this regard, we found that certain models introduced by the ITU could be a starting point suitable for research and further adaptation. An Informative Appendix I of Recommendation ITU-T Y.4000 [5] presents an example of identified business roles in the IoT ecosystem and their relationships. As this example does not represent all possible relevant roles in IoT business deployments, we intend to extend it to include the impact of new computing paradigms. We also consider Recommendation ITU-T Y. 4100 [6] because it provides common IoT requirements based on the general use cases of the IoT and IoT actors. In light of the expectation that the number of connected things will be so enormous that the IoT data will constitute a predominant part of the data carried by networks, Recommendation ITU-T Y.4114 [7] and presented key possible mappings from IoT business roles [5] to the IoT data roles are also taken into account in this research.

Business roles identified in the Informative Appendix of Recommendation ITU-T Y.4000 [5] are as follows:

- The device provider provides devices to the network provider and application provider;
- The network provider performs access and integration of other provider resources, provides IoT capabilities and their support and management of their infrastructure, and provides network capabilities and resources to different providers;
- The platform provider provides capabilities to application providers, such as data storage, data processing, device management, integration capabilities, and open interfaces;
- The application provider provides IoT applications to application customers while using capabilities or resources of the network provider, device provider, and platform provider;
- The application customer is the user of the IoT applications provided by the applications provider.

Applicable mappings between IoT actors described in ITU Recommendations [6] and [7] with IoT business roles described in Appendix I of [5] are presented in Table 3.

Table 3. Mappings between the IoT actors and business roles in selected International Telecommunication Union (ITU) recommendations.

IoT Actors Identified in ITU Recommendations ITU-T Y.4100 [6] and ITU-T Y.4114 [7]	Business Roles in Informative Appendix I of Recommendation ITU-T Y.4000 [5]
Data Manager is responsible for managing the capture, processing, storage, and transfer of IoT data to meet the IoT service provision requirements [6]. Data manager actor can be a human Data manager or a machine Data manager actor	Application provider Device provider
Service Provider provides services related to things, such as location tracking, monitoring, and service discovery [6].	Application provider, Platform provider, Network provider.
IoT User uses services related to things, such as location tracking, monitoring, and service discovery [6].	Application customer
IoT Data Provider collects data from things and injects the data processed within the IoT system as well as data from external sources and provides them via the IoT data carrier to the IoT data consumer [7].	Device provider, Network provider, Platform provider, Application provider
IoT Data Consumer consumes IoT data. Usage of the consumed data depends on application purposes [7].	Device provider, Network provider, Platform provider, Application provider, Application customer.
IoT Data Framework Provider provides general IoT data processing capabilities and related infrastructure (e.g., storage and computing resources, data processing run time environment) as required by the IoT data provider, IoT data carrier, IoT data application provider, and IoT data consumer for the support of data operations execution [7].	Network provider, Platform provider.
IoT Data Application Provider provides applications related to the execution of IoT data operations (e.g., applications for data analysis, data pre-processing, data visualization, and data query) [7].	Device provider, Network provider, Application provider.
IoT Data Carrier carries data among the IoT data provider, the IoT data framework provider, the IoT data application provider, and the IoT data consumer [7].	Network provider.

Analysing these mappings between the IoT actors and business roles in selected ITU recommendations, and taking into account the definitions for 'Data Manager' given in ITU-T Y.4100 [6], 'Device Provider' and 'Application Provider' given in ITU-T Y.4000 [5], identified as Data manager corresponding business roles, our view is that Data Manager as an IoT actor needs to be more granulated. With more granulation, the overall data flow and corresponding responsibilities become more understandable and clearer.

Data Manager actor corresponds to Device Provider when provided devices that involve some data management functionalities [6]. Depending on the provided device processing capabilities, the corresponding Data Manager actor needs to be granulated as Dew Data Manager, Mist Data Manager, Edge Data Manager, Fog Data Manager, and Cloud Data Manager.

Data Manager actor corresponds to Application Provider when the provided applications involve some data management functionalities [6]. As the Application Provider uses the resources or capabilities of the device provider, network provider, and platform provider, the corresponding Data Manager actor needs to be granulated as Network Data Manager, Platform Data Manager, and Application Data Manager.

A variety of device options and use cases, combined with a variety of IoT applications, makes the IoT value chain a complicated ecosystem that can have a countless number of partnerships between the participants.

Additionally, designation, manufacturing, and distributing of IoT devices can be done with incompatible standards in different jurisdictions. The situation is the same with IoT actors, i.e., many of them are outside the jurisdiction in which the IoT service is delivered.

Therefore, we will consider additional perspectives (regulatory, product development and technology consultancy companies, industry organizations, and others) on how the market players in the IoT value chain are understood, and these are summarised in Table 4.

Table 4. Mappings between the IoT actors identified from various perspectives.

IoT Actors	IoT Actors Identified from Various Perspectives
IoT Developer	IoT service developer [17] IoT application developer [17]
IoT Security Specialist	Security specialists [64]
IoT Data Protection and Privacy Specialist	Data protection officer [6]
IoT Data Manager	Data manager [6] Application provider + data management [5] Device provider + data management [5]
IoT Device Provider	Device provider [5,17,65] Device manufacturers, module manufacturers [66] Designers and producers of connected devices [67] IoT module providers [67] The designers and manufacturers of the objects [64] The manufacturers of the module components [64] Device, component, and chipset manufacturers [68] Device manufacturer/provider [69] Device manufacturers, component manufacturers [70]
IoT Network equipment Provider	Suppliers of the middleware [64] Network equipment providers [67] Infrastructure manufacturers [70] Network equipment manufacturers [64] Connectivity equipment developers and vendors [68]
IoT Platform Provider	Connectivity platforms [39] Device management platforms [39] Cloud platforms [39] Application enablement platforms [39] Advanced analytics platforms [39] IoT platform provider [67], [17] Platform vendors [70] Platform provider [65], [5]
IoT Connectivity Provider	Network provider [5,57,64] Infrastructure provider [17] Operators [65] Connectivity service provider [43] Connectivity provider (network developer) [71] Connectivity provider [17,66,67] Connectivity/network provider [69] Connectivity/mobile network operators [68] Middleware/analytics vendors (connectivity providers, service provider) [70]
IoT Service Provider	Service provider [6,66,69]. (application provider, platform provider, network provider) [5] IoT service provider [43] Service providers and data aggregators [64] Service (cloud service providers, IoT platforms) [68] Service enabler and service creator [71] Cloud computing companies [64], IoT cloud provider [67]
IoT Application Provider	Application provider [5,65,69]
IoT Integrator	IoT service integrator [17] Integrators [64], System integrator [65,66]
IoT User	IoT user [17,43] Application customer [5]
End User	End user [17,43,69] Markets-payers (consumer, end user, company, public sector) [68]

6. IoT Model

As the focus in the IoT ecosystem is on data, monitoring the flow of data is complex Multiple roles of IoT actors are possible, and some IoT actor could be IoT data market stakeholder at the same time or have relations with IoT data market stakeholders. Identifying all these IoT actors and clarifying their roles and responsibilities is of great importance regarding various aspects.

As the IoT deals with unlimited heterogeneous connected devices, there is a need for a flexible layered architecture. Keeping in mind the previously presented various perspectives, the diversity of IoT concepts and inconsistencies is evident. Although there is no all-encompassing IoT architecture in place, there are some key components and features that are shared in most IoT deployments. Therefore, the previously presented IoT concepts can be adapted to a new IoT model, presented in Figure 1.

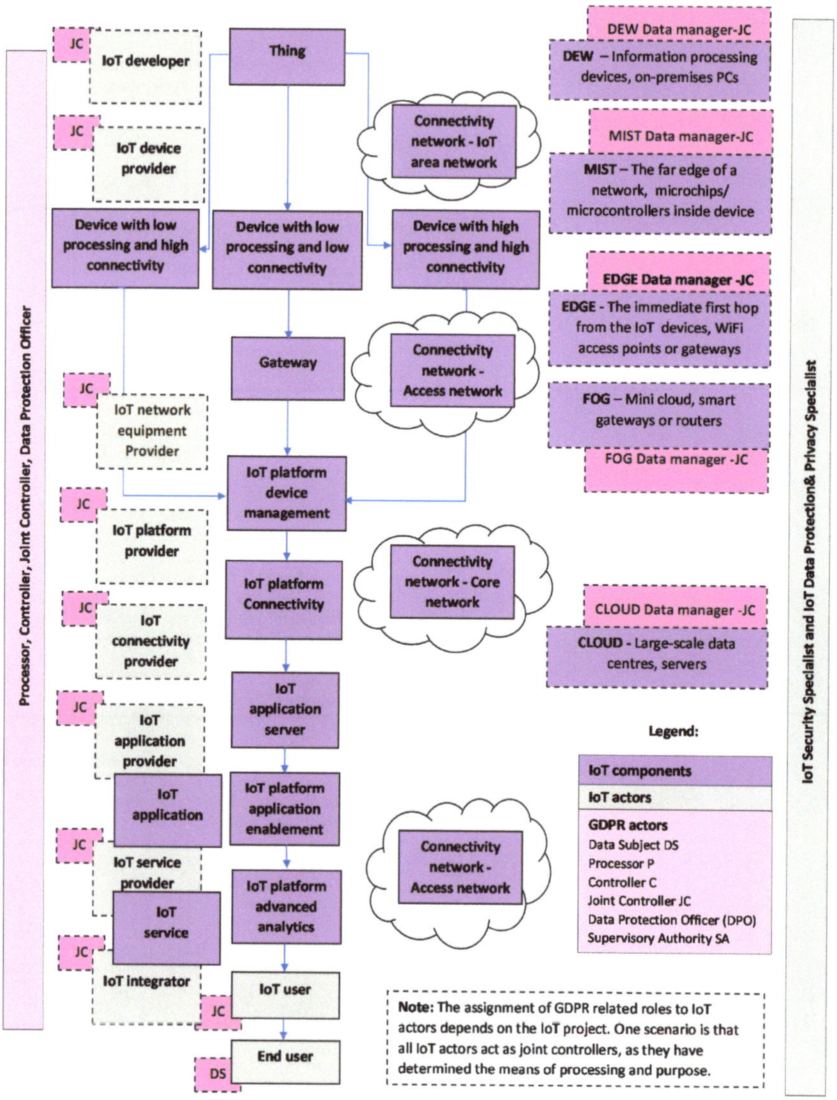

Figure 1. IoT Model.

6.1. Methodology for Identifying IoT Components and IoT Actors

We used the basic model of the network for the IoT, identified in Recommendation ITU-T Y.4113 as a starting point. This model consists of: Device, IoT area network, Gateway, Access network, Core network, IoT platform, and IoT application server [13].

We used all these components in making a new IoT model, but acknowledging that other networks may appear, we present an IoT area network, Access network and Core network as one Connectivity network. To make it easier to understand, we present all three networks in Figure 1.

We add Thing as an object of the physical world or the information world, capable of being identified and integrated into communication networks (for example, a human being is a Thing in the case of remote diagnostics and health monitoring)

Acknowledging the existence of a potentially unlimited number of diverse IoT Platforms, to make it easier to understand, we present the five most common IoT platforms, i.e., Connectivity platforms, Device management platforms, Cloud platforms, Application enablement platforms, and Advanced analytics platforms, in Figure 1.

The connectivity network is combined with IoT network equipment and the IoT Connectivity platform, and it is operated by an IoT Connectivity provider.

Device is identified by the ITU as a piece of equipment with the mandatory capabilities of communication and optional capabilities of sensing, actuation, data capture, data storage, and data processing [5].

The Device is delivered by the IoT Device provider while the IoT Device management platform unifies and simplifies the management of IoT devices, the provisioning of software updates to devices, and offers other functionalities. Network equipment is provided by the IoT network equipment provider.

IoT Security is mandatory in the IoT. Therefore, all related activities aiming to deliver security may be considered as IoT components.

IoT Data Protection and Privacy is mandatory in IoT. Therefore, all related activities aiming to ensure data protection and privacy may be considered as IoT components.

IoT user is the purchaser of an IoT service who incorporates the IoT service as one component in their own product and/or service (It could be a car manufacturer or electricity provider who includes a smart meter in their services).

End user is the actual user of the products (a car owner, user of applications and services).

IoT application can be referred to as the application provided by an IoT application provider, while IoT service can be referred to as the service provided by an IoT service provider.

IoT integrator is the IoT actor who delivers end-to-end solutions.

IoT developer is focusing primarily on the creation of software.

6.2. Methodology for Identifying Relevant GDPR Actors

In the EU, Article 29 Data Protection Working Party issued a specific Opinion 8/2014 on the Recent Developments of the Internet of Things on 16 September 2014, emphasizing that IoT stakeholders should ensure that data at every level is used for purposes known to the user and compatible with the original purpose of the processing. Accurate identification of the involved IoT stakeholders is necessary to qualify their legal status as data controllers who must comply with various obligations. It is stated that most device manufacturers collect and process personal data generated by the device, which qualified them as data controllers. Third-party application developers, unless the data is properly anonymised, must be considered as data controllers. Other third parties may use IoT devices to collect and process information about individuals, so they are also qualified as data controllers. IoT data platforms can also qualify as data controllers for processing activities for which they determine purposes and means, under EU data protection law [72].

In the case of the IoT determining controller or processor roles, this is always dependent on the characteristics of the actual IoT project [73].

Data Protection Officer (DPO) is generally necessary in IoT, because of the large scal of personal data processing. The DPO informs and advises the controller or the processo monitors compliance with the GDPR, provides advice when required, and acts as th contact point for the supervisory authority. In Figure 1, Data protection and privac specialist and DPO are separate, but this specialist may be appointed as DPO. DPO guide the company toward GDPR compliance. Large companies have departments to handl data protection and privacy related matters.

Data protection impact assessment (DPIA) is of great importance in the IoT, and wher the DPIA indicates a high risk, the controller must consult the supervisory authority.

Supervisory Authority is a public authority in EU Member States, and it is als typically referred to as Data Protection Authority or equivalent.

Lead Supervisory Authority is applicable in the case of multinational companie where the company may choose not to appoint a DPO for each country of operation.

Controller determines the purposes and means of the processing of personal dat while the processor processes personal data on behalf of the controller, according to th Data Processing Agreement signed with the controller.

It is often possible to have joint controllers where every IoT actor determines th purposes and means of the processing of personal data.

For example, an IoT integrator who delivers end-to-end solutions could be only on controller, while other IoT actors are processors who processes personal data on behal of the IoT integrator according to the Data Processing Agreement signed with the Io integrator as the controller.

7. Current Status and Collaborative Regulatory Approach

Every year, during ITU Global Symposiums for Regulators (GSR), the Best Practic Guidelines are adopted by the global community of ICT regulators. Recent GSR2020 Bes Practice Guidelines point out that IoT is one complex issues that is waiting to be addresse As new issues call for novel approaches, formal regulations should leave enough space fo self-regulation and hybrid and collaborative regulatory models [74].

A cross-sectoral IoT nature requires a cross-sectoral regulatory approach for maximiz ing the IoT benefits while minimizing the IoT risks. As ICT underpins almost every secto of the economy, traditional ICT sector regulations in the silo-style is not viable anymore This need has been demonstrated in practice based on the analysis of the main barrier to the adoption of smart city IoT projects that were identified while research was bein conducted on the assessment of more than 350 projects, which are funding, silos, an politics [75].

The ITU has developed the concepts of "collaborative regulation" and "fifth-generatio regulation", according to the concept of ICT regulation generations. Currently, the sta tus is far from satisfactory from the data on the G5 Benchmark (The Benchmark of Fift Generation Collaborative Regulation) in the Global ICT Regulatory Outlook 2020, issued by the ITU. The G5 Benchmark covers 80 economies from all regions on the glide path towards collaborative regulations, and it uses 2018–2019 data. It shows that nine countries out of every ten are still regulating the ICT sector as a separate economic sector, while only sixteen countries in total have a holistic and forward-looking regulatory framework. Eight indicators out of a total of twenty-five belong to the assessment of collaboration degree measured between the ICT regulator and the Competition Authority; the Consumer Protec tion Commission; the Data Protection Commission; the Spectrum Agency; the Broadcasting Regulator; the Financial Regulator; the Energy Regulator, and the Internet agency [8].

A broader picture of the current regulatory state can be obtained if the ICT Regulatory Tracker is also considered. This tracker, issued by the ITU, shows the evolution of the four generations of ICT regulation whereby, in 2019, only 32.6 % of 193 countries belonged to the fourth generation of ICT regulations [76].

The collaboration was recognized as a cornerstone of success in the annual IESE Cities in Motion index, which examines all aspects of quality of life and sustainability in 181 key global cities. According to the report [77], the best ranking cities fully understand that the challenges are too great to be addressed individually and indicate that collaboration is key for achieving long-term success.

Currently, involved authorities in the data economy are Competition authorities, Data protection authorities, Electronic Communications National Regulatory authorities (NRAs), Cybersecurity authorities, and Governmental offices promoting open data policies/information fairness. Data protection authorities are responsible for the application of the GDPR and, in some cases, the ePrivacy Directive, while NRAs are responsible for the regulation of the telecoms market and application of the ePrivacy Regulations [27].

8. Data Brokers, Gatekeepers, and Other Actors

As the number of devices increases, the amount of data collected by these devices also increases. According to the Statista report, the total data volume of connected IoT devices worldwide is projected to reach 79.4 zettabytes by 2025 [78].

Across the IoT, data is created by devices and sent to applications to be sent, consumed, and used. A new/old actor appears—a data broker exploits and sells personal data about individuals to third parties. According to the USA Federal Trade Commission (FTC) report released in May 2014, data brokers are companies that collect consumers' personal information and resell or share that information along with others. The data broker industry is complex, consisting of collecting consumer data, mostly without their knowledge, combining online and offline data, and analysing data about consumers to make visions of consumers. Commonly, multiple data brokers provide data to each other [79].

Despite the terms and conditions for privacy, as the opt-in-based agreement provided by the data brokers, the data providers (or data sources) still do not know how their data is being processed, delivered, and used. So far, IoT data markets have not been well-formed due to lack of transparency between providers and brokers/consumers [80]. Current IoT data markets are classified into two types of market, as privacy protection markets and privacy valuation markets.

Authors in Oh et al. [81] considered the following four major stakeholders for modelling the IoT data market:

- Data providers;
- Multiple data brokers who collect raw data from various source and sells big data;
- The data service provider who utilize big data from the data brokers;
- Service consumer.

The findings of Wolfie [82] emphasized that data brokers, online platforms, advertising technology providers, and business in industries can now monitor and analyse individuals in various aspects. As a result of the recent technology developments, we are talking about unprecedented new qualities of ubiquitous corporate surveillance with potential danger that could end in a society without privacy. Much of these activities occur in the background and remain blurred to most consumers as well as to policymakers. It is no secret that many companies use misleading and ambiguous language in their terms and conditions and privacy policies.

There is a lack of transparency in the practice of data brokers, and on the way from the source to the data product, data may change hands many times, and it is challenging to identify all actors in this data value chain.

According to The Vermont Statutes, a first-of-its-kind bill to regulate data brokers went into effect in January 2019. Data brokers, i.e., businesses collecting and selling data about Vermont, USA, residents are required to register and to share information with the public about how they operate. But the Vermont law only covers third-party data companies, while the first-party data holders that collect data directly from users, such as Google, Amazon, or Facebook, are not covered by this law. Despite the big list of firms registered, there is little clear information about what these firms are doing with the data and whether users can remove themselves from their database [83].

On 15 December 2020, the EU released drafts of the long-awaited Digital Services Act and Digital Markets Act, proposing measures to regulate online platforms to protect consumers and competition. The Digital Services Act includes rules for intermediary services (Internet access providers, domain name registrars), hosting services, online platforms, and very large online platforms. The Digital Markets Act includes rules for gatekeeper online platforms aimed at prohibiting unfair practices by them [84]. The gatekeeper may also have a role as device manufacturer and developers of operating systems. Consequently, with all these actors and their multiple roles, the complex IoT ecosystem becomes even more complex.

9. Conclusions

There is no single Internet of Things definition as it is still evolving, together with the IoT evolution. A similar situation can be seen in the case of IoT architecture, as there is no standard IoT architecture, while about a thousand IoT related standards are present.

In this regard, IoT complexity and numerous perspectives have led to different IoT models being proposed by many researchers, communities, and organizations. However, to our knowledge, there is a lack of research from a regulatory perspective.

Bearing in mind that the IoT ecosystem involves various IoT actors, regulatory challenges are significantly greater than before. Intending to identify IoT components, IoT actors and relationships among them, we made a comparison of various approaches and mappings between identified IoT actors. We believe that this mapping of IoT actors from various perspective, along with the presented IoT model, gives a clearer picture and better clarification of the blurred IoT actors' relationships.

IoT devices generate large amounts of data, so data management is one of the biggest challenges in the IoT. As the significance of data and data related activities are increasing, consequently the significance of data laws, regulations, and policies are also increasing. Here, the GDPR is of extreme importance, together with draft ePrivacy Regulations.

Because of the high degree of fragmentation between the many IoT actors, a high risk to data protection exist. In that way, keeping in mind existing complexity in defining controller, joint controller, and processor roles and the distribution of responsibilities among them, our contribution could help relevant authorities to better understand the data management layers. The situation is further complicated if we consider that joint controllers, according to the GDPR, are not obliged to share their responsibilities equally. Now, the real test for the GDPR is in its enforcement, and future challenges lie in clarifying how to apply the GDPR principles to technologies such as IoT, as stated in the European Commission first evaluation and review of GDPR.

Recent research suggests that the future of IoT lies in combining the advantages of multiple computing paradigms. Firstly, Cisco introduced a simplified model of three IoT computation stack and data management layers placed in the edge layer, in the fog layer, and in the cloud layer. Later, NIST presented a conceptual model of fog and mist computing aimed at facilitating meaningful conversations on the topic. Comparing these models, it can be noticed that Cisco's model does not identify the mist layer, while the NIST recommendations only give focus to fog and mist computing, emphasizing that fog computing is hierarchical, while edge computing is limited to a modest number of peripherals. Furthermore, a four-layer platform has also been evolved, namely the cloud-fog-edge-dew computing model. Compared to the Cisco model, this model introduces dew computing with the primary aim of enabling content when there is no Internet connectivity. Compared to the NIST model, this model does not include mist computing, while it does identify dew computing and edge computing.

Acknowledging that there are other similar computing paradigms and that some of these computing paradigms are a sub-set of others, we present the new five-layer IoT model, where a symbiosis of cloud-fog-edge-mist-dew computing paradigms exists. In this regard, as data controllers and data processors must set up appropriate technical and organizations measures to achieve the data protection principles required by the GDPR, our model is focuses on device processing capabilities and computing paradigms. From that perspective, we posit here the granulation of the Data Manager role in the IoT model in order to better understand where the responsibility for managing the capture, storage, transferring, and processing of IoT data begins. It is evident from our model that all IoT actors have their share of data protection responsibility, from IoT Developer to End user.

In future work, we plan to explore the relationships between the identified IoT actors, data brokers, and large online platforms.

Author Contributions: Conceptualization, S.H. and S.M.; methodology, S.H. and S.M.; validation, S.H.; formal analysis, S.H.; investigation, S.H.; writing—original draft preparation, S.H.; writing—review and editing, S.M. and M.R.; supervision, S.M. All authors have read and agreed to the published version of the manuscript.

Funding: This research received no external funding.

Institutional Review Board Statement: Not applicable.

Informed Consent Statement: Not applicable.

Data Availability Statement: Data sharing not applicable.

Acknowledgments: We thank to Sreenivasan Jayashree (Multimedia University, Faculty of Management, 6300 Cyberjaya, Malaysia) for proof reading and invaluable discussion.

Conflicts of Interest: The authors declare no conflict of interest.

References

1. Cisco Annual Internet Report (2018–2023) White Paper. Available online: https://www.cisco.com/c/en/us/solutions/collateral/executive-perspectives/annual-internet-report/white-paper-c11-741490.html (accessed on 1 March 2021).
2. Ibarra-Esquer, J.E.; González-Navarro, F.F.; Flores Rios, B.L.; Burtseva, L.; Astorga-Vargas, M.A. Tracking the Evolution of the Internet of Things Concept Across Different Application Domains. *Sensors* **2017**, *17*, 1379. [CrossRef] [PubMed]
3. IEEE Internet of Things: Contribute to the Ever-Changing Definition of IoT. Available online: https://iot.ieee.org/definition.html (accessed on 1 March 2021).
4. Jakobs, K. Standardizing the IoT and its Applications—Learning from the Past?! In *Internet of Things–Concepts, Technologies, Applications, and Implementations*; John Wiley & Sons: Hoboken, NJ, USA, 2018; pp. 191–218.
5. Y.4000: Overview of the Internet of Things. Available online: https://www.itu.int/rec/T-REC-Y.4000/en (accessed on 1 March 2021).
6. Y.4100: Common Requirements of the Internet of Things. Available online: https://www.itu.int/rec/T-REC-Y.4100/en (accessed on 1 March 2021).
7. Y.4114: Specific Requirements and Capabilities of the Internet of Things for Big Data. Available online: https://www.itu.int/rec/T-REC-Y.4114-201707-I (accessed on 1 March 2021).
8. Global ICT Regulatory Outlook 2020—Pointing the Way forward to Collaborative Regulation. Available online: https://www.itu.int/pub/D-PREF-BB.REG_OUT01 (accessed on 1 March 2021).
9. Regulation (EU) 2016/679 of the European Parliament and of the Council of 27 April 2016 on the Protection of Natural Persons with Regard to the Processing of Personal Data and on the Free Movement of Such Data, and Repealing Directive 95/46/EC (General Data Protection Regulation) (Text with EEA Relevance) OJ L 119, 4.5.2016. pp. 1–88. Available online: https://eur-lex.europa.eu/legal-content/EN/TXT/?uri=CELEX%3A02016R0679-20160504 (accessed on 1 March 2021).
10. European Commission (2017). Proposal for a Regulation of the European Parliament and of the Council Concerning the Respect for Private Life and the Protection of Personal Data in Electronic Communications and Repealing Directive 2002/58/EC (Regulation on Privacy and Electronic Communications) COM/2017/010 Final-2017/03 (COD). Available online: https://eur-lex.europa.eu/legal-content/EN/TXT/?uri=CELEX%3A52017PC0010 (accessed on 1 March 2021).
11. Alshohoumi, F.; Sarrab, M.; Alhamadani, A.; Al-Abri, D. Systematic Review of Existing IoT Architectures Security and Privacy Issues and Concerns. *Int. J. Adv. Comput. Sci. Appl.* **2019**, *10*. [CrossRef]
12. Lynn, T.; Endo, P.T.; Ribeiro, A.M.N.C.; Barbosa, G.B.N.; Rosati, P. The Internet of Things: Definitions, Key Concepts, and Reference Architectures. In *The Cloud-to-Thing Continuum*; Lynn, T., Mooney, J., Lee, B., Endo, P.T., Eds.; Palgrave Macmillan: London, UK, 2020. [CrossRef]
13. Y.4113: Requirements of the Network for the Internet of Things. Available online: https://www.itu.int/rec/T-REC-Y.4113/en (accessed on 1 March 2021).
14. Hanes, D.; Salgueiro, G.; Grossetete, P.; Barton, R.; Henry, J. *IoT Fundamentals: Networking Technologies, Protocols, and Use Cases for the Internet of Things*; Cisco Press: Indianapolis, IN, USA, 2017.

15. Iorgam, M.L.; Feldman, R.; Barton, M.J.; Martin, N.; Goren, C. Mahmoudi. Fog ComputingConceptual Model— Recommendation of the National Institute of Standards and Technology. Available online: https://www.nist.gov/publications/fog-computing-conceptual-model (accessed on 1 March 2021).
16. Y.4460: Architectural Reference Models of Devices for Internet of Things Applications. Available online: https://www.itu.int/rec/T-REC-Y.4460/en (accessed on 1 March 2021).
17. Weber, M.; Žarko, I.P. A Regulatory View on Smart City Services. *Sensors* **2019**, *19*, 415. [CrossRef] [PubMed]
18. Yousefpour, A.; Fung, C.; Nguyen, T.; Kadiyala, K.; Jalali, F.; Niakanlahiji, A.; Kong, J.; Jue, J.P. All one needs to know about fog computing and related edge computing paradigms: A complete survey. *J. Syst. Arch.* **2019**, *98*, 289–330. [CrossRef]
19. Ray, P.P. An Introduction to Dew Computing: Definition, Concept and Implications. *IEEE Access* **2018**, *6*, 723–737. [CrossRef]
20. Šojat, Z.; Skala, K. Views on the role and importance of dew computing in the service and control technology. In Proceedings of the 2016 39th International Convention on Information and Communication Technology, Electronics and Microelectronics (MIPRO), Opatija, Croatia, 30 May–3 June 2016; pp. 164–168.
21. Technical Specification D2.1—Data Processing and Management Framework for IoT and Smart Cities and Communities. Available online: http://handle.itu.int/11.1002/pub/813b0846-en (accessed on 1 March 2021).
22. European Data Protection Supervisor Guidelines on the Concepts of Controller, Processor and Joint Controllership under Regulation (EU) 2018/1725, 7 November 2019. Available online: https://edps.europa.eu/data-protection/our-work/our-work-by-type/guidelines_en (accessed on 1 March 2021).
23. European Data Protection Board Guidelines 07/2020 on the Concepts of Controller and Processor in the GDPR. Available online: https://edpb.europa.eu/our-work-tools/public-consultations-art-704/2020/guidelines-072020-concepts-controller-and-processor_en (accessed on 1 March 2021).
24. ETSI TR 103 591 V1.1.1 82019-10 SmartM2M; Privacy Study Report; Standards Landscape and Best Practices. Available online: https://www.etsi.org/deliver/etsi_tr/103500_103599/103591/01.01.01_60/tr_103591v010101p.pdf (accessed on 1 March 2021).
25. Information Commissioner's Office: What Needs to be Included in the Contract? Available online: https://ico.org.uk/for-organisations/guide-to-data-protection/guide-to-the-general-data-protection-regulation-gdpr/contracts-and-liabilities-between-controllers-and-processors-multi/what-needs-to-be-included-in-the-contract/#11 (accessed on 25 November 2020).
26. European Commission Communications (2020), Data Protection as a Pillar of Citizens' Empowerment and the EU's Approach to the Digital Transition—Two Years of Application of the General Data Protection Regulation COM/2020/264 Final. Available online: https://eur-lex.europa.eu/legal-content/EN/TXT/?uri=CELEX%3A52020DC0264 (accessed on 1 March 2021).
27. BEREC Report on the Data Economy. Available online: https://berec.europa.eu/eng/document_register/subject_matter/berec/reports/8599-berec-report-on-the-data-economy (accessed on 1 March 2021).
28. Zantalis, F.; Koulouras, G.; Karabetsos, S.; Kandris, D. A Review of Machine Learning and IoT in Smart Transportation. *Future Internet* **2019**, *11*, 94. [CrossRef]
29. Mackenzie, M.; Rebbeck, T. What Is the IoT Value Chain and Why Is It Important? Available online: https://www.analysysmason.com/research/content/comments/iot-value-chain-rdme0/ (accessed on 1 March 2021).
30. Synergy Consulting Group. Enabling the IoT Ecosystem with Policy and Regulation How Policymakers and Regulators Can Encourage Sustainable Market Growth. Available online: https://www.synergyconsulting.ae/insights/enabling-iot-ecosystem-policy-regulation/ (accessed on 1 March 2021).
31. Paradis, I. International Carriers' Path to the IoT Gold Mine, The Mobile Century. Available online: http://themobilecentury.com/international-carriers-path-to-the-iot-gold-mine/ (accessed on 1 March 2021).
32. Cicciari, M. The internet of word-things. *IEEE Spectr.* **2016**, *53*, 25. [CrossRef]
33. Y.2222: Sensor Control Networks and Related Applications in a Next Generation Network Environment. Available online: https://www.itu.int/rec/T-REC-Y.2222-201304-I/en (accessed on 1 March 2021).
34. IoT Device Certification Landscape. Available online: https://www.gsma.com/iot/resources/iot-device-certification-landscape/ (accessed on 1 March 2021).
35. Elizalde, D. How Does an IoT Device Work? Available online: https://danielelizalde.com/iot-hw-blocks/ (accessed on 1 March 2021).
36. Y.4101: Common Requirements and Capabilities of a Gateway for Internet of Things Applications. Available online: https://www.itu.int/rec/T-REC-Y.4101-201710-I/en (accessed on 1 March 2021).
37. IoT 101: An Introduction to the Internet of Things, First Edition. Available online: https://www.leverege.com/ebooks/iot-intro-ebook (accessed on 1 March 2021).
38. IoT Analytics: IoT Platforms Company Landscape 2020. Available online: https://iot-analytics.com/product/iot-platforms-landscape-database-2020/ (accessed on 1 March 2021).
39. JT Solutions IoT Blog—The 5 Types of IoT Platforms. Available online: https://blog.jtiot.com/the-5-types-of-iot-platforms (accessed on 1 March 2021).
40. Lucero, S. IoT Platforms: Enabling the Internet of Things, White Paper, IHS Technology. Available online: https://cdn.ihs.com/www/pdf/enabling-IOT.pdf (accessed on 1 March 2021).
41. Y.4208: Internet of Things Requirements for Support of Edge Computing. Available online: https://www.itu.int/rec/T-REC-Y.4208/en (accessed on 1 March 2021).

42. Y.101: Global Information Infrastructure Terminology: Terms and Definitions. Available online: https://www.itu.int/rec/T-REC-Y.101-200003-I/en (accessed on 1 March 2021).
43. BEREC Report on Enabling the Internet of Things. Available online: https://berec.europa.eu/eng/document_register/subject_matter/berec/reports/5755-berec-report-on-enabling-the-internet-of-things (accessed on 1 March 2021).
44. M.3050.1: Enhanced Telecom Operations Map (eTOM)—The Business Process Framework. Available online: https://www.itu.int/rec/T-REC-M.3050.1/en (accessed on 1 March 2021).
45. Internet of Things Security is More Challenging Than Cybersecurity, White Paper, Wind River Systems, Inc. Rev. Available online: https://www.windriver.com/whitepapers/security/iot-security-is-more-challenging-than-cybersecurity (accessed on 1 March 2021).
46. Fagan, M.; Fagan, M.; Megas, K.N.; Scarfone, K.; Smith, M. Foundational Cybersecurity Activities for IoT Device Manufacturers—The National Institute of Standards and Technology (NIST) U.S. Department of Commerce. Available online: https://csrc.nist.gov/publications/detail/nistir/8259/final (accessed on 1 March 2021).
47. Fagan, M.; Megas, K.N.; Scarfone, K.; Smith, M. IoT Device Cybersecurity Capability Core Baseline—The National Institute of Standards and Technology (NIST). Available online: https://csrc.nist.gov/publications/detail/nistir/8259a/final (accessed on 1 March 2021).
48. EUROSMART. The Voice of The Digital Security Industry. Available online: https://www.eurosmart.com/who-we-are/#members (accessed on 1 March 2021).
49. CYBER—Cyber Security for Consumer Internet of Things: Baseline Requirements. Available online: https://www.etsi.org/deliver/etsi_en/303600_303699/303645/02.01.00_30/en_303645v020100v.pdf (accessed on 1 March 2021).
50. ENISA Advisory Group—Opinion Consumers and IoT Security. Available online: https://www.enisa.europa.eu/about-enisa/structure-organization/advisory-group/ag-publications/final-opinion-enisa-ag-consumer-iot-perspective-09.2019 (accessed on 1 March 2021).
51. Silva, R.; Silva, J.S.; Boavida, F. Opportunistic fog computing: Feasibility assessment and architectural proposal. In Proceedings of the 2017 IFIP/IEEE Symposium on Integrated Network and Service Management (IM), Lisbon, Portugal, 8–12 May 2017; pp. 510–516. [CrossRef]
52. Jiang, J.; Li, Z.; Tian, Y.; Al-Nabhan, N. A Review of Techniques and Methods for IoT Applications in Collaborative Cloud-Fog Environment. *Secur. Commun. Networks* **2020**, *2020*, 1–15. [CrossRef]
53. Li, Z.; Wang, Y. An Introduction and Comparison of the Application of Cloud and Fog in IoT. In *Lecture Notes of the Institute for Computer Sciences, Social Informatics and Telecommunications Engineering*; Springer Nature: Cham, Switzerland, 2020; pp. 63–75.
54. Bonomi, F.; Milito, R.; Zhu, J.; Addepalli, S. Fog computing and its role in the Internet of Things. In Proceedings of the first edition of the MCC workshop on Mobile cloud computing, Helsinki, Finland, 13–17 August 2012; pp. 13–16.
55. Cisco—What is Edge Computing. Available online: https://www.cisco.com/c/en/us/solutions/computing/what-is-edge-computing.html (accessed on 25 November 2020).
56. Yingwei, W.; Skala, K.; Rindos, A.; Gusev, M.; Shuhui, Y.; Yi, P. Dew Computing and Transition of Internet Computing Paradigms, ZTE Communications, No.4 2017, No.59. Available online: https://www.zte.com.cn/global/about/magazine/zte-communications/2017/4/en_222/466311.html (accessed on 1 March 2021).
57. Naveen, J. Fog vs Edge vs Mist Computing: Which One Is the Most Suitable for Your Business? Available online: https://www.allerin.com/blog/fog-vs-edge-vs-mist-computing-which-one-is-the-most-suitable-for-your-business (accessed on 1 March 2021).
58. Rahman, A.U.; Afsana, F.; Mahmud, M.; Kaiser, M.S.; Ahmed, M.R.; Kaiwartya, O.; James-Taylor, A. Toward a Heterogeneous Mist, Fog, and Cloud-Based Framework for the Internet of Healthcare Things. *IEEE Internet Things J.* **2019**, *6*, 4049–4062. [CrossRef]
59. Dolui, K.; Datta, S.K. Comparison of edge computing implementations: Fog computing, cloudlet and mobile edge computing. *2017 Global Internet of Things Summit (GIoTS)* **2017**, 1–6. [CrossRef]
60. Varghese, B.; Wang, N.; Nikolopoulos, D.S.; Buyya, R. Feasibility of Fog Computing. Available online: https://www.academia.edu/31011430/Feasibility_of_Fog_Computing (accessed on 1 March 2021).
61. Jagannath, U.R.; Saravanan, S.; Suguna, S.K. Applications of the Internet of Things with the Cloud Computing Technologies: A Review. In *Edge Computing. EAI/Springer Innovations in Communication and Computing*; Al-Turjman, F., Ed.; Springer Nature: Cham, Switzerland, 2018. [CrossRef]
62. Mahesa, R. How Cloud, Fog, and Mist Computing Can Work Together. Available online: https://ibmcode-staging.us-east.containers.mybluemix.net/articles/how-cloud-fog-and-mist-computing-can-work-together/ (accessed on 1 March 2021).
63. Roberts, P. IBM and Samsung Bet on Bitcoin Tech to Save the Internet of Things, The Security Ledger. Available online: https://securityledger.com/2015/01/ibm-and-samsung-bet-on-bitcoin-to-save-iot/ (accessed on 1 March 2021).
64. Document, No. 1: A Mapping of Arcep and Its Partners' Understanding of the Issues Pertaining to the Internet of Things. Available online: https://en.arcep.fr/news/press-releases/p/n/arcep-and-its-partners-publish-a-white-paper-to-prepare-for-the-internet-of-things-revolution-to-en.html (accessed on 1 March 2021).
65. Kar, S.; Chakravorty, B.; Sinha, S.; Gupta, M.P. Analysis of stakeholders within IoT ecosystem. In *Digital India. Advances in Theory and Practice of Emerging Markets*; Kar, A., Sinha, S., Gupta, M., Eds.; Springer Nature: Cham, Switzerland; ISBN 978-3-319-78377-2.
66. GSMA—What's Unique about the IoT? Available online: https://www.gsma.com/iot/whats-unique-iot/ (accessed on 1 March 2021).

67. OECD—IoT Measurement and Applications. OECD Digital Economy Papers 2018. Available online: https://www.oecd-ilibrary.org/science-and-technology/iot-measurement-and-applications_35209dbf-en (accessed on 1 March 2021).
68. Winchcomb, T.; Massey, S.; Beastall, P. Review of Latest Developments in the Internet of Things, A Report for Ofcom by Cambridge. Available online: https://www.ofcom.org.uk/research-and-data/telecoms-research/general/review-of-latest-developments-in-the-internet-of-things (accessed on 1 March 2021).
69. Asia-Pacific Telecommunity, SATRC Report on ICT Regulatory Framework for M2M Communications and IOT for the SATRC Countries. Available online: https://www.apt.int/sites/default/files/SATRC-SAPVI-02_M2M_Report.docx (accessed on 1 March 2021).
70. Moss, J.; Barnett, G. OVUM—The Internet of Things: Understanding the Evolving Value Chain. Available online: https://silo.tips/download/the-internet-of-things-understanding-the-evolving-value-chain (accessed on 1 March 2021).
71. Ericsson—Exploring IoT Strategies Insights on IoT Value Chain Positioning from Leading Telecom Service Providers. Available online: https://www.ericsson.com/en/press-releases/2018/4/exploring-iot-strategies-telecom-service-providers-pursue-multiple-paths-to-iot-revenue (accessed on 1 March 2021).
72. Article 29 Data Protection Working Party (2014). Opinion 8/2014 on the Recent Developments on the Internet of Things. Available online: https://www.pdpjournals.com/docs/88440.pdf (accessed on 1 March 2021).
73. Advisera, Expert Advice Community. Available online: https://community.advisera.com/topic/joint-controllers-share-of-responsibilities-in-iot/#comment-15806 (accessed on 1 March 2021).
74. GSR—Global Symposium for Regulators (GSR) 2020 Best Practice Guidelines The Gold Standard for Digital Regulation. Available online: https://www.itu.int/en/ITU-D/Conferences/GSR/2020/Pages/default.aspx (accessed on 1 March 2021).
75. IoT Analytics—5 Things to Know about the LPWAN Market in 2020, Internet Analytics Market Insights for the Internet of Things. Available online: https://iot-analytics.com/5-things-to-know-about-the-lpwan-market-in-2020/ (accessed on 1 March 2021).
76. ITU—ICT Regulatory Tracker 2019. Available online: https://www.itu.int/net4/itu-d/irt/#/tracker-by-country/regulatory-tracker/2019 (accessed on 1 March 2021).
77. IESE Cities in Motion IESE Business School, University of Navarra. Available online: https://blog.iese.edu/cities-challenges-and-management/2019/05/10/iese-cities-in-motion-index-2019/ (accessed on 1 March 2021).
78. STATISTA—Data Volume of Internet of Things (IoT) Connections Worldwide in 2019 and 2025. Available online: https://www.statista.com/statistics/1017863/worldwide-iot-connected-devices-data-size/ (accessed on 1 March 2021).
79. FTC—Data Brokers A Call for Transparency and Accountability, May 2014, Federal Trade Commission. Available online: https://www.ftc.gov/system/files/documents/reports/data-brokers-call-transparency-accountability-report-federal-trade-commission-may-2014/140527databrokerreport.pdf (accessed on 1 March 2021).
80. Oh, H.; Park, S.; Lee, G.M.; Heo, H.; Choi, J.K. Personal Data Trading Scheme for Data Brokers in IoT Data Marketplaces. *IEEE Access* **2019**, *7*, 40120–40132. [CrossRef]
81. Oh, H.; Park, S.; Lee, G.M.; Choi, J.K.; Noh, S. Competitive Data Trading Model with Privacy Valuation for Multiple Stakeholders in IoT Data Markets. *IEEE Internet Things J.* **2020**, *7*, 3623–3639. [CrossRef]
82. Wolfie, C. Corporate Surveillance in Everyday Life—How Companies Collect, Combine, Analyze, Trade, and Use Personal Data on Billions. A Report by Cracked Labs. Available online: https://crackedlabs.org/dl/CrackedLabs_Christl_CorporateSurveillance.pdf (accessed on 1 March 2021).
83. MacMillan, D. Data Brokers Are Selling Your Secrets. How States Are Trying to Stop Them, Washington Post. Available online: https://www.washingtonpost.com/business/2019/06/24/data-brokers-are-getting-rich-by-selling-your-secrets-how-states-are-trying-stop-them/ (accessed on 1 March 2021).
84. European Commission—The Digital Services Act Package. Available online: https://ec.europa.eu/digital-single-market/en/digital-services-act-package (accessed on 1 March 2021).

sensors

MDPI

Article

Hyper-Angle Exploitative Searching for Enabling Multi-Objective Optimization of Fog Computing

Taj-Aldeen Naser Abdali [1], Rosilah Hassan [1,*], Azana Hafizah Mohd Aman [1], Quang Ngoc Nguyen [2] and Ahmed Salih Al-Khaleefa [1]

1 Centre for Cyber Security, Faculty of Information Science and Technology (FTSM), Universiti Kebangsaan Malaysia, UKM, Bangi 43600, Malaysia; P94546@siswa.ukm.edu.my (T.-A.N.A.); azana@ukm.edu.my (A.H.M.A.); ahmed.salih89@siswa.ukm.edu.my (A.S.A.-K.)
2 Department of Communications and Computer Engineering, Faculty of Science and Engineering, Waseda University, Tokyo 169-8050, Japan; quang.nguyen@aoni.waseda.jp
* Correspondence: rosilah@ukm.edu.my

Abstract: Fog computing is an emerging technology. It has the potential of enabling various wireless networks to offer computational services based on certain requirements given by the user. Typically, the users give their computing tasks to the network manager that has the responsibility of allocating needed fog nodes optimally for conducting the computation effectively. The optimal allocation of nodes with respect to various metrics is essential for fast execution and stable, energy-efficient, balanced, and cost-effective allocation. This article aims to optimize multiple objectives using fog computing by developing multi-objective optimization with high exploitive searching. The developed algorithm is an evolutionary genetic type designated as Hyper Angle Exploitative Searching (HAES). It uses hyper angle along with crowding distance for prioritizing solutions within the same rank and selecting the highest priority solutions. The approach was evaluated on multi-objective mathematical problems and its superiority was revealed by comparing its performance with benchmark approaches. A framework of multi-criteria optimization for fog computing was proposed, the Fog Computing Closed Loop Model (FCCL). Results have shown that HAES outperforms other relevant benchmarks in terms of non-domination and optimality metrics with over 70% confidence of the t-test for rejecting the null-hypothesis of non-superiority in terms of the domination metric set coverage.

Keywords: fog computing; task allocation; multi-objective optimization; evolutionary genetics; hyper-angle; crowding distance

Citation: Naser Abdali, T.-A.; Hassan, R.; Mohd Aman, A.H.; Nguyen, Q.N.; Al-Khaleefa, A.S. Hyper-Angle Exploitative Searching for Enabling Multi-Objective Optimization of Fog Computing. *Sensors* **2021**, *21*, 558. https://doi.org/10.3390/s21020558

Received: 7 December 2020
Accepted: 8 January 2021
Published: 14 January 2021

1. Introduction

Internet of Things (IoT) has been used in several fields such as health care, environmental engineering, transportation, and safety [1,2]. The idea behind IoT is to connect physical items to the virtual world, so they can be controlled remotely and act as physical access points to Internet services [3]. These devices increased rapidly around the world and generate a huge amount of data, termed Big Data (BD) [4,5]. One of the fundamental challenges in IoT is the data transmissions [6,7] to the Cloud Computing (CC), which indicate to the infrastructure where both data storage and processing operate outside of the IoT devices [8,9].

CC data center is far from end-user, then causes high latency and affects the actual time constraints in many applications [10]. Therefore, CISCO [11] suggests the new paradigm Fog Computing (FC) to ensure reliable sending and receiving data between the Cloud and IoT devices [12]. Figure 1 gives a conceptual elaboration of the architecture of IoT, CC, and FC. The first layer is the IoT environment, this layer close to the user and the physical environment. It contains several devices such as mobile phones, sensors, smart cards, readers, and smart vehicles. The second layer fog layer this layer is located on the edge of the network means between IoT and cloud computing. This layer contains a huge number of fog nodes which generally including routers, gateways, switches, access points, base stations, and specific fog servers. The third layer is the cloud computing layer and consists of several effective servers and storage devices and provides various application services for smart homes, smart transportation, smart factories, and so on.

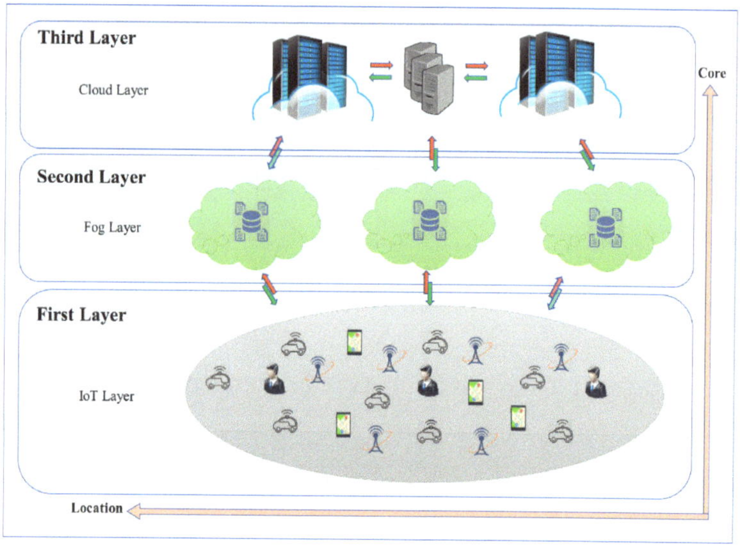

Figure 1. A basic conceptual framework of IoT, cloud computing, and fog computing.

The distributed nature of FC and the relatively limited computation, energy, and communication power of its nodes have motivated researchers to assure its load balancing aspect when various applications are required to be executed in FC. The load balancing of fog computing is accomplished by a set of methodological approaches named Task Allocation (TA) [13] in the literature. The term TA indicates allocating various network nodes optimally to execute a given task or application while maintaining various objectives. In the context of TA for FC, we are interested in dividing the given task into a set of sub-tasks with the independency aspect and dividing them on the network nodes with matching various constraints. Next, they will be presented with a mathematical model for calculating the various fog measures, including energy efficiency, cost-effectiveness, time latency, stability, and reliability. Having the ability to evaluate the candidate solution from the optimization and to provide its objectives values, we call Fog Computing Closed Loop (FCCL). This type of problem is regarded as a Non-Deterministic Polynomial Hard Problem (NP-hard) [14], which makes it a challenging optimization problem. This is due to the huge number of combinations of nodes' task allocation and the various conditions of the nodes and the tasks. Typical approaches for solving such a problem use a meta-heuristic family of optimization algorithms, and more specifically, the multi-objective type of meta-heuristic was enhanced to apply for fog computing to hold the huge number of tasks and set them based on their priority.

Multi-Objective Optimization (MOO) algorithms [15,16] aim at optimizing many objectives' functions using heuristic random searching in order to find a set of non-dominated solutions [17]. There is a high similarity between single objective [18] and multi-objective meta-heuristics [19,20] in the aspect of relying on a random pool of generated solutions, evaluating them, and selecting the best among them to generate off-spring. However, the essential difference between the single objective and multi-objective heuristic searching is the means of evaluating solutions. More specifically, in the multi-objective searching, the solutions are evaluated based on ranks that include a sub-set of non-dominated solutions instead of simple fitness value as in the single-objective optimization. Consequently, the goal of the MOO algorithm is to explore the solution space for finding maximum coverage of non-dominated solutions.

The goal of this article is to develop an optimization framework for computational fog computing. We aim to enable non-dominated optimization for fog computing by assuring high domination of the resulted decisions in terms of various performance metrics, which gives the decision-maker more flexibility as well as high achieved performance. Specifically, the integration of a novel hyper-angle exploitive searching optimization with the crowding distance of Non-Dominated Sorting Genetic Algorithm II (NSGA-II) in the context of fog computing optimization assists in providing more dominant solutions in terms of the fog measures that the decision-maker aims at optimizing. The article presents the following contributions.

- Proposing a fog computing optimization framework with multi-criteria perspectives. The multi-criteria cover the following metrics: Time Latency, Energy Consumption, Energy Distribution, Renting Cost, and Stability.
- Developing a novel optimization algorithm based on meta-heuristic genetic. The developed algorithm supports exploitive searching based on the hyper-angle indicator. We designate it as Hyper-Angle Exploitive Searching (HAES).
- Formulating a novel Fog Computing Closed Loop (FCCL) mathematical function and using HAES for optimizing it after discretization.
- Designing an adaptive objective partitioning by activating the sub-set of objectives at each iteration out of the entire objectives.
- Evaluating the developed HAES based on multi-objective optimization performance metrics and benchmarking mathematical functions and evaluating the optimized FCCL based on HAES, then analyzing its performance in comparison with other relevant optimization benchmarking algorithms.

2. Background and Literature Review

The article is focusing on multi-objective optimization for FC. Hence, the literature contains two phases. Firstly, the related work of the MOO algorithms is presented in Section 2.1, and Section 2.2 provides the related works of MOO fog computing optimization in.

2.1. MOO Algorithms

The studies on meta-heuristic-based MOO in the literature contain various approaches. Different criteria and techniques are used to generate the dominant Pareto Front (PF) and provide extensive exploration. In [21], a fitting function or interpolation method was applied from a finite set of objective values to calculate PF by selecting the individuals that have the shortest distance to the reference points based on the error matrix. The two algorithms, called MOGA/fitting and MOGA/interpolation, dealt with MOO without focusing on attaining the optimal solutions. Bao et al. [22] proposed Hierarchical NDS (HNDS), which focuses on reducing the number of comparisons in the search. HNDS initially sorts all the candidate solutions in ascending order, depending on their first objective. Next, HNDS compares the first solution with the rest of the candidate solutions, one by one, to make a speedy distinction by realizing different superiority solutions and then avoid the high number of unnecessary comparisons.

Other notable studies have extended the existing single-objective searching algorithms to multi-objective ones by introducing the concept of NSGA-II, which is fast NDS with crowding distance. This extension applies to Multi-Objective Vortex Searching (MOVS), which was proposed in [23]. MOVS uses the inverse incomplete gamma function with a parameter ranging from 0 to 1 to spread solutions over the PF. improved NSGA-II to make it more efficient and have better diversity by presenting a more efficient implementation of NDS, namely the dominance degree approach for NDS. Part and Select Algorithm (PSA) was also proposed to maintain diversity, and the entire algorithm after being integrated into NSGA-II was called Diversity DNSGA2–PSA. Additionally, several researchers have added a local search strategy to NSGA-II [24]. For example, the study in [25] proposed Heavy Perturbation (HP)-based NSGA-II. Two objectives, the size and total weight of a clique were considered. In particular, the larger the size of a clique in terms of set inclusion is and the higher the total weight is, the better a solution is. HP-NSGA-II is then dedicated to the clique problem of a weighted graph with weights of vertices in which the perturbation is conducted by either improving a selected elite with a local search procedure or swapping its left and right parts.

Several types of research work also developed nature-inspired models for MOO. For instance, an improved method of GA based on an evolutionary computational model namely the Physarum-Inspired Computational Model (PCM), was proposed in [26]. The initialization of the population used prior knowledge of PCM. Hill climbing was also used to improve the diversity of solutions, and the traveling salesman problem, which is one of the most classical NP-hard problems in combinatorial optimization, was utilized. Apart from improving the optimization of found solutions, several researchers have aimed at improving the searching speed. In the same context, [27] proposed an algorithm for MOO and compared it with four other competing algorithms on three different datasets to reduce the optimization complexity for a large number of objectives from $O\left(N \log^{M-1} N\right)$ to $O(MN \log N + MN^2)$, where M denotes the number of objectives and N denotes the number of solutions. The algorithm removes unnecessary comparisons among solutions to improve the running time.

The work in [28] added the angle concept to crowd distance searching to balance the searching procedure among all angles. Other researchers have also used the framework of NSGA-II with different extensions. For example, [29] used a set of reference points while searching to maintain diversity. Then, from previous approaches, the concept of crowd distance, when combined with angle searching, achieves the extensive scope of the search. Specifically, authors in [30] have used range angle as a criterion to balance the search, then using it in finding criterion solutions as the goal of the study.

Overall, the previous research works that have focused on meta-heuristic for multi-objective aimed at incorporating various criteria for accomplishing exploration as well as exploitation. The crowding distance of NSGA-II is effective for exploration, while the angle searching was used in MOGA-AQCD as an additional base for the crowd-distance exploration. However, the angle usage for exploitation has not been explicitly considered and performed by the existing studies. This article then aims at tackling this aspect by proposing a novel MOO searching that incorporates angle searching for exploitation.

Particularly, the present paper proposes a MOO searching algorithm that uses crowding distance for exploration and angle searching for exploitation. The proposal optimizes the exploitation by selecting solutions from angular sectors that have the maximum found solutions. The crowding distance is also used for exploration; however, we aim at avoiding redundant operators for exploration. This goal is achieved by considering angle searching for exploitation, provided that the crowding distance has successfully played its role in the exploration process. To our knowledge, this is the first meta-heuristic searching algorithm for MOO that jointly considers and optimizes the angle criterion for exploitation and crowding distance for exploration at the same time. In the next section, we present the system models and the research background.

2.2. Fog Computing Optimization

Solving IoT challenges of data processing within real-time constraints have created the need to not rely on cloud network for processing. As a result, the concept of Fog Computing was first introduced by Cisco in 2012. However, congested networks, high latency in service delivery, and poor Quality of Service (QoS), non-stability, and increased cost have been experienced [31]. Such challenges have motivated researchers to focus on fog computing optimization.

The literature contains a significant amount of algorithmic works for fog computing optimization. Each work has focused on certain aspects of the fog network and followed a certain approach for optimization. While some work has tried to include more practical aspects of fog computing needs and nature, others were more simplified and ignored some crucial matters. In the work of [32], the authors have represented the fog computing optimization as a scheduling problem, where the algorithm has to assign tasks to nodes with assuring two objectives the stability and speed. Their model ignores energy and cost matters, which are considered to be crucial aspects of fog computing. On the other side, they used classical multi-objective optimization NSGA-II to solve their model without significant changes to explore the solution space more efficiently and find more dominant solutions. We find that other models have considered energy and cost like the work of [33]; however, there is no consideration of stability or reliability for finishing the work. Similarly, the work of [34] has included energy and latency while ignoring cost and reliability, while the work of [35] has included time latency and cost as objectives and it ignored energy and reliability.

A summary of the covered objectives of each model is given in Table 1. To the best of our knowledge, there is no developed model for fog computing optimization including four objectives: time latency, energy, cost, and reliability at the same time. Such inclusion implies more challenging multi-objective optimization. On the other side, all the previous works have applied NSGA-II and other similar non-dominated searching optimization without development in the searching aspect, which is needed because of the non-convex nature of the problem and a huge number of constraints resulting in the optimization surface non-linear and non-convex with NP-hard nature.

Table 1. Summary of the covered objectives in the fog computing model in the literature.

Authors/Objectives	Energy Consumption	Renting Cost	Stability	Time Latency	Energy Distribution
[32]	✗	✗	✓	✓	✗
[33]	✗	✗	✓	✗	✓
[34]	✓	✗	✗	✓	✗
[35]	✗	✓	✗	✓	✗
Proposed Model	✓	✓	✓	✓	✓

3. Proposed Methodology

This section presents the developed method for accomplishing the goal of the article. It starts with presenting the problem formulation of optimization and fog computing framework was provided in Section 3.1. Next, in Section 3.2, we provide the algorithm named hyper-angle exploitive searching. The fog computing closed-loop model is given in Section 3.3. Table 2 elaborates on the mathematical terms used in the article.

3.1. Problem Formulation of Optimization and Fog Framework

Assume that we have a tuple $x = (x_1, x_2, \ldots x_n) \in X$, where $X \subseteq R^n$ and a tuple $y = (y_1, y_2, \ldots y_m) \in Y$ where $Y \subseteq R^m$ in which the following constraints are held:

$$y_1 = f_1(x_1, x_2, \ldots x_n) \tag{1}$$

$$y_2 = f_2(x_1, x_2, \ldots x_n) \tag{2}$$

$$y_m = f_m(x_1, x_2, \ldots x_n) \tag{3}$$

In such a scenario: x is called the decision vectors; y is the objective vector. X is the solution space, and Y is the objective space to model a minimization problem, with two vectors a and b. We call b dominates a, denoted as $a \prec b\ iff$:

$$\begin{cases} \forall\, i \in \{1, 2, \ldots m\} : f_i(a) \le f_i(b) \\ \exists\, j \in \{1, 2, \ldots m\} : f_j(a) < f_j(b) \end{cases} \tag{4}$$

The domination of b over a is applied when b is superior over a with at least one of the objectives j, and b is not worse than a in the remaining objectives i.

Table 2. Terms and symbols used for presenting the mathematical models.

Symbol	Meaning
$DG(V_t, E_t)$	Graph of tasks.
$V = \{t_1, t_2, \ldots t_m\}$	Tasks to be executed in the fog network.
$E = \{e_1, e_2, \ldots e_k\}$	The dependency relation between the tasks.
$e_i = (t_{m1}, t_{m2})$	A connection between task t_{m1} and t_{m2}.
$P = \{P_1, P_2, \ldots P_m\}$	Computation load of the task.
$L = \{L_1, L_2 \ldots . L_m\}$	Communication loads of the task.
$G = \{G_1, G_2, \ldots G_n\}$	Subsets of independent graphs of tasks (a task in any graph can be executed with any order comparing with other tasks in the same graph).
$V = \{v_1, v_2, \ldots v_n\}$	Speed of CPU of nodes in the network.
$UDG(V_n, E_n)$	Graph of nodes.
$V = \{n_1, n_2 \ldots n_n\}$	Nodes are available for service in the fog network.
$RC = \{r_1, r_2 \ldots r_n\}$	Renting cost of nodes.
$RR = \{rr_1, rr_2 \ldots rr_n\}$	Reliability of nodes.
E_{comp}	Energy consumption because of the computational load.
E_{comm}	Energy consumption because of communication.
B	The bandwidth of the connection's links between nodes that participate in executing the task.
E_σ	Energy balance is represented by the standard deviation of the energy
C	The cost, which is represented by the total rental cost.
S	The stability term is a measure of the reliability of the nodes that execute the task.
$d = \left[d_{ij}\right] = [d\left(n_i, n_j\right)]$	The distance information between every two nodes
$e = [e_i]$	The energy consumption rate of nodes in the network
P_0	The maximum computational load that can be given to a certain node
L_0	The maximum communication load that can be given to a certain node

3.2. *Hyper-Angle Exploitive Searching HAES*

This section presents a hyper angle exploitive searching HAES algorithm. Firstly, we present its working principle and the difference between HAES and MOGA-AQCD [30] in Section 3.2.1. Secondly, we present the objective partitioning in Section 3.2.2. Lastly, the algorithm of HAES in Section 3.2.3.

3.2.1. Working Principle and the Difference between HAES and MOGA-AQCD

Both the proposed HAES and MOGA-AQCD use the concept of angle quantization for searching, which is based on dividing the space into equal-angle sectors and building a histogram that calculates the number of solutions selected for each sector. However, HAES behaves differently from MOGA-AQCD in terms of the selection of the new solutions. MOGA-AQCD favors solutions located in the least angular sector in terms of the previously selected solutions when two solutions are non-dominated with each other. In contrast HAES favors solutions located in the maximum angular sector in terms of the previously selected solutions. Typically, the MOGA-AQCD concept is to perform extensive exploration to yield substantial optimal solutions, whereas the HAES concept is that sectors that cover suitable solutions in the past are also likely to be rich in the future. We then provide an example to explain the critical difference between HAES and MOGA-AQCD regarding the searching concept.

The concept of HAES is depicted in Figure 2. The solution space is decomposed into a set of angular sectors. Each angular sector contains a set of solutions. The already found solutions are marked with black bullets and the candidate solutions are represented with white bullets. HAES selects the solutions that are located in the highest angular sector with respect to the number of solutions. We mark the selected solutions with yellow bullets and the ignored solutions with blue bullets.

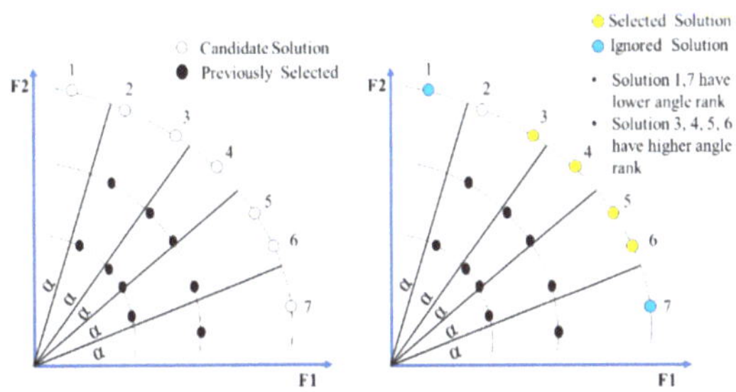

Figure 2. The selected solution in solution space by HAES.

3.2.2. Objectives Partitioning

The multi-objective optimization when working on a high number of objectives requires searching within a wide objective space, which makes it challenging to converge toward the boundary of the objective space. Hence, we do boundary searching mechanisms by activating the sub-set of objectives at each iteration out of the entire objectives. We name it objective partitioning; its role is to reach the boundary of the solution space with respect to the activated objectives. We select at each iteration of the optimization size $k < m$, where m denotes the number of objectives, and we use it for evaluating the solutions, sorting them, and selecting non-dominated ones. The sub-set of objectives is selected randomly at each iteration using a uniform distribution.

3.2.3. Algorithm of HAES

The general algorithm of HAES is presented in Algorithm 1. The algorithm takes the number of generations $NGen$, the number of solutions $NSol$, the sector range value $SectorRange$, and the set of objectives SoB as inputs, the size of objectives partitioning. The output of the algorithm is the Pareto front $ParetoFront$. As can be seen in Algorithm 1, the algorithm starts with the initialization of the first population in line 10, keeping it as a previous population in line 11, initialization of the counter of the population in line 12, initialization of the angle range rank in line 13, and initialization of crowding distance in line 14. Next, an iterative while loop is performed until the number of generations is finished. The loop is composed of calling for the evaluation of the solutions in the previous generation using the objective partitioning in function $selectSubSet$ (line 15) and the objective function calculation in the function $evaluate$ (line 16), updating the crowding distance using the function $updateCrowdingDistance$ (line 17), updating the ranges using the function $updateRanges$ (line 18), selecting the elites that are responsible for generating the off-spring using $selectElites$ (line 20), generating the off-spring using the function $geneticOperations$ (line 22), and the concatenation of the parents and their off-spring using the concatenation operator $||$ (line 23), and finally the new population is selected again from the resulted concatenated using the $electElites$ one more time (line 25). This process is repeated until the total iterations are finished, then the Pareto front of the last generated solution is the result of the algorithm, as presented in line 26.

The algorithm calls three essential functions: $updateCrowdingDistance()$, $updateRanges()$, and $selectElites()$. We provide the details of each of them in Algorithm 2, Algorithm 3, and Algorithm 4, respectively. For the $updateCrowdingDistance()$, the algorithm (detailed in Algorithm 2) takes the number of solutions $NSolutions$ and the objective values $objectiveValues$ as input, and provides the set of crowding distance $crowdingDistance$. The algorithm starts with the initialization of the set of the crowding distance with the size of solutions $NSolutions$ (line 7). Next, the two extreme solutions are assigned the value of infinity (line 8). Afterward, the algorithm sorts the solutions as the separated lists according to their objective values (line 9). Then, the algorithm updates the crowding distance in an accumulated way, corresponding to the difference between each objective of a solution and the value of its next solution in the sorted list (line 11).

Algorithm 1 Pseudocode of the HAES Algorithm

```
Input:
    N_Gen                                   //Number of Generations
        N_Sol                               //Number of Solutions
    SectorRange                             //Sector Range
    SoB = f_i, where i = 1, 2, ..., m;      //Set of Objectives
    K                                       //size of objectives partitioning
Output:
    ParetoFront                             //Found Pareto Front
Start:
        P_0 = InitiateFirstPopulation N_Sol;        //generate first population randomly
        populationPrevious = P_0;           //first population is the previous population
        counterOfGeneration = 1;
        angleRangeRank = zeros (1, 2π/SectorRange) //initialize the angle range rank
        while (CounterofGeneration < N_Gen)
          SSoB=selectSubSet(SoB, k)
          [solutionsRanks,objectiveValues] = evaluate (populationPrevious,SSoB)
          [crowdingDistance] = updateCrowdingDistance (populationPrevious,objectiveValues)
          [angleRangeRank] = updateRanges (populationPrevious,solutionsRanks,
          SectorRange,angleRangeRank, SoB) //select N_Sol from the previous solutions
          selected Elites = selectElites
          (P_0,solutionsRanks,angleRangeRank,crowdingDistance, N_Sol)
          offSpring = geneticOperations (selected Elites)
          combinedPop = selectedElites | | offSpring sortedCombinedPop =
    NonNominatedSorting (combinedPop)
    P_New = selectElites (sortedCombinedPop, angleRangeRank,NSol)
    populationPrevious = P_New;
    CounterofGeneration++;
    end while
End
```

Algorithim 2 Pseudocode of calculating the crowding distance

```
Input:
    N_Sol
    objectiveValues
Output:
      CrowdingDistance
Start:
      crowdingDistance = zeros (N_Sol);
      crowdingDistance (1) = crowdingDistance (N_Sol) = ∞
      for (each i objective of objectiveValues) sortedSolutions = sort (N_Sol,i);
        for (solution j from 2 to N_Sol)
            crowdingDistance (j) = crowdingDistance(j) + objectiveValues(i) − objectiveValues(i − 1);
        end for
      end for
End
```

The *updateRanges*() function is provided in Algorithm 3. It takes three variables: Solutions, *SectorRange*, and *SoB*, as input. Additionally, it gives *angleRangeRank* as output. The approach of obtaining *angleRangeRank* is based on performing an iterated loop in the input Solutions and updating the counter of each sector in the *SectorRange* that contains the solution, as presented in the for loop from line 10 to line 13.

Algorithim 3 Pseudocode of updating the angle range rank

```
Input:
        Solutions
        SectorRange
        SoB
Output:
        angleRangeRank
Start
            L = length (Solutions)
            angleRangeRank = zeros (360/SectorRange)
            for (i = 1 to L)
                A_i = angle (solution(i))//angle of solution i
                angleRangeRank (j) = map (A_i, SectorRange) + angleRangeRank (j)
            end for
            return angleRangeRank
End
```

The final procedure receives the pool of solutions Pool of Solutions, the rank of solution Rank, the angle range rank *AngleRangeRank*, the array of the crowding distance *CrowdingDistance*, and the number of solutions to be selected N as input and provides selected solutions (Algorithm 4). The procedure performs an iterated loop for N times, where it selects two solutions in each time and calculates three measures for each solution: rank, angle range rank, and crowding distance. Next, the selection function determines which one has a better rank (line 17), better angle range rank (line 19), and better crowding distance (line 21). Then, the selection process is applied by checking the condition (line 22–24) to identify which favors a solution that has a better rank. In the case that two solutions have the same rank, then the solution with better angle range rank is selected. If the two solutions both have the same values of rank and angle range rank, then the approach will select the solution that has better crowding distance. In addition, the definition of "better" is provided for rank in line 17, for angle range rank in line 19, and for crowding distance in line 21. The detail of the algorithm for selecting the elites is shown in Algorithm 4.

3.3. Fog Computing Closed Loop Model (FCCL)

This section presents our developed integrated objectives fog computing model FCCL. It is composed of five main sections. Section: 3.3.1 explains the first layer which is the fog interface. Section 3.3.2 is an overview of the task decomposer and task model. Next, Section 3.3.3 the task dispatcher. Then, Section 3.3.4 contains the network model, and lastly, Section 3.3.5 contains the optimization objectives.

From a fog computing perspective, our problem is formulated similarly. The fog has an interface that receives from the user a request of executing a computational task with the needed criterion for optimization. Next, it calls an optimization algorithm that provides a set of non-dominated solutions with respect to the provided criteria. The user will make a decision for selecting one among them. The criteria are denoted by vectors $y = (y_1, y_2, \ldots y_m)$, where $\{y_i\}$ denotes a criterion for fog computing optimization. Without loss of generality, we consider five criteria, namely, Energy Consumption, Energy Distribution, Renting Cost, and Stability.

$y = (\textit{energy consumption, energy distribution, renting cost, and stability})$. The solutions that are provided to the user gives the selected fog nodes for the execution of the request; we are represented by vector $x = (x_1, x_2, \ldots x_n)$. The goal is to maximize the domination aspect of the provided solutions and their diversity. This gives the user more variety of choices. To elaborate more, we present Figure 3, which elaborates the user giving a request to the user interface and waiting for a set of non-dominated solutions to select one. The fog interface communicates with the task decomposer that decomposes the task that is requested by the user to execute in the fog network. The role of the task decomposer is to partition the task into subsets of independent subtasks; we call each subset a group. Each group is executable independently on the other task.

Algorithim 4 Pseudocode of selecting the elites

```
Input:
    Pool of Solutions
    Rank
    AngleRangeRank
    CrowdingDistance
    N                                        //number of the selected solutions
Output:
    selected solutions
Start:
for (solution = 1 to N)                      //number of the selected solutions
    Select two individuals A, B randomly for an individual
    Compute Non-domination rank (rank)
    Compute Crowding distance (distance)
    Compute Angle rank level (angle Range Rank)

            //Compare Solutions
    betterRank = A_rank < B_rank
    sameRank = A_rank == B_rank
    betterAngleRangeRank = A_angleRangeRank > B_angleRangeRank
    sameAngleRangeRank = A_angleRangeRank == B_angleRangeRank
    betterCrowdingDiandstance = A_distance > B_distance
    if (betterRank)
      or (sameRank and betterAngleRangeRank)
      or (sameRank and sameAngleRangeRank and betterCrowdingDistance)
    then
            add A to the selected solutions
    else
            add B to the selected solutions
    end if
  end for
End
```

This aspect enables shorter execution time, which is one of the metrics to be optimized. The task decomposer communicates with the task dispatcher that is responsible for calling the mathematical functions of the fog criterion for calculating the objective function for any candidate solution. Obviously, the task dispatcher receives the needed information from the fog network and the task decomposition and specification before carrying the optimization. The optimization is carried using a multi-objective optimization algorithm named HAES.

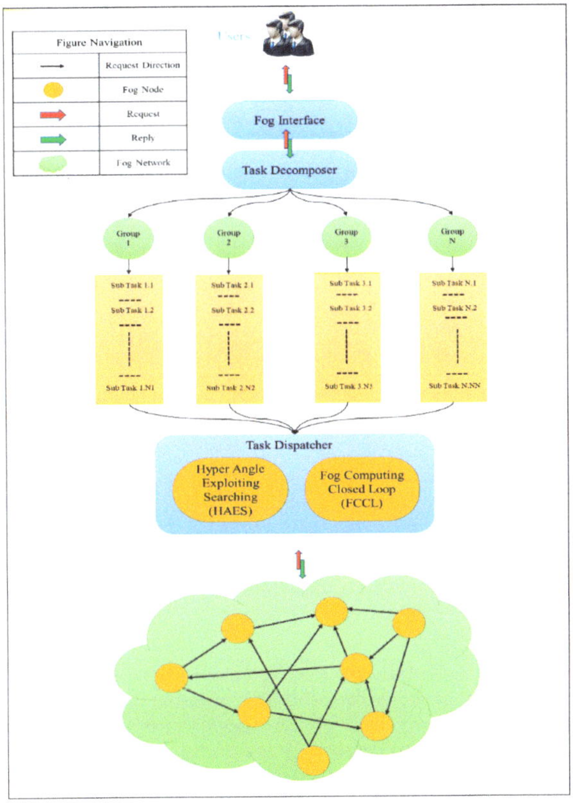

Figure 3. The framework of multi-criteria optimization for fog computing.

3.3.1. Fog Interface

The fog interface will accept from the user two inputs. The first one is the task, and the second one is the preference vector of the various objectives for optimizing the task. The vector of preference between the five objectives is the five components vector, given as $pre = [pr_1\ pr_2\ pr_3\ pr_4\ pr_5]$ with the constraint $\sum_{i=1}^{5} pr_i = 1$. The second input is the configuration input, which is also given by a vector named $conf = [itMax\ popSize]$, where $itMax$ denotes the maximum number of iterations, and $popSize$ denotes the size of the population. Assuming that there is more interest in the time execution (makespan) and stability, the second interest is in the cost, and the third interest in the energy consumption and the energy balance, then the value of $pre = \left(1 \times pr, 1 \times pr, \frac{1}{2} \times \mathrm{pr}, \frac{1}{3} \times \mathrm{pr}, \frac{1}{3} \times \mathrm{pr}\right)$. This implies, $1 \times pr + 1 \times pr + \frac{1}{2} \times \mathrm{pr} + \frac{1}{3} \times \mathrm{pr} + \frac{1}{3} \times \mathrm{pr} = 1$. Then, $pr = 6/19$.

3.3.2. Task Decomposer and Task Model

The logical decomposition of data fusion tasks is a fundamental process in the design of systems aiming at combining multiple and heterogeneous cues collected by sensors. In recent years, a relevant body of research has focused on formalizing logical models for multi-sensor data fusion in order to propose appropriate and general task decomposition. Therefore, we suggest a task decomposer, which is elaborated in Figure 4, to decompose the data and classify based on priority. The role of the task's decomposer is to decompose the tasks into a set of independent tasks; we denote them into groups $G = \{G_1, G_2, \ldots G_N\}$. Example 1 went particularly into decomposing and classifying the tasks.

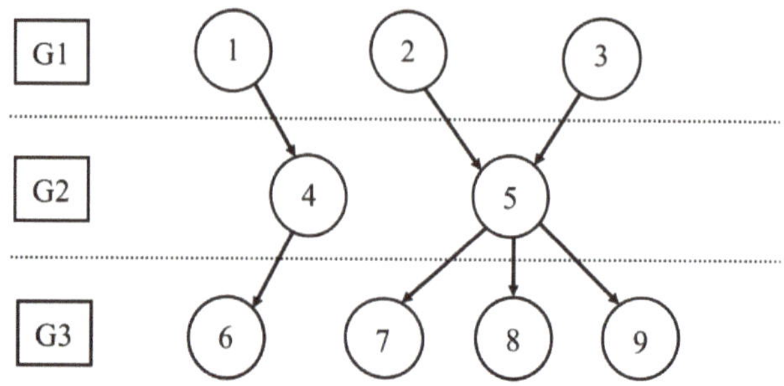

Figure 4. Task Decomposer.

This component has the role of accepting the task from the user. The task itself is modeled as a directed graph $DG(V,E)$, where $V_t = \{t_1, t_2, \ldots t_m\}$, $E = \{e_1, e_2, \ldots e_k\}$ where m denotes the number of tasks in the graph and k denotes the number of directed edges. Where each edge $e_i = (t_{m1}, t_{m2})$, it denotes that t_{m2} is dependent on t_{m1}. Another piece of information that is related to the task and has to be provided by the interface is the workload of tasks in terms of both computation and communication, where the computation is described by set $P = \{P_1, P_2, \ldots P_m\}$, where each P_i denotes the computation that is the number of a clock for the task P_i, and the communication load is described by the set $L = \{L_1, L_2 \ldots . L_m\}$, where L_i represents the communication loads, which describes the total length of data to be exchanged among selected nodes for executing the task.

Example 1:

Task decomposer will classify the nodes in the network into groups, and each group depends on the number of nodes in the fog network. In addition, its direct graph, which is the fog nodes, will forward the request to the next node. The result of the task decomposer is set of three groups as G_1 = {1,2,3}, G_2 = {4,5}, and G_3 = {6,7,8,9}. As we see, the tasks in each group are independent of each other, and they can be processed in any order.

3.3.3. Task Dispatcher

The task dispatcher is responsible for allocating certain nodes in the fog network for the execution of the sub-tasks that result from the task decomposer. It contains the optimization algorithm HAES, which was presented in Section 3.2.3. The fog computing closed loop is presented in Section 3.3.

3.3.4. Network Model

We assume that the network is an undirected graph $UDG(V_n, E_n)$, where $V = \{n_1,\ n_2 \ldots n_n\}$, where n denotes the number of nodes in the network. $E = \{(n_i, n_j)\}$ where $n_i, n_j \in V$. Assuming that the nodes have wireless connections between each other, then we are interested in the distance between every two nodes. Each node i has a rate of computational energy consumption e_i $[\frac{j}{sec}]$, and each two nodes $n_i, n_j \in V$, have distance between them, which is given as $d_{ij} = d(n_i, n_j)$. In addition, we assume that each node n_i has a speed for execution v_i. Furthermore, we assume that each node has a maximum capacity for executing computational load p_0 and maximum capacity for executing communication load l_0.

3.3.5. Optimization Objectives

We present in this section the equations of the optimization objectives. Our model has the aspect of integrating five objectives at the same time, which makes it distinguished from other models in the literature.

A. Time Latency

Time latency is an expression of how much time it takes for a packet of data to get from one designated point to another. It is sometimes measured as the time required for a packet to be returned to its sender, which is calculated by the following formula.

$$T = \sum_{i=1}^{m}\sum_{j=1}^{n} t_{ij} \tag{5}$$

$$t_{ij} = {t_{ij}}^{1} + {t_{ij}}^{2} \tag{6}$$

$${t_{ij}}^{1} = \frac{P_{ij}}{v_i} \text{ computation time} \tag{7}$$

$${t_{ij}}^{2} = \frac{l_{ij}}{B} + {t_{ij}}^{queue} \text{ communication time} \tag{8}$$

where ${t_{ij}}^{queue}$ denotes the queue waiting time. P_{ij} denotes the task computational load that is assigned to node i. The speed is v_i of the node i. l_{ij} denotes the communication load between i and j. Lastly, B denotes the bandwidth.

B. Energy Consumption

In order to send number packets from node A until node B, where the distance between the two nodes is $d(A,B) = d$, we calculate the consumed energy as Equation (9).

$$e(A,B) = e(d) = \begin{cases} (e_{elec} + \varepsilon_{amp} d^2) l(A,B) \; for \; transmit \\ e_{elec} l(A,B) \; for \; receive \end{cases} \tag{9}$$

where e_{elec} denotes energy consumption for operating the radio model for each bit in the data. d denotes distance between the two nodes A, B. The coefficient of transmit amplifier given by ε_{amp}. $l(A,B)$ denotes the number of bits to be sent from node A to node B.

Based on the term $e(A,B) = e_{A,B}$, $l(A,B) = l_{A,B}$, we can calculate the total energy consumption based on terms E_{comp}, E_{comm}, which represent the computation energy consumption and communication energy, respectively. The total energy is given in Equation (10), the computation energy is given in Equation (11), and the communication energy is given in Equation (12).

$$E = E_{comp} + E_{comm} \tag{10}$$

$$E_{comp} = \sum_{i=1}^{n} e_i t_i \tag{11}$$

where e_i denotes to the energy consumption because of execution in node i, t_i denotes to the time allocation of the node i, $e_{i,j}$ denotes the energy consumption because of communication between nodes i and j, and $l(i,j)$ number of bits transferred between nodes i and j.

$$E_{comm} = \sum_{i,j,\, i \neq j}^{m} e_{i,j} l_{i,j} \tag{12}$$

C. Energy Distribution

This term indicates the differences among the nodes in terms of the energy levels. The term is calculated as the standard deviation of the node's energy as it is given in Equation (13).

$$E_{\sigma} = \sqrt{\frac{\sum_{i=1}^{n} \left(E_i - \overline{E}\right)^2}{n-1}} \tag{13}$$

where E_i denotes the consumed energy of node i; $\overline{E}$ denotes the average consumed energy of all nodes. n denotes the total number of nodes.

D. Renting Cost

The renting cost is defined as the total cost of rent, which is the summation of node rental rate r_i multiplied by the time of allocating the node according to Equation (14).

$$C = \sum_{i=1}^{n} t_i r_i \quad (14)$$

where r_i denotes the renting rate of the node i. t_i denotes the time of allocating the node.

E. Stability

This term indicates the total stability of the task execution. It is calculated as the summation of the reliability percentage of a certain node rr_i multiplied by the time of allocating the nodes. The calculation is depicted in Equation (15).

$$S = \sum_{i=1}^{n} t_i rr_i \quad (15)$$

where rr_i denotes to the reliability rate of the node n_i, and t_i denotes the time of allocating the fog node i.

F. Constraints

Before assigning any given solution to the fog network, it is needed to assure that it meets the constraints. Basically, there are two types of constraints that should be satisfied. The first one is the connectivity constraint, which states that any sub-network is assigned an execution of a task; it should be connected in order to execute the task that is assigned to the sub-network. The second constraint is named the load constraint. It states that for a task T with computational load P and communication load L, it should be allocated at least N_0 for execution. The value N_0 is calculated based on Equation (16).

$$\begin{cases} N_0 = Max\left(\frac{L}{L_0}, \frac{P}{P_0}\right) \\ N \geq N_0 \end{cases} \quad (16)$$

4. Experimental Design and Parameters Setup

This section comprises three categories for presenting the evaluation of the proposed model and base benchmarks used in the evaluation. The first category, in Section 4.1, is the evaluation metrics of HAES and FCCL models. This section talks specifically about the most common and standard evaluation measures, which are hyper-volume, non-dominated solution, generational distance measure, inverse relative generational distance measure, delta metric measure, and set coverage measure. In addition, the parameters for HAES mode with base models. The second category, Section 4.2, is a dedicated section that presents the multi-objective mathematical functions that will test HAES and compare it with state-of-the-art approaches. The third, Section 4.3, presents the parameters for the FCCL model.

4.1. Evaluation Metrics of HAES and FCCL

This section presents the evaluation metrics that are used for evaluating our developed approaches, which are HAES and FCCL. Fog computing evaluation metrics are the same objectives that are used for optimization. We present the hyper-volume in sub-section A. Next, we present the number of non-dominated solutions in sub-section B. Afterward, the generational distance is presented in sub-section C. Next, the inverse relative generational distance measure in sub-section D, and the delta metric is provided in sub-section E. Lastly, set coverage is giving in sub-section F.

A. Hyper Volume (HV) Measure

The hyper volume (HV) metric is widely used in evolutionary MOO to evaluate the performance of the searching algorithm [36]. It computes the volume of the dominated portion of the objective space related to the worst solution. This region is the union of the hypercube, with its diagonal as the distance between the reference point and a solution X from the Pareto Set (PS). High values of this measure present the desirable solutions. HV is presented by the following (Equation (17)):

$$HV = volume(\cup_{x \in P_s} Hyper\ Cube\ (x)). \tag{17}$$

B. Number of Non-Dominated Solutions (NDS)

The number of non-dominated solutions (NDS), which expresses the effectiveness of the optimization algorithm [37], can be calculated as the cardinality of PS as (18):

$$NDS(N) = |P_s|. \tag{18}$$

C. Generational Distance Measure (GDM)

This metric, also called the GD metric [38], is a measure to evaluate the performance of a found Pareto Set (PS) compared with a reference point set (a true Pareto set (PS)). This measure is based on the distance among obtained solutions and reference points, which is calculated as follows (Equation (19)):

$$GD(P_s, P_T) = \frac{\left(\sum_{i=1}^{|P_s|} d_i^2\right)^{\frac{1}{2}}}{|P_s|}. \tag{19}$$

D. Inverse Relative Generational Distance Measure (IRGD)

Inverse Relative Generational Distance Measure (IRGD)

Another metric that is used is the inverse Relative Generational Distance or IRGD, and it is given in Equation (20).

$$IRGD(P_s, P_T) = \frac{|P_s|}{\left(\sum_{i=1}^{|P_s|} d_i^2\right)^{\frac{1}{2}}}. \tag{20}$$

E. Delta Metric Measure

The delta or diversity metric Δ shows the extent to which it achieves the spread [14]. The delta measure receives the non-dominated set of solutions and provides the diversity metric, and can be computed according to the following equation:

$$\Delta = \frac{d_f + d_l + \sum_{i=1}^{N-1} |d_i - d^-|}{d_f + d_l(N-1)d^-} \tag{21}$$

where N is the number of solutions, d_f and d_i, the Euclidean detachments between the extreme and border solutions, and d is all the consecutive distances, d_i (i = 1, 2, . . . , N − 1). This measure is required to be slight and maintained to be less, because this measure indicates uniform distribution. In addition, it provides various selections to the decision-maker.

F. Set Coverage Measure

Set coverage measure [37], also called C metric, compares the Pareto sets *Ps1* and *Ps2* and can be identified by (22):

$$C(P_{s1}, P_{s2}) = \frac{|\{y \in P_{s2} | \exists x \in P_{s1} : y \prec x\}|}{P_{s2}} \tag{22}$$

C equals the ratio of nondominated solutions in P_{s2} dominated by non-dominated solutions in P_{s1} to the number of solutions in P_{s2}. Thus, when evaluating a set Ps, the value of $C(X;\ Ps)$ must be minimized for all Pareto sets X.

4.2. Multi-Objective Mathematical Functions

The algorithms are evaluated based on various relevant MOO mathematical functions. The formulas, optimization range, and true PF of each mathematical function are provided in Table 3. They have been used in most of the existing studies on MOO optimization as the benchmarking functions. The convexity is different for each function. Table 3 shows the bounds of the variables and the optimal solutions or PFs. In this way, our proposed approach can be validated against critical MMO measures. We selected three approaches NSGA-II, NSGA-III, and MOGA-AQCD, which were presented in the background section as the three relevant benchmarks to evaluate HAES.

To make the study quantitative, ten experiments are performed for each function using different seeds. This study also refers to the previous studies so that the same methodology of evaluation as of Multi-Objective Evolutionary Algorithms (MOEAs) is performed. The test function is chosen based on the well-known studies, including Fleming's study (FON) [39], Kursawe's study (KUR) [40], Poloni's study (POL) [41], and Schaffer's study (SCH) [42]. We then followed those guidelines and suggested six test problems, in which five of them are presented in Table 3, call ZDT1, ZDT2, ZDT3, ZDT4, and ZDT6. All problems have two objective functions, and none of these problems has any constraint. In addition, the number of variables, the bounds, the Pareto-optimal solutions, and the nature of the Pareto-optimal front for each problem.

Table 3. Mathematical functions for evaluating MOO measures.

Problem	n	Variable Bounds	Objective Function	Optimal Solution	Remark
FON	3	$[-4, 4]$	$f1\,(x) = 1 - \exp\left(-\sum_{i=1}^{3}\left(xi - \frac{1}{\frac{1}{\sqrt{3}}}\right)^2\right)$ $f2\,(x) = 1 - \exp\left(-\sum_{i=1}^{3}\left(xi - \frac{1}{\frac{1}{\sqrt{3}}}\right)^2\right)$	$x_1 = x_2 = x_3$	Non-convex
KUR	3	$[-5, 5]$	$f_1(x) = \sum_{i=1}^{n-1}\left(-10\,exp\left(-0.2\sqrt{x_i^2 + x_i^2 + 1}\right)\right)$ $f(x) = \sum_{i=1}^{n}\left(\lvert x_i\rvert^{0.8} + 5\sin x_i^3\right)$	[43]	Non-convex
POL	2	$[-\pi, \pi]$	$f_1(x) = \left[1 + (A_1 - B_1)^2 + (A_2 - B_2)^2\right]$ $f_2(x) = \left[(x_1 + 3)^2 + (x_2 + 1)^2\right]$	[43]	Non-convexDisconnected
SCH	1	$[10^{-3},\ 10^3]$	$f_1(x) = x^2$ $f_2(x) = (x-2)^2$	$X \in [0,\ 2]$	Convex
ZDT1	30	$[0, 1]$	$f_1(x) = \sum_{i=1}^{n-1}(-10\,\exp(-0.2\sqrt{x_i^2 + x_{i+1}^2}$ $f_2\,(x) = \sum_{i=1}^{n}(\,\lvert x_i\rvert^{0.8} + 5\sin x_i^3)$	$x_1 \in [0,\ 1]$ $x_i = 0$ $i = 2,\ 3, \ldots, n$	Convex
ZDT2	30	$[0, 1]$	$f_1(x) = x_1$ $f_2(x) = g(x)\left[1 - (x_1/g_x)^2\right]$ $g(x) = 1 + 9\left(\sum_{i=2}^{n} x_i\right)/(n-1)$	$x_1 \in [0,\ 1]$ $x_i = 0$ $i = 2,\ 3, \ldots, n$	Non-convex
ZDT3	30	$[0, 1]$	$f_1(x) = x_1$ $f_2(x) = g(x)\left[1 - \sqrt{\frac{x_1}{g(x)}} - \frac{x_1}{g(x)}\sin(10\,\pi\,x_1)\right]$ $g(x) = 1 + 9\left(\sum_{i=2}^{n} x_i\right)/(n-1)$	$x_1 \in [0,\ 1]$ $x_i = 0$ $i = 2,\ 3, \ldots, n$	Convex, Disconnected

Table 3. *Cont.*

Prob-lem	n	Variable Bounds	Objective Function	Optimal Solution	Remark
ZDT4	10	[0, 1] [−5, 5]	$f_1 = x_1$ $f_2 = g(x)\left[1 - (\frac{f_1}{g})^{0.5}\right]$ $g = 1 + 10(N-1) + \sum_{i=2}^{N}(x_i^2 - 10\cos(4\pi x_i))$		Non-convex
ZDT6	10	[0, 1]	$f_1(x) = 1 - \exp(-4x_1)\sin^6(6\pi x_1)$ $f_2(x) = g(x)\left[1 - (\frac{f_1(x)}{g(x)})^2\right]$ $g(x) = 1 + 9\left[\left(\sum_{i=2}^{n} x_i\right)/(n-1)\right]^{0.25}$	$x_1 \in [0, 1]$ $x_i = 0$ $i = 2, 3, \ldots, n$	Convex, non-uniformly spaced

The implementation is conducted using MATLAB 2019b. The parameters for NSGA-II, NSGA-III, MOGA-AQCD, and HAES are given in Tables 4 and 5. The same number of solutions and generation was used for all the algorithms in order to have a fair comparison. An increase in the number of solutions and generations typically yields better performance results. The numbers of the population and generations are selected to be (100) and (500), respectively. The parameters of the crossover are determined based on two parts: fraction and ratio. The fraction is selected to be $2/n$, where n denotes the solution length and the ratio is selected to be 1,2. For the scale of the mutation, we selected the value of 0,1. These values are the default ones that are used by the MATLAB optimization package.

Table 4. Parameters of NSGA-II, MOGA-AQCD, and HAES.

Parameters		NSGA-II	MOGA-AQCD	HAES
No. of solution		100	100	100
No. of generation		500	500	500
Crossover option	Fraction	2/n	2/n	2/n
	Ratio	1,2	1,2	1,2
Mutation option	Fraction	2/n	2/n	2/n
	Scale	0,1	0,1	0,10
	Shrink	0.5	0.5	0.5
Quantification of angle space (α)		N/A	10^{-7} for all test except KUR 5×10^{-7}	10^{-7} for all test except KUR 5×10^{-7}

Table 5. Parameters of NSGA-III.

Parameters		NSGA-III
No. of Solution		100
No. of Generation		500
Crossover Percentage		0.5
Mutation Option	Mutation Percentage	0.5
	Mutation Rate	0.02
Number of Divisions		10

These selected numbers are to obtain the PF within a balanced time. However, increasing both or one of them yields highly dominated solutions, given extensive exploration will be conducted in the searching space.

4.3. HAES Evaluation Based on FCCL Model

The evaluation is done based on population size 200 and number of generations 200. We run the model on 10 experiments. Each experiment is conducted on a different value of the quantization, $\alpha = \{20, 23, 25, 28, 30, 33, 35, 37, 40, 45\}$. In addition, each experiment i repeated 10 times with different values of seed, which are given in Table 6. The results ar decomposed into two sub-sections. The first one is the presentation of the results of th multi-objective mathematical functions, and the second one is the results of the evaluatio of the fog computing closed-loop model.

Table 6. Table of parameters used for evaluation FCCL Model.

Parameter	Value
Population size	200
Number of generations	200
Number of random experiments	10
α	$\{20, 23, 25, 28, 30, 33, 35, 37, 40, 45\}$
Number of nodes	30
Number of tasks	6
Number of objectives	5
Crossover	1.2
Mutation	0.5, 1.5

5. Evolution and Enhanced Model Results

This part presents the results of the two models, HAES and FCCL, and discus the experiment results comparing to the other models and their differences. Section 5.1 elaborates on the first phase which is the optimization of HAES with three benchmark as follows NSGA-II, NSGA-III, and MOGA-AQCD; Section 5.2, the second phase, is th model of FCCL and the comparison of our model with the same benchmarks for phase on

5.1. HAES Experimental Investigation and Results

The evaluation of the HAES algorithm is performed firstly based on mathematica functions with a challenging MOO nature as follows: FON, KUR, POL, SCH, ZDT1, ZDT2 ZDT3, ZDT4, and ZDT6. It presents the Pareto front, average hyper volume metric, averag non-dominated solutions metric, an average of delta metric, and the average of generationa distance metric, respectively, in each figure for HAES and other three benchmarks. As w observe in Figure 5, the Pareto front is plotted with two axes figures, because each of the mathematical functions has two objectives. Considering that HAES has an exploiting natur that enables the algorithm to each more dominant solution even if the regions of exploratio were less, this has made it more capable of minimizing the values of the objectives.

In order to present this clearly, we show for each mathematical function two scales: the first one shows the general Pareto at the top and the second one shows the area of solutions found by HAES at the bottom. The Pareto front was lower for the functions FON, POL, SCH ZDT1, ZDT2, ZDT3, and ZDT4, which is more domination with respect to these functions The only function that has not achieved lower values of the Pareto front is KUR. However HAES has achieved a more diverse Pareto front for KUR compared with the benchmarks Figure 5 elaborate on the results for mathematical functions for each metric particularly.

In order to identify the superiority in terms of domination, we provide two tables the first one is showing the domination of the benchmarks over HAES in Table 7, and the second shows the domination of HAES over the benchmarks in Table 8. As can be seen, the values in Table 7 are higher than their corresponding values in Table 8, which means that HAES is more dominant over the MOGA-AQCD, NSGA-III, and NSGA-II.

Table 7. Average set coverage values of HAES compared to those of MOGA-AQCD, NSGA-III, and NSGA-II.

Functions	MOGA-AQCD	NSGA-III	NSGA-II
FON	1.100×10^{-2}	3.000×10^{-2}	1.500×10^{-2}
KUR	3.100×10^{-2}	2.290×10^{-1}	2.700×10^{-2}
POL	8.000×10^{-3}	1.000×10^{-3}	4.000×10^{-2}
SCH	2.000×10^{-3}	6.880×10^{-1}	2.000×10^{-3}
ZDT1	0.000×10^{-0}	0.000×10^{-0}	1.500×10^{-2}
ZDT2	0.000×10^{-0}	0.000×10^{-0}	0.000×10^{-0}
ZDT3	6.000×10^{-3}	0.000×10^{-0}	1.500×10^{-2}
ZDT4	4.815×10^{-2}	6.000×10^{-1}	9.000×10^{-2}
ZDT6	2.750×10^{-1}	0.000×10^{-0}	2.710×10^{-1}

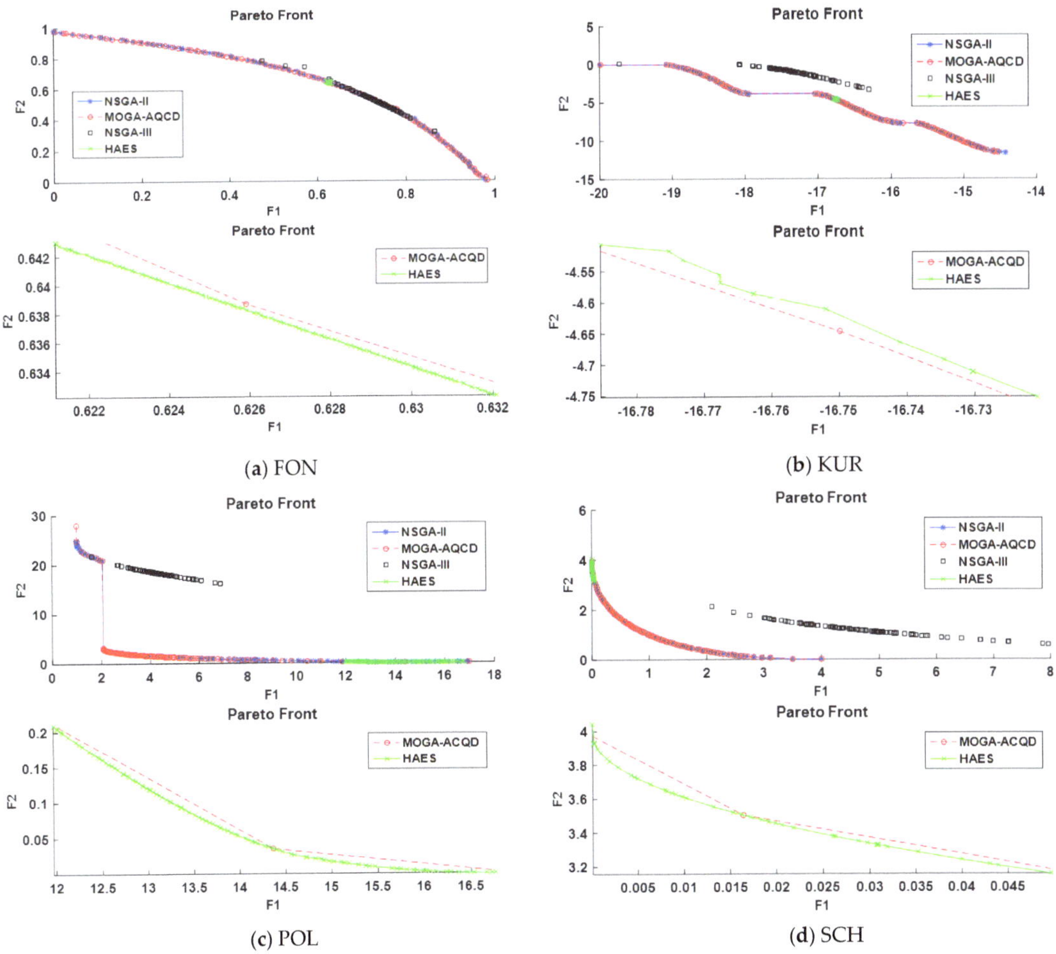

(**a**) FON (**b**) KUR (**c**) POL (**d**) SCH

Figure 5. *Cont.*

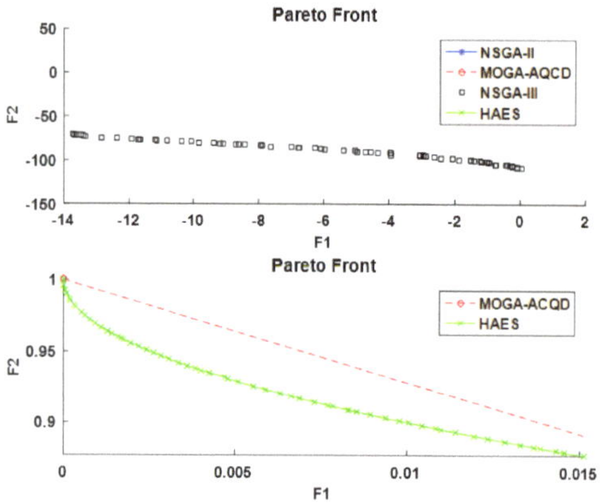

(**e**) ZDT1

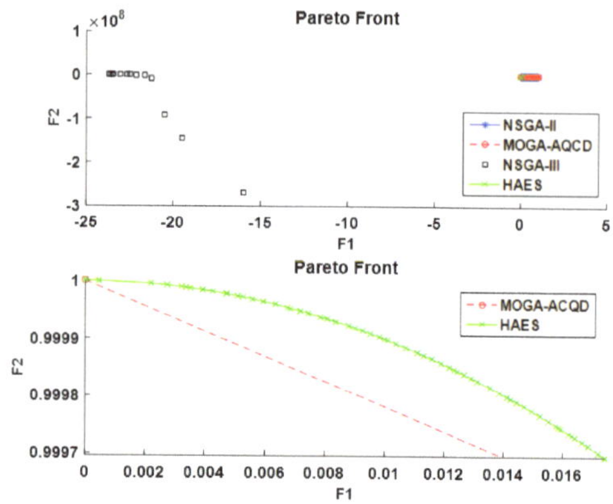

(**f**) ZDT2

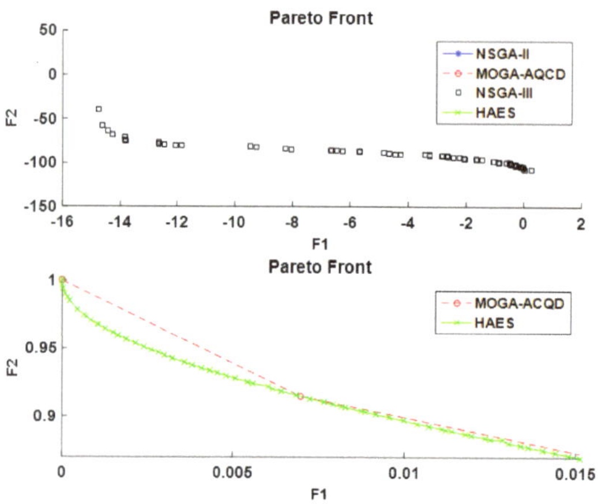

(**g**) ZDT3

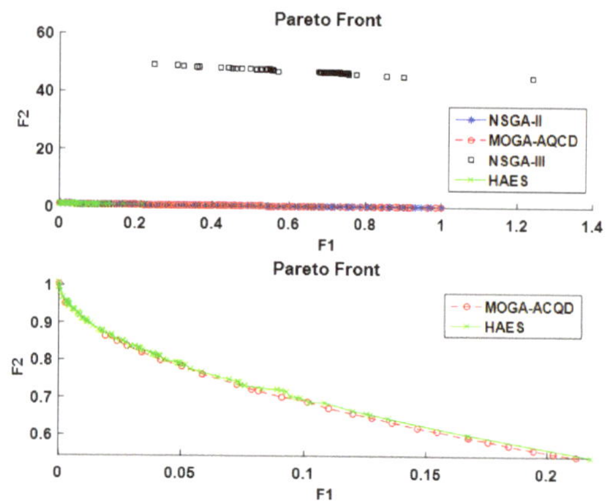

(**h**) ZDT4

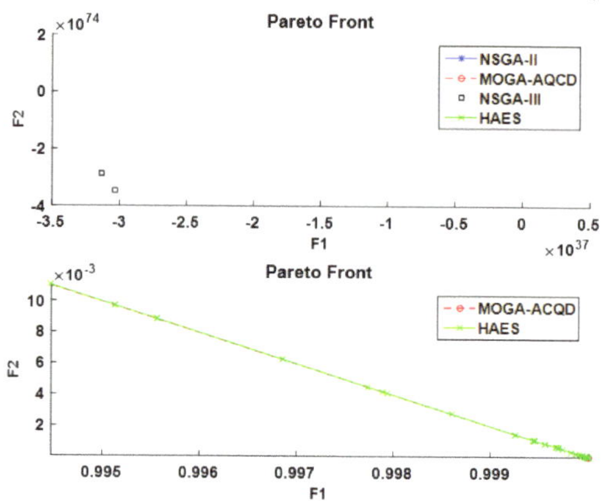

(**i**) ZDT6

Figure 5. Pareto front with two scales: sub-figure HAES with MOGA-AQCD for FON, KUR, POL, SCH, ZDT1, ZDT2, ZDT3, ZDT4, and ZDT6.

Table 8. Average set coverage of MOGA-AQCD, NSGA-III, and NSGA-II compared to that of HAES.

Functions	MOGA-AQCD	NSGA-III	NSGA-II
FON	0.000×10^{-0}	0.000×10^{-0}	0.000×10^{-0}
KUR	8.140×10^{-3}	7.488×10^{-3}	1.279×10^{-2}
POL	1.000×10^{-3}	0.000×10^{-0}	1.300×10^{-2}
SCH	2.000×10^{-3}	0.000×10^{-0}	2.000×10^{-3}
ZDT1	0.000×10^{-0}	1.000×10^{-0}	3.000×10^{-3}
ZDT2	0.000×10^{-0}	1.000×10^{-0}	0.000×10^{-0}
ZDT3	0.000×10^{-0}	1.000×10^{-0}	0.000×10^{-0}
ZDT4	0.0481209	0	0.1139833
ZDT6	0	1	0

In order to assess the performance of HEAS in terms of the richness of the found solutions compared with the benchmarks, we present the hyper-volume. As it is shown in Table 9, ZDT6 has accomplished high hyper-volume only for KUR and ZDT6, while it was less for the other functions. This is interpreted as more domination of solutions that was accomplished for HAES compared with the benchmarks. This makes it more challenging to obtain high hyper-volume compared with MOGA-AQCD, NSGA-II, and NSGA-III, which has generated a lower dominant Pareto front.

Table 9. Average of MOO metrics for benchmarking mathematical functions.

Problems	Evaluation Measure	HAES	MOGA-AQCD	NSGA-III	NSGA-II
FON	Average of Hyper Volume	**5.685**	0.298	0.089	0.297
	Average Non-Dominated Solutions	100	100	100	100
	Delta Metric	0.991	**0.196**	1.011	0.281
	Average Generational Distance	**0.00109**	0.001199	0.001483	0.001199
KUR	Average of Hyper Volume	15.85	25.66	2.316	**25.67**
	Average Non-Dominated Solutions	61.8	**100**	**100**	**100**
	Delta Metric	0.8695	**0.3695**	1.035	0.4129
	Average Generational Distance	0.01893	0.006606	0.07131	**0.006420**
POL	Average of Hyper Volume	0.4963	368.2	17.45	**369.1**
	Average Non-Dominated Solutions	100	100	100	100
	Delta Metric	**0.9289**	1.308	1.026	0.9444
	Average Generational Distance	**0.001193**	0.007846	0.204	0.008936
SCH	Average of Hyper Volume	0.02784	13.26	**17.45**	13.26
	Average Non-Dominated Solutions	100	100	100	100
	Delta Metric	1.057	**0.6812**	1.021	0.6812
	Average Generational Distance	0.001227	**0.0008915**	1.15	**0.0008915**
ZDT1	Average of Hyper Volume	0.0012	0.6591	**187.1**	0.6579
	Average Non-Dominated Solutions	**100**	**100**	66	**100**
	Delta Metric	0.9863	**0.4984**	0.9223	0.6562
	Average Generational Distance	7.92×10^{-4}	**4.18×10^{-4}**	10.9096	5.02×10^{-4}
ZDT2	Average of Hyper Volume	1.6993	0.3274	0.3247	**2.1159**
	Averages Non-Dominated Solutions	**100**	**100**	13.8	**100**
	Delta Metric	0.9985	**0.3258**	1.295	0.6794
	Average Generational Distance	0.0011	**5.06×10^{-4}**	2.31×10^{11}	5.31×10^{-4}
ZDT3	Average of Hyper Volume	0.0012	0.7763	341.5	0.7771
	Average Non-Dominated Solutions	**100**	**100**	39.1	**100**
	Delta Metric	0.9915	0.7661	0.9718	**0.7541**
	Average Generational Distance	**5.55×10^{-4}**	6.81×10^{-4}	14.3872	6.60×10^{-4}

Table 9. *Cont.*

Problems	Evaluation Measure	HAES	MOGA-AQCD	NSGA-III	NSGA-II
ZDT4	Average of Hyper Volume	0.2211	0.6407	**0.829**	0.6119
	Average Non-Dominated Solutions	87.3	**100**	67.6	**100**
	Delta Metric	1.014	0.4384	1.013	**0.3854**
	Average Generational Distance	$\mathbf{9.05} \times 10^{-4}$	0.0012	7.9171	09.05×10^{-4}
ZDT6	Average of Hyper Volume	**0.4746**	0.2646	0	0.2636
	Average Non-Dominated Solutions	59.8	**100**	1.4	**100**
	Delta Metric	1.214	**0.635**	0.9666	**0.7989**
	Average Generational Distance	0.0363	3.35×10^{-4}	3.47×10^{85}	$\mathbf{3.20} \times 10^{-4}$

In addition to hyper-volume, we generated an NDS measure that indicates the numbe of found solutions in the Pareto front. A higher value of NDS is equivalent to bette performance in general. However, it is important to read NDS as a secondary metric afte domination. We observe that HAES has accomplished competing values of NDS to th benchmarks for FON, POL, ZDT1, ZDT3, ZDT4, and ZDT6. Hence, it is considered a goo performing algorithm from the perspective of not only domination, but also NDS.

The delta metric shows how much the solutions were equally distributed on th resultant Pareto front. A lower value of the delta metric implies a more equal distributio of the found solutions on the Pareto front. Considering that HAES's focus is to search in a exploiting way, it provides lower distributed solutions in the Pareto front, which makes it value higher compared with the benchmarks and in general closer in order to the value o delta metric of NSGA-III. On the other side, we observe that NSGA-II and MOGA-AQCD have lower values of delta metric.

Another metric that is used to evaluate the performance of MOO is GD, which i preferred to be lower. It shows that HAES has accomplished lower GD for FON, POL, SCH ZDT1, ZDT2, ZDT4, and ZDT6. We also observe that NSGA-III has suffered from relatively higher values of GD compared with the other approaches. It is important to point out tha GD is not always correlated with the percentage of domination due to the change of scales between one objective and the other.

5.2. FCCL Investigation and Results

This section presents the evaluation of implementing HAES on the fog computing closed-loop model. Three main measures are presented for each of the provided configu rations in the experimental design, namely, IRGD, which represented the inverse of the relative generational distance, HV, which represents the hyper volume, and NDS, which denotes the number of non-dominated solutions. The evaluation measures are presented with the different configurations in Figure 6. Looking at the figure, we observe that HAES was capable of accomplishing full IRGD and NDS for configurations 23, 25, 33, and 45 Additionally, we observe that HAES' different configuration was not able to bring HV to its maximum value.

For a more quantitative comparison of the difference in the performance between HAES and other benchmarks, we generated the results of the t-test in Figure 7 for three metrics: IRGD, HV, and set coverage. Their values reveal that HAES has outperformed other benchmarks with respect to set coverage with a confidence of more than 70%, and with respect to IRGD with a confidence of more than 90%. However, HAES was less superior with respect to HV, with a confidence of more than 90%.

Looking at the hyper-volume as a secondary measure after the domination and considering that reaching more optimal solutions might limit their spread in the objective space, we interpret that hyper-volume of HAES has not outperformed the hyper-volume of the benchmarks. However, we could have accomplished more optimal solutions with HAES compared with the benchmarks, as both IRGD and set-coverage of HAES have outperformed their corresponding values in the benchmarks.

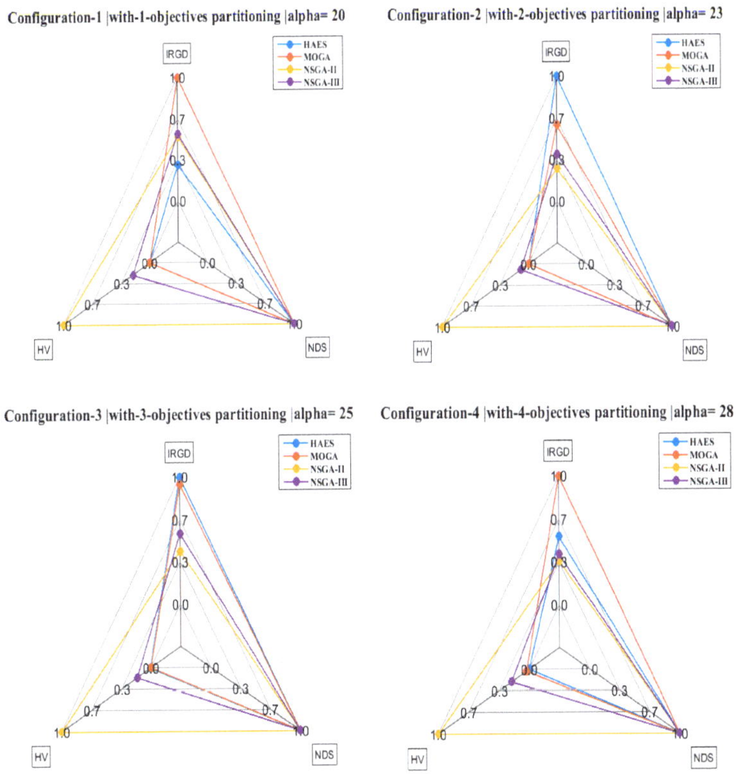

Figure 6. *Cont.*

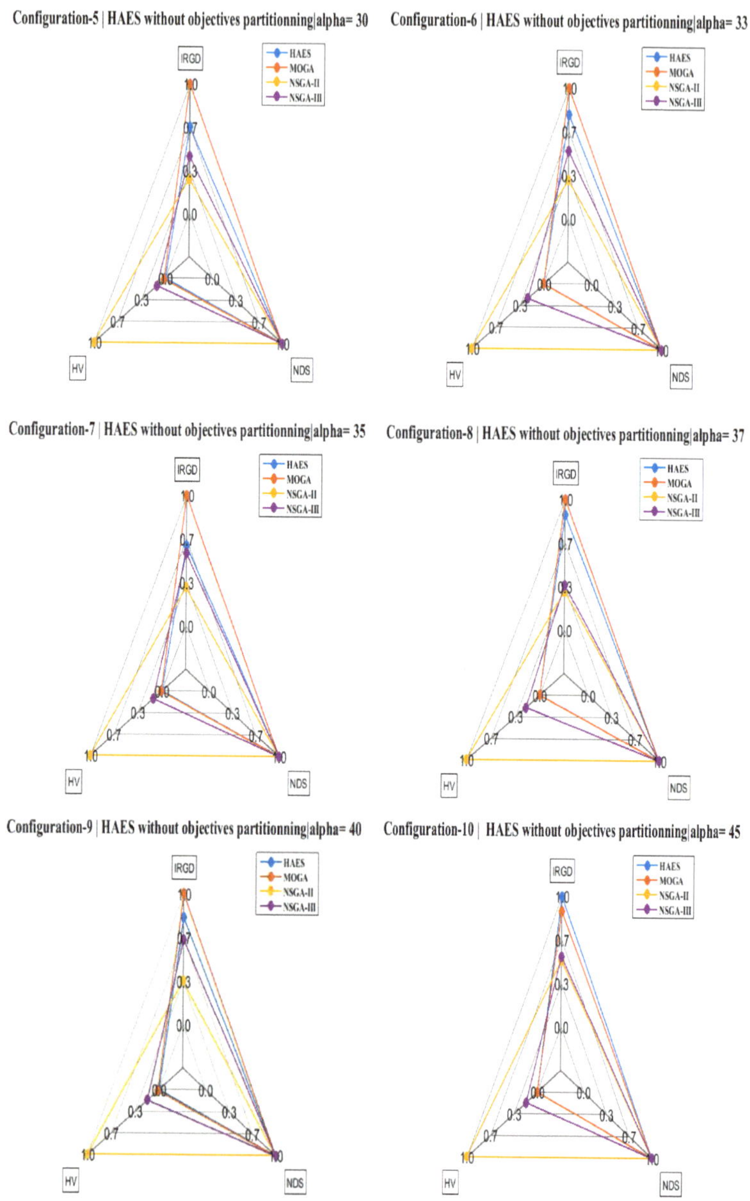

Figure 6. Comparison between HAES different configurations in terms of alpha and the other algorithms.

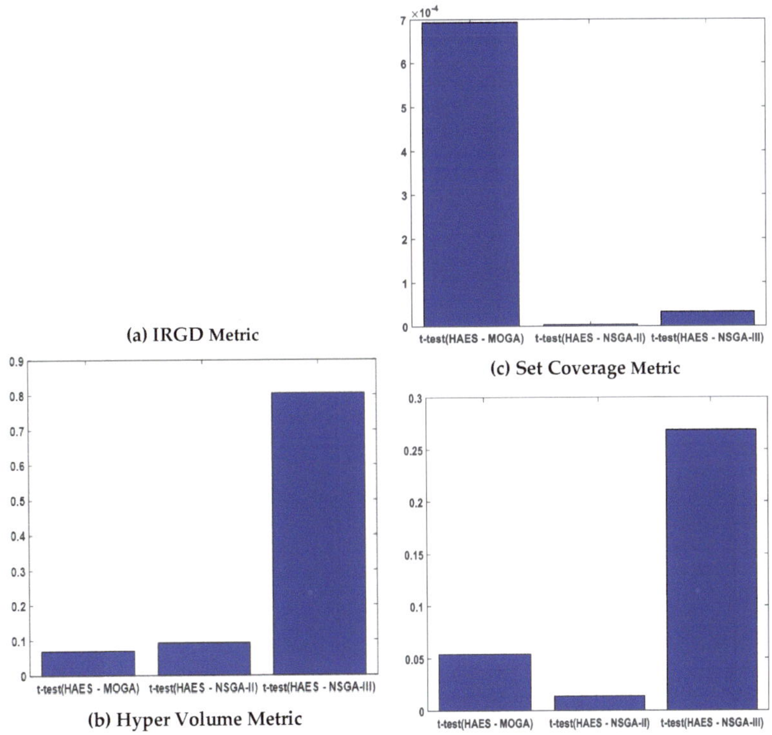

Figure 7. *t*-test to compare the performance of HAES and MOGA-AQCD, NSGA-II, and NSGA-III.

6. Conclusions and Future Works

This article has presented a novel formulation of the problem of fog computing optimization with a multi-objective perspective. The covered objectives are the time latency, the energy consumption with the energy distribution, the renting cost, and stability. The multi-objective and the conflicting nature of the problem require adopting meta-heuristic searching for solving it. However, due to the relatively high number of objectives, different from the relevant existing studies in literature, this research has proposed a novel hyper-angle genetic optimization. The role of the hyper angle is to prioritize solutions within the same rank based on their best-accomplishing rank, which gives the algorithm more exploitive capability. In addition, the article has adopted the concept of objective decomposition by evaluating the approach on various sizes of sub-set of objectives for the objective's decomposition. Objective decomposition enables exploring the boundary of the objective space before going to the intermediate region while searching. Such an approach is crucial for the relatively large number of objectives. Furthermore, various values of angle resolutions were used for the evaluation. It was found that the number of sub-set of objectives while performing the objectives decomposition as well as the value of the angle play an important role in the overall performance. The approach is limited in its dependency on static parameters for both. Hence, our planned future work is to enable an adaptive number of objectives, in which the value of the angle is investigated.

Author Contributions: Supervision: R.H.; validation: A.H.M.A.; visualization and writing—original draft: T.-A.N.A.; review and editing: A.S.A.-K. and Q.N.N. All authors have read and agreed to the published version of the manuscript.

Funding: This paper is supported under grant Fundamental Research Grant Scheme FRGS/1/2018/TK04/UKM/02/17 and Dana Impak Perdana UKM DIP-2018-040.

Data Availability Statement: The data that support the findings of this study are available from th corresponding author, [RH], upon reasonable request.

Acknowledgments: The authors would like to acknowledge the support provided by the Networ and Communication Technology (NCT) Research Groups, FTSM, UKM in providing facilities througl out this paper. The authors would also like to thank the Editor and the anonymous reviewers fo their valuable comments and suggestions.

Conflicts of Interest: The authors declare no conflict of interest.

References

1. Badii, C.; Bellini, P.; Difino, A.; Nesi, P. Sii-Mobility: An IoT/IoE architecture to enhance smart city mobility and transportatio services. *Sensors* **2019**, *19*, 1. [CrossRef]
2. Wu, F.; Wu, T.; Yuce, M.R. An internet-of-things (IoT) network system for connected safety and health monitoring application *Sensors* **2019**, *19*, 21. [CrossRef] [PubMed]
3. Ibrahim, M.Z.; Hassan, R. The Implementation of Internet of Things using Test Bed in the UKMnet Environment. *Asia Pac. J. In Technol. Multimed* **2019**, *8*, 1–17. [CrossRef]
4. Mohammadi, M.; Al-Fuqaha, A.; Sorour, S.; Guizani, M. Deep learning for IoT big data and streaming analytics: A survey. *Iee Commun. Surv. Tutor.* **2018**, *20*, 2923–2960. [CrossRef]
5. Sadeq, A.S.; Hassan, R.; Al-rawi, S.S.; Jubair, A.M.; Aman, A.H.M. A Qos Approach For Internet Of Things (Iot) Environmen Using Mqtt Protocol. In Proceedings of the 2019 International Conference on Cybersecurity (ICoCSec), Negeri Sembilan, Malaysia 25–26 September 2019; pp. 59–63.
6. Jia, M.; Yin, Z.; Li, D.; Guo, Q.; Gu, X. Toward improved offloading efficiency of data transmission in the IoT-cloud by leveragin secure truncating OFDM. *Ieee Internet Things J.* **2018**, *6*, 4252–4261. [CrossRef]
7. Aman, A.H.M.; Yadegaridehkordi, E.; Attarbashi, Z.S.; Hassan, R.; Park, Y.-J. A survey on trend and classification of internet o things reviews. *Ieee Access* **2020**, *8*, 111763–111782. [CrossRef]
8. Stergiou, C.; Psannis, K.E.; Kim, B.-G.; Gupta, B. Secure integration of IoT and cloud computing. *Future Gener. Comput. Syst.* **2018** *78*, 964–975. [CrossRef]
9. Hassan, R.; Jubair, A.M.; Azmi, K.; Bakar, A. Adaptive congestion control mechanism in CoAP application protocol for internet o things (IoT). In Proceedings of the 2016 International Conference on Signal Processing and Communication (ICSC), Noida, India 26–28 December 2016; pp. 121–125.
10. Iyer, G.N. Evolutionary games for cloud, fog and edge computing—A comprehensive study. In *Computational Intelligence in Data Mining*; Springer: Berlin/Heidelberg, Germany, 2020; pp. 299–309.
11. Cisco Systems. *Fog Computing and the Internet of Things: Extend the Cloud to Where the Things Are*; White paper; Cisco Systems Cisco San Jose, CA, USA, 2015.
12. Shabisha, P.; Braeken, A.; Steenhaut, K. Symmetric Key-Based Secure Storage and Retrieval of IoT Data on a Semi-trusted Cloud Server. *Wirel. Pers. Commun.* **2020**, *113*, 1–17. [CrossRef]
13. Zhu, C.; Tao, J.; Pastor, G.; Xiao, Y.; Ji, Y.; Zhou, Q.; Li, Y.; Ylä-Jääski, A. Folo: Latency and quality optimized task allocation in vehicular fog computing. *Ieee Internet Things J.* **2018**, *6*, 4150–4161. [CrossRef]
14. Bjerkevik, H.B.; Botnan, M.B.; Kerber, M. Computing the interleaving distance is NP-hard. *Found. Comput. Math.* **2019**, *20*, 1–35 [CrossRef]
15. Wang, H.; Olhofer, M.; Jin, Y. A mini-review on preference modeling and articulation in multi-objective optimization: Current status and challenges. *Complex Intell. Syst.* **2017**, *3*, 233–245. [CrossRef]
16. Han, D.; Li, Y.; Song, T.; Liu, Z. Multi-Objective Optimization of Loop Closure Detection Parameters for Indoor 2D Simultaneous Localization and Mapping. *Sensors* **2020**, *20*, 1906. [CrossRef] [PubMed]
17. Mayer, M.J.; Szilágyi, A.; Gróf, G. Environmental and economic multi-objective optimization of a household level hybrid renewable energy system by genetic algorithm. *Appl. Energy* **2020**, *269*, 115058. [CrossRef]
18. Albadr, M.A.; Tiun, S.; Ayob, M.; AL-Dhief, F. Genetic Algorithm Based on Natural Selection Theory for Optimization Problems *Symmetry* **2020**, *12*, 1758. [CrossRef]
19. Abdali, T.-A.N.; Hassan, R.; Muniyandi, R.C.; Mohd Aman, A.H.; Nguyen, Q.N.; Al-Khaleefa, A.S. Optimized Particle Swarm Optimization Algorithm for the Realization of an Enhanced Energy-Aware Location-Aided Routing Protocol in MANET *Information* **2020**, *11*, 529. [CrossRef]
20. Mai, Y.; Shi, H.; Liao, Q.; Sheng, Z.; Zhao, S.; Ni, Q.; Zhang, W. Using the Decomposition-Based Multi-Objective Evolutionary Algorithm with Adaptive Neighborhood Sizes and Dynamic Constraint Strategies to Retrieve Atmospheric Ducts. *Sensors* **2020** *20*, 2230. [CrossRef]
21. Han, C.; Wang, L.; Zhang, Z.; Xie, J.; Xing, Z. A multi-objective genetic algorithm based on fitting and interpolation. *Ieee Access* **2018**, *6*, 22920–22929. [CrossRef]
22. Bao, C.; Xu, L.; Goodman, E.D.; Cao, L. A novel non-dominated sorting algorithm for evolutionary multi-objective optimization *J. Comput. Sci.* **2017**, *23*, 31–43. [CrossRef]

23. Arslan, H.D.; Özer, G.; Özkiş, A. Evaluation of Final Product Integrated with Intelligent Systems in Architectural Education Studios. *Online J. Art Des.* **2017**, *5*, 119.
24. Qu, D.; Ding, X.; Wang, H. An improved multiobjective algorithm: DNSGA2-PSA. *J. Robot.* **2018**, *2018*, 9697104. [CrossRef]
25. Cai, D.; Gao, Y.; Yin, M. NSGAII with local search based heavy perturbation for bi-objective weighted clique problem. *Ieee Access* **2018**, *6*, 51253–51261. [CrossRef]
26. Chen, X.; Liu, Y.; Li, X.; Wang, Z.; Wang, S.; Gao, C. A new evolutionary multiobjective model for traveling salesman problem. *Ieee Access* **2019**, *7*, 66964–66979. [CrossRef]
27. Roy, P.C.; Islam, M.M.; Deb, K. Best order sort: A new algorithm to non-dominated sorting for evolutionary multi-objective optimization. In Proceedings of the 2016 on Genetic and Evolutionary Computation Conference Companion, Denver, CO, USA, 20–24 July 2016; pp. 1113–1120.
28. Deb, K.; Pratap, A.; Agarwal, S.; Meyarivan, T. A fast and elitist multiobjective genetic algorithm: NSGA-II. *Ieee Trans. Evol.. Comput.* **2002**, *6*, 182–197. [CrossRef]
29. Mkaouer, W.; Kessentini, M.; Shaout, A.; Koligheu, P.; Bechikh, S.; Deb, K.; Ouni, A. Many-objective software remodularization using NSGA-III. *Acm Trans. Softw. Eng. Methodol. (Tosem)* **2015**, *24*, 1–45. [CrossRef]
30. Metiaf, A.; Wu, Q.; Aljeroudi, Y. Searching with direction awareness: Multi-objective genetic algorithm based on angle quantization and crowding distance MOGA-AQCD. *Ieee Access* **2019**, *7*, 10196–10207. [CrossRef]
31. Mahmud, R.; Kotagiri, R.; Buyya, R. Fog computing: A taxonomy, survey and future directions. In *Internet of Everything*; Springer: Berlin/Heidelberg, Germany, 2018; pp. 103–130.
32. Sun, Y.; Lin, F.; Xu, H. Multi-objective optimization of resource scheduling in fog computing using an improved NSGA-II. *Wirel. Pers. Commun.* **2018**, *102*, 1369–1385. [CrossRef]
33. Liu, L.; Chang, Z.; Guo, X.; Mao, S.; Ristaniemi, T. Multiobjective optimization for computation offloading in fog computing. *Ieee Internet Things J.* **2017**, *5*, 283–294. [CrossRef]
34. Cui, L.; Xu, C.; Yang, S.; Huang, J.Z.; Li, J.; Wang, X.; Ming, Z.; Lu, N. Joint optimization of energy consumption and latency in mobile edge computing for Internet of Things. *Ieee Internet Things J.* **2018**, *6*, 4791–4803. [CrossRef]
35. Zahoor, S.; Javaid, S.; Javaid, N.; Ashraf, M.; Ishmanov, F.; Afzal, M.K. Cloud–fog–based smart grid model for efficient resource management. *Sustainability* **2018**, *10*, 2079. [CrossRef]
36. Rakshit, P. Memory based self-adaptive sampling for noisy multi-objective optimization. *Inf. Sci.* **2020**, *511*, 243–264. [CrossRef]
37. Zitzler, E.; Thiele, L. Multiobjective evolutionary algorithms: A comparative case study and the strength Pareto approach. *Ieee Trans. Evol. Comput.* **1999**, *3*, 257–271. [CrossRef]
38. Kleeman, M.P.; Seibert, B.A.; Lamont, G.B.; Hopkinson, K.M.; Graham, S.R. Solving multicommodity capacitated network design problems using multiobjective evolutionary algorithms. *Ieee Trans. Evol. Comput.* **2012**, *16*, 449–471. [CrossRef]
39. Fonseca, C.M.; Fleming, P.J. Multiobjective optimization and multiple constraint handling with evolutionary algorithms. I. A unified formulation. *Ieee Trans. Syst. Man Cybern.-Part A Syst. Hum.* **1998**, *28*, 26–37. [CrossRef]
40. Kursawe, F. A variant of evolution strategies for vector optimization. In Proceedings of the International Conference on Parallel Problem Solving from Nature, Edinburgh, Scotland, 17–21 September 2016; pp. 193–197.
41. Poloni, C. *Hybrid GA for Multi Objective Aerodynamic Shape Optimisation*; Genetic Algorithms in Engineering and Computer Science; Winter, G., Periaux, J., Galan, M., Cuesta, P., Eds.; 1997; pp. 397–414.
42. Lin, J.C.-W.; Zhang, Y.; Zhang, B.; Fournier-Viger, P.; Djenouri, Y. Hiding sensitive itemsets with multiple objective optimization. *Soft Comput.* **2019**, *23*, 12779–12797. [CrossRef]
43. Deb, K. *Multi-Objective Optimization using Evolutionary Algorithms*; John Wiley & Sons: Hoboken, NJ, USA, 2001; Volume 16.

sensors

MDPI

Article

Horizontal Pod Autoscaling in Kubernetes for Elastic Container Orchestration

Thanh-Tung Nguyen [1], Yu-Jin Yeom [1], Taehong Kim [1,*], Dae-Heon Park [2,*] and Sehan Kim [2]

1 School of Information and Communication Engineering, Chungbuk National University, Cheongju, Chungbuk 28644, Korea; tungnt@cbnu.ac.kr (T.-T.N.); yujinyeom@cbnu.ac.kr (Y.-J.Y.)
2 Electronics and Telecommunications Research Institute, Daejeon 34129, Korea; shkim72@etri.re.kr
* Correspondence: taehongkim@cbnu.ac.kr (T.K.); dhpark82@etri.re.kr (D.-H.P.)

Received: 21 July 2020; Accepted: 14 August 2020; Published: 17 August 2020

Abstract: Kubernetes, an open-source container orchestration platform, enables high availability and scalability through diverse autoscaling mechanisms such as Horizontal Pod Autoscaler (HPA), Vertical Pod Autoscaler and Cluster Autoscaler. Amongst them, HPA helps provide seamless service by dynamically scaling up and down the number of resource units, called pods, without having to restart the whole system. Kubernetes monitors default Resource Metrics including CPU and memory usage of host machines and their pods. On the other hand, Custom Metrics, provided by external software such as Prometheus, are customizable to monitor a wide collection of metrics. In this paper, we investigate HPA through diverse experiments to provide critical knowledge on its operational behaviors. We also discuss the essential difference between Kubernetes Resource Metrics (KRM) and Prometheus Custom Metrics (PCM) and how they affect HPA's performance. Lastly, we provide deeper insights and lessons on how to optimize the performance of HPA for researchers, developers, and system administrators working with Kubernetes in the future.

Keywords: cloud computing; container orchestration; custom metrics; Docker; edge computing; Horizontal Pod Autoscaling (HPA); Kubernetes; Prometheus; resource metrics

1. Introduction

In recent years, with the rapid emergence of cloud computing and later edge computing, virtualization techniques have become a sensational topic for both academic research and industrial development as they enable Cloud platforms such as Amazon Web Services (AWS) [1], Google Cloud Platform (GCP) [2], Microsoft Azure [3] to achieve elasticity on a large scale [4]. One of the emerging virtualization techniques is containerization technology, in which a lightweight operating system (OS) equipped with ready-to-deploy application components is packaged into a self-sufficient container ready to run on a host machine that supports multi-tenancy [4,5]. In the host system, different containers run together on the same host OS and in the same Kernel, which helps reduce storage requirements and allows them to achieve near-native performance compared to the host OS [6].

As containers can be deployed on a large scale [7], there is a tremendous need for container orchestration platforms that are highly automatic in terms of deployment, scaling, and management. Amongst various orchestration platforms including Docker Swarm [8], Amazon Elastic Container Service (Amazon ECS) [9], Red Hat OpenShift Container Platform (Red Hat OCP) [10], Kubernetes [11] has become the de facto standard for its popularity. It is an open-source platform, on which it is easy to package and run containerized applications, workloads, and services, and provides a framework for operating scalable distributed systems. Moreover, the containerized applications have portability to be run on any type of OSs and cloud infrastructures [12]. Kubernetes uses Docker as a base environment to run portable and self-sufficient containers, whose instantiations is known as Docker images. It provides a control plane to manage and schedule those containers to run on its cluster of host machines, called nodes, based on their available resources and each container's specific requirements.

In Kubernetes, one of the most important features is autoscaling because it allows containerized applications and services to run resiliently without the necessity of human intervention. There are three types of autoscalers provided by Kubernetes.

- Horizontal Pod Autoscaler (HPA) supports high availability by adjusting the number of execution and resource units, known as pods [11], based on various requirements. When triggered, HPA creates new pods to share the workloads without affecting the existing ones currently running inside the cluster.
- Vertical Pod Autoscaler (VPA) [11] directly changes the specifications, such as requested resources, of pods and maintains the number of working pods. Therefore, it requires restarting these pods and thus disrupts the continuity of applications and services.
- Cluster Autoscaler (CA) [11] increases the number of nodes when it is no longer possible to schedule pods on the existing ones. Currently, CA only works on commercial cloud platforms such as GCP [2] and AWS [1].

To support autoscaling, Kubernetes monitors pods, applications, host machines, and cluster statistics, called metrics. Autoscalers are triggered when these metrics reach certain thresholds. While Kubernetes provides Resource Metrics by default, their monitoring targets are limited to CPU and memory usage of pods and host machines. Therefore, Customizable (or simply Custom) Metrics can be added with the assist of external software to improve the performance and flexibility of HPA. In this paper, we consider Custom Metrics provided by Prometheus [13], an open-source project run by Cloud Native Computing Foundation (CNCF) [14].

There have been several works that aimed to improve the performance of Kubernetes autoscalers. For example, techniques in References [15,16] are proposed to improve CA and VPA and in References [17,18] focused on improving the performance of HPA and resource monitoring. However, they have not delved into the fundamental knowledge. For instance, questions such as "How does Kubernetes HPA react to different types of metrics?", "What are the effects of different scraping periods of metrics on Kubernetes?" or "Is monitoring only CPU and memory usage enough for HPA?" need to be addressed. Moreover, documents on Kubernetes and its autoscalers can be found on the official website and several other sources on the Internet [11], but they are written on a functional point of view or simply provide only tutorials on how to install and run Kubernetes. There is a lack of comprehensive and fundamental analysis of Kubernetes's operational behaviors. Therefore, in this paper, we focus on HPA and seek to improve knowledge on the subject with the following contributions:

- Firstly, we evaluate HPA on diverse aspects such as scaling tendency, metric collection, request processing, cluster size, scraping time, and latency with various experiments on our testbed. Our comprehensive analysis of the results provides knowledge and insights that are not available on the official website and other sources.
- Secondly, besides Kubernetes's default Resource Metrics, we also evaluate HPA using Prometheus Custom Metrics. By understanding the difference between two types of metrics, readers can have a much firmer grasp on HPA's operational behaviors.
- Lastly, we provide practical lessons obtained from the experiments and analysis. They could serve as fundamental knowledge so that researchers, developers, and system administrators can make informed decisions to optimize the performance of HPA as well as the quality of services in Kubernetes clusters.

The rest of this paper is organized as follows. Section 2 discusses the existing literature regarding Kubernetes and autoscaling research. Section 3 analyzes Kubernetes's architecture while Section 4 thoroughly discusses HPA, different metrics, and the use of Readiness Probe, which helps the readers have a better understanding of Kubernetes and HPA before going into details with performance analysis. Section 5 derives lessons on the performance of HPA from the results of diverse experimental scenarios. Lastly, Section 6 concludes the paper.

2. Related Work

Kubernetes was originally developed by Google and later transferred to the CNCF [14] as a solution to deploy, manage, and scale containerized applications efficiently in cloud data centers. However, since it is an open-source project, Kubernetes can be configured and modified to be a solid foundation, on which other platforms that meet specific demands can be built and developed. The authors of Reference [19] argue that the current version of Kubernetes scheduler only takes into consideration the virtualizable physical infrastructure including CPU and memory usage, which makes only logical sense. However, from a corporation's point of view, to improve the efficiency of data centers, other conditions such as geographic location, power infrastructure, and business processes also need to be considered. Thus, the authors propose an enhanced scheduler called Edgetic, which forecasts the optimal placement of pods in terms of performance and power consumption.

In Reference [15], Thurgood and Lennon discuss a number of scenarios including a smart-home environment where there is a great number of input devices while the number of users such as family members frequently fluctuates. This presents the need for HPA and later CA when all the existing nodes are busy. However, currently, default Kubernetes CA is only provided by cloud platform providers such as GCP [2], thus they propose an elastic CA solution for Kubernetes called Free/Open-source Software (FOSS). This solution employs VMware ESXi hosts as nodes and VM tools including vCenter, Foreman for CA operations. Specifically, the vCenter server creates a VM alarm when a CPU or memory threshold is reached at any node, which then executes a bash script creating new VM nodes through Foreman.

In Reference [16], the authors propose a non-disruptive VPA solution called Resource Utilization Based Autoscaling System (RUBAS) incorporating container migration. They argue that resource can be overestimated, which leads to poor utilization rate. Thus, RUBAS calculates the actually required resource for VPA. Moreover, the authors try to tackle the issue of having to restart pods and containers in VPA by creating a checkpoint image using Checkpoint Restore in Userspace (CRIU). Rossi [20] proposes a reinforcement learning model for both horizontal and vertical autoscaling. It aims to ensure the required response times of applications. The authors of Reference [21] develop a hybrid adaptive autoscaler, Libra. It also considers the optimal resource allocation for applications incorporating conventional HPA. Libra is essentially a control loop of VPA and HPA. In the first phase, Libra calculates the appropriate CPU limit with a canary application and based on this new CPU limit, adjusts the number of pods for production applications. After that, if the load reaches the limit, the loop is repeated.

In References [22,23], the authors argue that Kubernetes is currently using relative metrics collected from */cgroup* virtual file system through cAdvisor. These metrics can be different from the actual CPU usage in the processors, which can be collected from the */proc* file system. This dissimilarity can cause underestimation of required resources. Therefore, the authors propose a correlation model between relative and absolute metrics for CPU-intensive applications, which is employed to correct relative metrics collected by Kubernetes to improve the performance of HPA. With a similar aim in mind, the authors of Reference [24] propose several influencing factors such as the conservative constant, which practically creates a buffer zone for metric fluctuation. Only when the metric value is out of this zone, do HPA actions happen. Another factor is the adaptation interval between successive scaling actions. This reduces unnecessary scaling when the metric value fluctuates.

In References [25,26], the authors apply Kubernetes into resource provisioning for containerized fog computing applications. A network-aware scheduling algorithm that takes into account the network infrastructure such as nodes' CPU and RAM capacities, device types, and geographic locations to make provisioning decisions is proposed. This algorithm, for example, can consider the round trip delay when scheduling instances of time-critical applications. Another Kubernetes-based fog computing platform that manages geographically distributed containers is proposed in Reference [27]. In this paper, the authors design a service called Autoscaling Broker (AS Broker) to get raw metrics and calculate the optimal number of replicas for HPA based on both CPU and memory usage while reducing applications' response time. In Reference [12], Chang et al. introduce a Kubernetes-based cloud monitoring platform that provides a dynamic resource provisioning algorithm based on resource utilization as well as application QoS metrics. On this platform, resource metrics are collected and displayed using Heapster v.0.19.1 [28], InfluxDB v.0.9.4.1 [29] and Grafana v.2.1.0 [30], while applications' response time is calculated with Apache JMeter [31]. This data is aggregated and input into a provisioning algorithm that essentially calculates and fetches the number of pods. Jin-Gang et al. propose a predictive HPA algorithm for a unified communication server using Docker in Reference [18]. The reactive scaling part of this algorithm is the same as Kubernetes's current algorithm. On the other hand, the algorithm also employs an auto-regressive integrated and moving average (ARIMA) model to predict the future workload, or the number of HTTP Requests [32] to trigger HPA and up-scales beforehand.

Another use-case for Kubernetes HPA is on API gateway systems as proposed in Reference [17]. This system is aimed to simplify internal connection to backend services. Since both the frontend and backend services' pods are subjects to horizontal autoscaling when necessary, their interconnections can also increase significantly leading to the need for scaling of the gateway system as well. In this work, the authors employ Prometheus custom metrics for HPA operations. However, which metrics and how they are used are not mentioned. In much the same way, Dickel, Podolskiy and Gerndt [33] propose applying Kubernetes HPA on stateful IoT gateways. While stateless applications such as HTTP can be horizontally-scaled easily, stateful ones including WebSocket and MQTT require more attention. For example, after HPA, there will be multiple gateway instances in the cluster. In the publish-subscribe model over information-centric IoT networks, clients (subscribers) and servers (publishers) are required to be connected through the same gateway. Thus, the authors design a framework for IoT gateways using WebSocket and MQTT protocols focusing on establishing and monitoring active connections between clients and servers. The paper also mentions utilizing Prometheus Operator's custom metrics [34] for HPA. However, similarly to the previously mentioned work, it does not give specific information on how these metrics are collected, calculated, and fetched to HPA-controlling entities.

It is important to note that while the majority of the aforementioned works study Kubernetes and its feature HPA, all of them failed to describe in detail the working principles of HPA and its behaviors when being used with different types of metrics or under various scaling configurations such as the scraping time. This is important for efficient development and management of containerized applications in Kubernetes. Thus, in this paper, we first discuss the architecture of Kubernetes, its components and their inter-communications to establish a solid foundation on the subject, which will be helpful for the readers to have a firm grasp of HPA-related concepts, such as methods of collecting different types of metrics, which will be explained subsequently. As far as we are aware, our paper is the first to achieve such a task. Lastly, we rigorously experiment on a wide range of scenarios to evaluate and analyze diverse aspects of the performance of HPA. Based on the analysis, we provide deep insights and make suggestions on how to optimize Kubernetes HPA to help researchers, developers and sysadmins make informed decisions.

3. Architecture of Kubernetes

In this section, we first describe the architecture of a Kubernetes cluster—the main components and their intercommunication inside the cluster. Then, we take a closer look into how Kubernetes's services enable applications currently running on pods within the cluster to work as a network service.

3.1. Kubernetes Cluster

As shown in Figure 1a, each Kubernetes cluster consists of at least one master node and several worker nodes. In practice, it is possible to have a cluster with multiple master nodes [11] to ensure high availability of the cluster by replicating the master, so in cases where one of the masters fails, a quorum still exists to run the cluster.

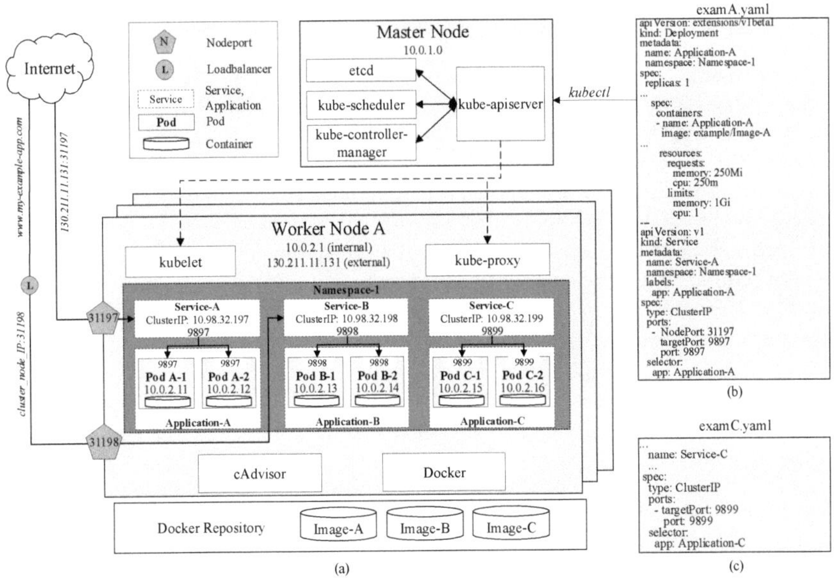

Figure 1. (**a**) The architecture of Kubernetes. (**b**,**c**) Examples of YAML code for application deployment in Kubernetes.

The most basic execution and resource unit in Kubernetes is called a pod, which contains a container or a group of containers and instructions on how these containers should be operated. Each pod represents an instance of an application and always belongs to a namespace. Furthermore, pods that belong to the same application are identical and have the same specifications. In this sense, a pod can be referred to as a replica as well. Upon the deployment of an application, the desired number of replicas, as well as the amount of requested resource, need to be specified. Figure 1b shows the application is created under the name Application-A in Namespace-1 and requests for each of its pods 250Mi and 250m, of memory and CPU available. ''Mi'' denotes ''Mebibyte'' and ''m'' denotes ''millicore''—a unique unit equal to 1/1000 of a CPU core. It is defined by Kubernetes as a granular way to measure the CPU resource so that more than one pod can share a CPU core.

Moreover, each pod is assigned with a unique IP address [11] within the cluster as shown in Figure 1a. This design allows Kubernetes to scale applications horizontally. For example, when an application requires more computational resources, instead of having to adjust the specifications of the existing pods, users can simply create another identical pod to share the load. This additional pod's IP address will then be included in the application's service that routes incoming traffic to the new pod as well as the existing ones. This will be discussed again in more detail.

3.1.1. Master Node

The master node has total control over the cluster through four main components of the Control Plane, namely kube-apiserver, kube-controller-manager, kube-scheduler, and etcd [11] as shown in Figure 1a.

- kube-controller-manager watches over and ensures that the cluster is running in the desired state. For instance, an application is running with 4 pods; however, one of which is evicted or missing, kube-controller-manager has to ensure that a new replica is created.
- kube-scheduler looks for newly created and unscheduled pods to assign them to nodes. It has to consider several factors including nodes' resource availability and affinity specifications.

In the previous example, when the new pod has been created and currently unscheduled, kube-scheduler searches for a node inside the cluster that satisfies the requirements and assigns the pod to run on that node.

- etcd is the back storage that has all the configuration data of the cluster.
- kube-apiserver is the foundational management component that can communicate with all other components and every change to the cluster's state has to go through it. kube-apiserver is also able to interact with worker nodes through kubelet, which will be discussed subsequently. Moreover, users can manage the cluster through the master by passing kubectl commands to kube-apiserver. In Figure 1, after running the command *kubectl apply -f examA.yaml*, the specifications in this file are passed through kube-apiserver to kube-controller-manager for replica-controlling and to kube-scheduler for scheduling pods on specific nodes. They will reply to kube-apiserver who will then signal to these nodes to create and run the pods. These configurations are stored in etcd as well.

3.1.2. Worker Node

Worker nodes allocate computing resources in the form of pods and run them according to the instructions from the master.

- kubelet is a local agent that operates the pods as instructed by the master node's kube-apiserver and keeps them healthy and alive.
- kube-proxy (KP) allows external and internal communication to pods of the cluster. As mentioned earlier, each pod is assigned a unique IP address upon creation. This IP addresses are used by KP to forward traffic from within and outside of the cluster to pods.
- Container Run-time: Kubernetes can be thought of as a specialized orchestration platform for containerized applications and thus requires container runtimes in all nodes including the master to actually run the containers. It can run on various runtimes including Docker, CRI-O and Containerd. Amongst those, Docker [8] is considered the most common one for Kubernetes. By packaging containers into lightweight images, Docker allows users to automate the deployment of containerized applications.
- CAdvisor (or Container Advisor) [35] is a tool that provides statistical running data of the local host or the containers such as resource usage. This data can be exported to kubelet or managing tools such as Prometheus for monitoring purposes. CAdvisor has native support for Docker and is installed in all nodes along with Docker to be able to monitor all nodes inside the cluster.

3.2. Kubernetes Service

In Kubernetes, it is possible to access each pod internally because it has a unique IP address that can be accessed inside the cluster. However, since pods can be created and die at any moment, using individual pods' IP addresses is not a plausible solution. Additionally, these IP addresses cannot be accessed from outside the cluster, which renders user requests or communications between applications deployed in different clusters impossible.

A solution to these cases is Kubernetes Service [11], which is an abstract object that exposes a set of pods to be easily accessed both internally and externally. There are three types of Kubernetes Service:

- ClusterIP is assigned to a service upon creation and stays constant throughout the lifetime of this service. ClusterIP can only be accessed internally. In Figure 1a, services A, B, and C are assigned with three different internal IP addresses and expose three service ports 9897, 9898, and 9899, respectively. For example, when the address 10.98.32.199:9899, which consists of the cluster IP and exposed port of Service-C, is hit within the cluster, traffic is automatically redirected to targetPort 9899 on containers of pods of Application-C as specified in the YAML file by the keyword selector in Figure 1c. The exact destination pod is chosen according to the selected strategy.

- NodePort is a reserved port by the service on each node that is running pods belonging to that service. In the example in Figure 1b, NodePort 31197 and service port 9897 are virtually coupled together. When traffic arrives at NodePort 31197 on node A, it is routed to Service A on port 9897. Then, similarly to the previous example, the traffic is, in turn, routed to pods A-1 and A-2 on targetPort 9897. This enables pods to be accessed from even outside the cluster. For instance, if node A's external IP address 130.211.11.131 is accessible from the internet, by hitting the address 130.211.11.131:31197, users are actually sending requests to pods A-1 and A-2. However, it is obvious that directly accessing nodes' IP addresses is not an efficient strategy.
- LoadBalanceris provided by specific cloud service providers. When the cluster is deployed on a cloud platform such as GCP [2], Azure [3] or AWS [1], it is provided with a load balancer that can be easily accessed externally with a URL (www.my-example-app.com). All traffic to this URL will be forwarded to nodes of the cluster on NodePort 31198 in a similar manner to the previous example as illustrated in Figure 1a.

4. Horizontal Pod Autoscaling

In Kubernetes, HPA is a powerful feature that automatically raises the number of pods, to increase the application's overall computational and processing power, without having to stop the application's currently running instances [11]. Once successfully created, these new pods are able to share the incoming load with the existing ones. From a technical point of view, HPA is a control loop implemented by kube-controller-manager. By default, every period of 15 s, also known as sync cycle, kube-controller-manager compares the collected metrics against their thresholds specified in HPA configurations. Figure 2a shows the configurations of an HPA. "minReplicas" and "maxReplicas" refer to the minimum and maximum numbers of pods that should be running in the cluster. In the example, minReplicas and maxReplicas are 2 and 4, respectively. The Replication Controller, which is a component of kube-controller-manager, keeps track of the replica set and ensures that no less than 2 and no more than 4 pods of this application run in the cluster at all times. The metric used for this HPA is CPU usage. Once the average value of CPU utilization reaches a preset threshold, HPA automatically increases the number of pods as by calculating the following

$$desiredReplicas = \left\lceil currentReplicas * \frac{currentMetricValue}{desiredMetricValue} \right\rceil, \tag{1}$$

where *desiredReplicas* is the number of pods after scaling, *currentReplicas* is the number of pods currently running, *currentMetricValue* is the latest collected metric value, *desiredMetricValue* is the target threshold. *desiredMetricValue* is actually the threshold *targetAverageValue* in Figure 2a.

In the example, when the *currentMetricValue*, which in this case is CPU usage, hits 150 m, which is higher than the threshold *desireMetricValue* of 60 m, the *desiredReplicas* equals to $\lceil 2 \times (150/60) \rceil$, or 5. However, as the maximum number of replicas is only 4, kube-controller-manager only signals to kube-apiserver to increase 2 more replicas. After this, if the average CPU usage declines to 40 m, the *desiredReplicas* is $\lceil 4 \times (40/60) \rceil$, or 3. Therefore, one of the newly created pods will be removed. However, it is worth noting that to avoid thrashing from creating and removing pods repeatedly as the metrics can fluctuate significantly, each newly created pod is kept running for at least a downscale delay period before it can be removed from the cluster. This period is set at 5 min [11]. Furthermore, it is possible for HPA to use several metrics, each of which has its own threshold. When any of these metrics reaches its threshold, HPA scales up the cluster in the above-mentioned manner. However, for scaling-down operations, all of these metrics are required to be below their thresholds.

The above-mentioned metrics used by kube-controller-manager for horizontal autoscaling are either Kubernetes's default Resource Metrics or external Custom Metrics provided by Prometheus [13], Microsoft Azure [3], et cetera. In the scope of this paper, we only discuss Kubernetes Resource Metrics and Prometheus Custom Metrics, which are the most popular for HPA.

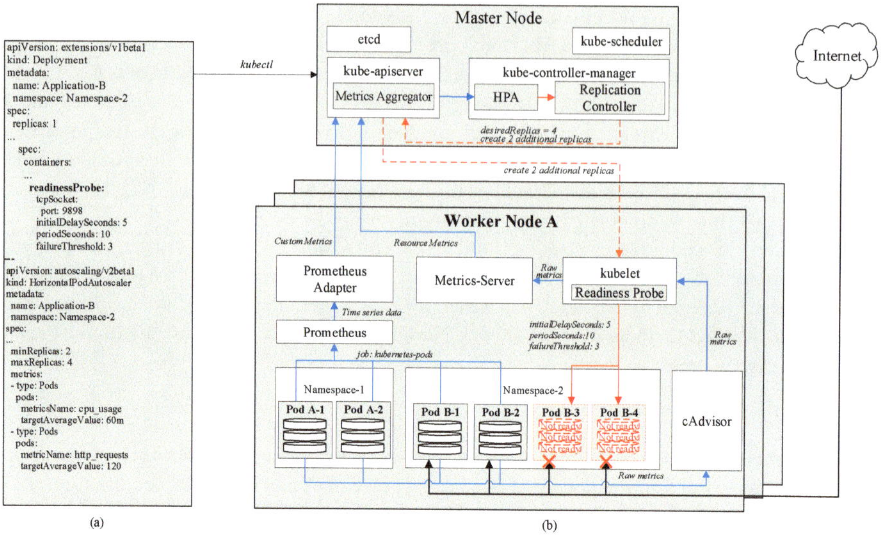

Figure 2. (**a**) Examples of YAML code for configuring HPA and Readiness Probe in Kubernetes. (**b**) Horizontal Pod Autoscaling's (HPA) architecture.

4.1. Kubernetes Resource Metrics

As shown in Figure 2b, cAdvisor acts as a monitoring agent to collect core metrics, such as CPU, memory usage, of the host machines and running pods, and publish these metrics through an HTTP port. For example, in Figure 2b, cAdvisor is currently monitoring 4 existing pods A-1, A-2, B-1, and B-2. With the help of kubelet, the Metrics-Server scrapes these metrics periodically. The default scraping period is 60 s and can be adjusted by changing Metrics-Server's deployment configurations. Then, the Metrics-Server exposes them to the Metrics Aggregator in kube-apiserver as CPU and memory usage of individual pods and nodes, whose average values will be calculated and fetched to HPA. This is referred to as the resource metrics pipeline [11]. Kubernetes Resource Metrics can be checked manually by passing commands kubectl top pod and kubectl top node to kube-apiserver.

4.2. Prometheus Custom Metrics

Prometheus [13] allows flexible monitoring as it exposes monitored targets as endpoints and periodically pulls their metrics through an HTTP server. It can monitor a wide range of targets including nodes, pods, services, or even itself. The monitoring operation for each of these targets is called a job. While Prometheus has its default global scraping period of 60 s, its jobs can have their own scraping periods. For jobs that are too short-lived to be scraped, Prometheus has component called Pushgateway into which these jobs can directly push their metrics once they exit. Then, by exposing the Pushgateway as an endpoint as well, Prometheus can scrape these metrics later even after the jobs have been terminated. As shown in Figure 2b, Prometheus scrapes metrics from existing pods A-1, A-2, B-1, and B-2 with a job called kubernetes-pods.

The scraped data is then stored in the form of time series in the Time Series Database (TSDB), which is exposed to the Prometheus Adapter [36], which is written in PromQL (Prometheus Query Language)—a functional query language [13] and has several queries to actually process raw time series metrics. For example, query *rate(http_requests_total[1m])* returns the per-second average rate of the time series collected within 1 min. The number of the time series depends on the scraping period. A 15-s scraping period would result in a total of 4 time series for the above query.

Once the queries are finished, the resulted metrics will be sent to the Metrics Aggregator in the kube-apiserver and by which fetched to the HPA. While the Metrics-Server can monitor only CPU and memory usage, Prometheus can offer a variety of custom metrics. The metric in the previous example can be employed as the average arrival rate of HTTP requests, which can be added to the HPA in Figure 2a. Therefore, in this case, where there are two input metrics, HPA scales up whenever either one of the metrics has reached its threshold and scales down when both metrics are below their thresholds.

4.3. Readiness Probe

In most cases, pods are not ready to serve traffic immediately after their creation as they may have to load data or configurations. Therefore, if traffic is sent to these newly created pods during their startup time, the requests will obviously be failed. As a solution, Kubernetes provides a feature named Readiness Probe [11], which checks the statuses of new pods and only allow traffic to them once they are Ready. In Figure 2a, the initialDelaySeconds, defines the amount of time between the pods' creation and when they are first checked for readiness. If, after the check, the pods are still not ready, Kubernetes will check again periodically every periodSeconds. In this example, once pods B-3 and B-4 are created, Kubernetes gives it 5 s to startup and get ready. After the first readiness check, if the pod's status is ready, it will start serving incoming traffic immediately. On the other hand, if it is not, Kubernetes checks every 10 s for 3 more times, which is defined by failureThreshold, before giving up and deciding to reset or deem the pod Unready based on preset configurations.

5. Performance Evaluations

In this section, we describe our experimental setup before showcasing and discussing evaluation results in detail to confirm our understanding of Kubernetes and its HPA feature. We also provide analysis and deep insights on how to optimize HPA.

5.1. Experimental Setups

We set up a Kubernetes cluster of 5 nodes, consisting of 1 master node and 4 worker nodes, inside a physical machine that runs on Intel(R) Core (TM) i7-8700 @ 3.20Ghz * 12. Each node of the cluster runs a virtual machine with Ubuntu 18.04.3 LTS operating system, Docker version 18.09.7, Kubernetes version 1.15.1. Regarding computing capabilities, the master is allocated 4 core processors and 8GB of RAM, compared to 2 core processors and 2GB of RAM for each of the worker nodes. Moreover, Gatling open-source version 3.3.1 [37] is employed as the load generator that sends HTTP requests to our application through a designated NodePort on each worker node.

Our application is designed to be CPU-extensive. In other words, once it successfully receives a HTTP request, it uses CPU resource until sending back a response to the source. The CPU request and limit of each replica are 100 m and 200 m, respectively. The number of replicas ranges from the minimum of 4 (average of 1 replica/node) to the maximum 24 (average of 6 replicas/node).

All experiments are run for 300 s. During the first 100 s, the average incoming request rate sent by Gatling is approximately 1800 requests/s, while it is roughly 600 requests/s for the next 100 s, which generates a total of 240,000 requests. We define these two periods as high traffic period (HTP) and low traffic period (LTP), respectively. The rest of the simulation time is used for observing the decrease of metrics while there are no arriving requests. In this paper, we test Kubernetes HPA's performances in 7 different experiments. Each experiment is repeated 10 times to ensure its accuracy.

5.2. Experimental Results

5.2.1. HPA Performances with Default Kubernetes Resource Metrics

Setup. Metrics' scraping period is set at the default value of 60 s.

Goal. We aim to evaluate the performance of HPA using default Kubernetes Resource Metrics (KRM) in terms of CPU usage, numbers of replicas and failed requests under the default scraping period.

Looking into Figure 3a, the average CPU usage increases to the limit because of the high rate of requests. After that, it decreases as the number of replicas is raised. It can be observed that the most striking point is that the metric value of CPU usage changes every scraping period (60 s). This is because Kubelet only scrapes the raw metrics from cAdvisor at the beginning of a scraping cycle. Then, the metrics are reported without any modification to the Metrics-Aggregator through the Metrics-Server. In other words, the reported values of the metrics are exactly equal to the scraped values.

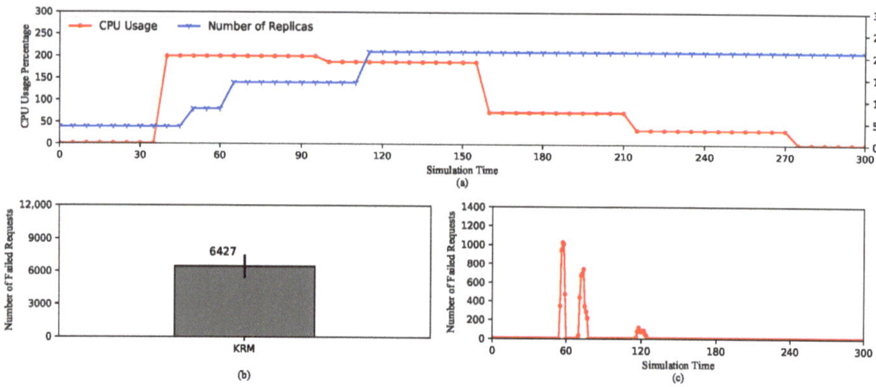

Figure 3. HPA using default Kubernetes Resource Metrics (KRM). (**a**) The average CPU usage and the scaling of the replica set. (**b**) The total number of the failed requests. (**c**) The timeline of the failed requests.

The first scaling action expanding the replica set to 8 happens around 50th second as a result of the increase in the metric value between 35th and 40th second. After that, as the metric value reaches the maximum of 200%, the replica set is expanded again to 15 at 65th second. However, the third scaling action happens at 110th s, which is 55 s later, even though the CPU usage is still high. This is due to a very important characteristic of HPA. By default, HPA checks the metric value every 15 s. At the time of checking, if there is no change in the value compared to the last check, it is deemed unnecessary to adjust the number of replicas. We can see that from 40th to 100th second, which is exactly one scraping cycle, the CPU usage does not change. After this, it decreases to about 180%, which actually triggers the third scaling action to 21 replicas. Here, the metric value again remains stable until 160th s, where it begins to drop and does not cause any more scaling-ups. The number of replica remains stable until the end of the experiment because HPA has to wait for 5 min from the last scaling-up to scale down. This design aims to avoid thrashing caused by continuous scaling actions.

Figure 3b,c show the number of failed requests and the time of failures. These requests are refused during the scaling operations, because newly created pods are not ready to serve traffic. Any requests routed to them during this time will be failed. Furthermore, we can see that the first scaling-ups cause the majority of failed requests because it happens during the HTP. The high request rate causes more requests to be refused. As opposed to this, the third scaling-up causes only a small number of failures.

In conclusion, it is important to note that Kubernetes HPA is designed to check the metric value periodically and decision-making on scaling depends on whether the metric value is changed compared to the last check. In addition, since KRM's value is only changed every scraping period, the number of created replicas depends on this factor. Therefore, the next experiment analyzes the effect of the length of the scraping period on HPA's performance.

5.2.2. HPA Performances with Default Kubernetes Resource Metrics and Different Scraping Periods

Setup. KRM's scraping period is adjusted to 15 s, 30 s, and 60 s, respectively.

Goal. We aim to investigate the effect of different scarping periods on HPA's performance.

Figure 4 shows the performance of HPA using KRM with three different scraping periods of 15 s, 30 s, and 60 s. In Figure 4a–c, we can see that the trend of metric values still follows the pattern described earlier. The values change every scraping period for all three cases. In other words, as the period is lengthened, the metric values remain at the same level for a longer amount of time.

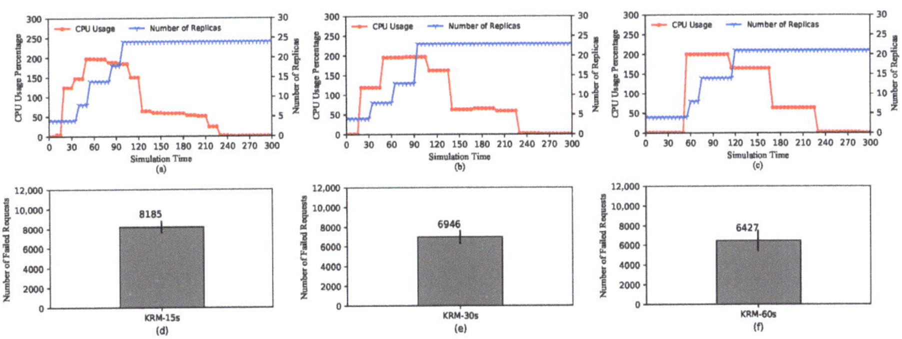

Figure 4. HPA using default Kubernetes Resource Metrics (KRM). (**a–c**) The average CPU usage and the scaling of the replica set for scraping periods of 15 s, 30 s, and 60 s, respectively. (**d–f**) The total number of the failed requests for scraping periods of 15 s, 30 s, and 60 s, respectively.

However, the maximum number of replicas tends to decrease as the scraping period increases. In Figure 4a,b, the maximum numbers of replicas are 24 and 23, respectively. On the other hand, for the case of the 60-second scraping period, it is only 21. It is because if the metric value does not change, it would not trigger scaling actions, as stated previously. In this case, because the value changes more frequently for the short scraping periods, the number of replicas is also raised more frequently. Moreover, Figure 4d–f show the numbers of failed requests in three cases. Similar to the maximum number of replicas, that of failed requests tends to decrease with longer scraping periods. This is because the more unready pods are getting incoming traffic, the more requests will be refused. Note that the failure of requests coming to unready pods can be solved with the Readiness Probe, whose effect will be analyzed in Section 5.2.7.

With the above explanations, it can be concluded that when under the same load for the same amount of time, a longer scraping period causes HPA to raise a smaller number of replicas. This comes with pros and cons. On the one hand, a longer scraping period may result in efficient resource allocation by triggering scaling actions relatively slowly adding a smaller number of replicas. On the other hand, it can lead to a lack of required resources if the incoming load becomes too high.

5.2.3. HPA Performances with Prometheus Custom Metrics

Setup. Prometheus Custom Metrics' scraping period is set at the default value of 60 s.

Goal. We aim to evaluate the performance of HPA using Prometheus Custom Metrics in terms of CPU usage, numbers of replicas and failed requests to compare with HPA using KRM.

Looking at Figure 5a, it can be seen that PCM's metric value changes very frequently, every time it is queried. This is completely contrary to the case of KRM. The reason for this lies in the way PCM is collected. Even though Prometheus also scrapes pods' metrics according to the scraping period, these metrics have to go through the *rate()* function at Prometheus Adapter, which calculates their per-second average of increase [13]. Moreover, it is important to note that the *rate()* function also performs extrapolation based on the current trend of metrics when necessary such as in cases where time-series data points are missing. Therefore, the CPU usage, in this case, changes every query period. These frequent changes, in turn, cause HPA to increase the number of replicas quickly to the maximum of 24 as opposed to only 21 in the case of KRM. Therefore, PCM offers the advantage of responsiveness to frequent changes in the metric value. Quickly increasing the number of replicas, or overall computational power, allows HPA to deal with surges of incoming loads. However, a disadvantage is a higher number of failed requests as shown in Figure 5b,c.

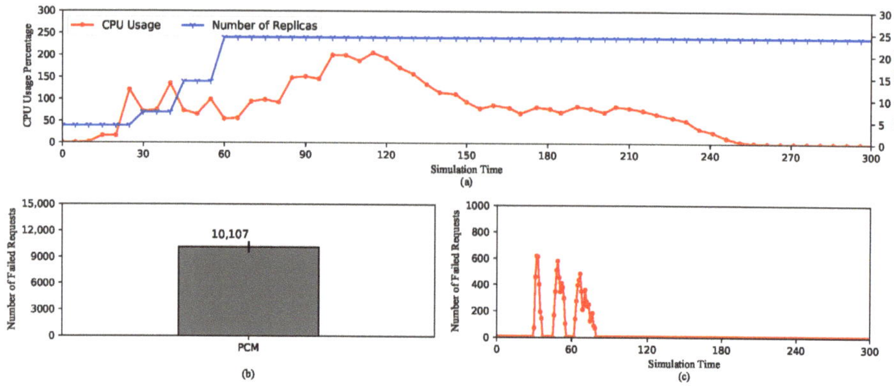

Figure 5. HPA using Prometheus Custom Metrics (PCM). (**a**) The average CPU usage and the scaling of the replica set. (**b**) The total number of the failed requests. (**c**) The timeline of the failed requests.

5.2.4. HPA Performances with Prometheus Custom Metrics and Different Scraping Periods

Setup. Metric scraping period is set 15 s, 30 s and 60 s.

Goal. We aim to investigate the effect of different scarping periods on the performance of HPA using PCM.

Figure 6 shows the performance of HPA in terms of CPU usage and the number of replicas with three different scraping periods of 15 s, 30 s, and 60 s, respectively. It can be seen that in all three cases, the trends of the graphs are very similar. As explained earlier, the reason is that Prometheus Adapter, the entity responsible for transforming raw metrics into Custom Metrics, can perform extrapolation to provide the metrics when raw data points are missing. If the moment of querying is in the middle of a scraping cycle, Prometheus Adapter, the *rate*() in particular, calculates the metrics based on the previously collected data points. As a result of the similarity in metric values, the numbers of replicas of three cases increase in a similar pattern during the same period.

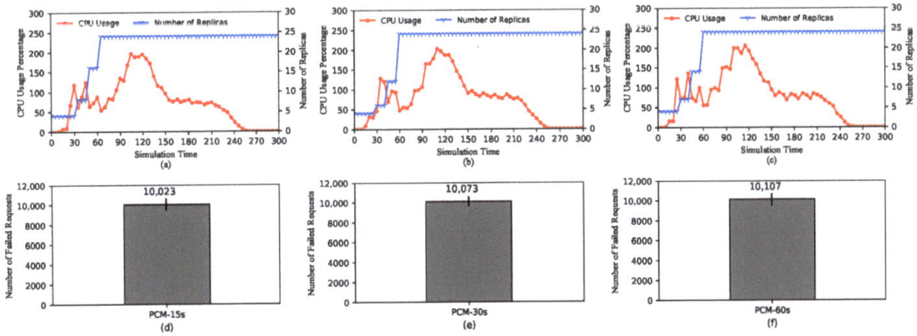

Figure 6. HPA using Prometheus Custom Metrics (PCM). (**a**–**c**) The average CPU usage and the scaling of the replica set for scraping periods of 15 s, 30 s, and 60 s, respectively. (**d**–**f**) The total number of the failed requests for scraping periods of 15 s, 30 s, and 60 s, respectively.

We can conclude that PCM is not as strongly affected by adjustments to the scraping period as KRM. A longer scraping period can be chosen as it will reduce the amount of computational and internal communicational resources required to collect and pull the metrics. However, it is worth noting that longer scraping periods mean there are fewer data points, which can reduce the precision of the *rate* function.

5.2.5. Comparison of HPA Performances in a 2-Worker Cluster and a 4-Worker Cluster

Setup. Two clusters of 2 and 4 worker nodes are set up. Worker nodes are identical and the maximum and minimum numbers of replicas are the same for both cases.

Goal. We aim to investigate and compare the performances of HPA in these two cases.

Figure 7 shows the performance comparison between HPAs using PCM in a 2-worker cluster and a 4-worker cluster (PCM-2W and PCM-4W). From Figure 7a, it can be concluded that the general trends of two CPU usage values are largely similar. The only noticeable differences in the CPU usage appear between 30th and 70th s. This is a result of the differences in the increases of replicas as shown in Figure 7b, where PCM-2W increases the number of replicas slower than PCM-4W does. This is because while the maximum number of replicas is 24 for both HPAs, the average numbers of pods on each node are 6 and 12 for the 4-worker and 2-worker clusters, respectively. It indicates that a node in the 2-worker cluster has to process a higher computation load when scaling, which makes creating and allocating resources for additional pods slower compared to a node in the 4-worker cluster.

Moreover, Figure 7c shows that the number of failed requests of PCM-2W is higher than that of PCM-4W. This is because newly created pods take longer to get ready in the case of PCM-2W. However, since the trends of increases in two cases are generally resembling each other, the difference is relatively moderate.

Furthermore, it is understandable that when the requests reach the 2-worker cluster's nodes, they would have to stand in a much longer queue than they would in the case of the 4-worker cluster. This is the main reason why the response time of PCM-2W is remarkably longer than that of PCM-4W in Figure 7d. In addition, the difference in the speeds of pod creation in two cases also contributes to this contrast.

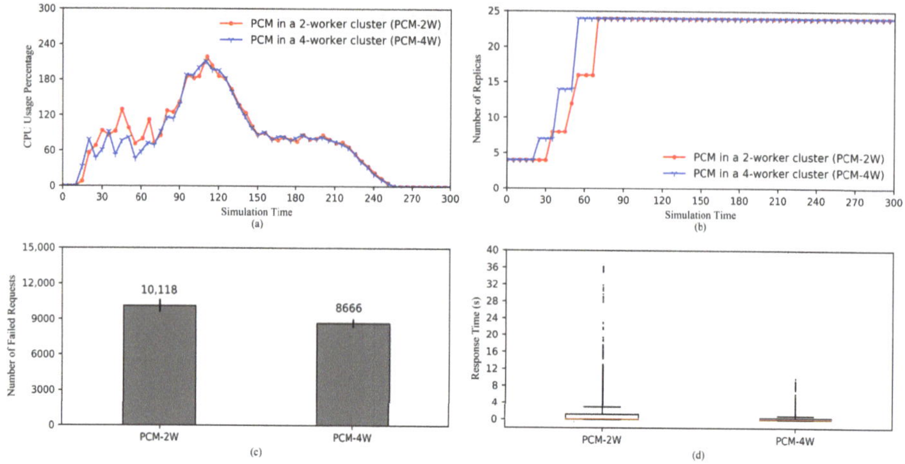

Figure 7. Comparisons of HPA performances in 2-worker and 4-worker clusters. (**a**) The average CPU usage. (**b**) The scaling of the replica set. (**c**) The total number of the failed requests. (**d**) Response time of the successful requests.

5.2.6. Comparison of HPA Performances with Different Custom Metrics

Setup. As opposed to KRM, which can monitor only CPU and memory usage, PCM can monitor other metrics such as the rate of incoming HTTP requests. In the first case, only the request rate is used for HPA. On the other hand, in the second case, it is combined with CPU usage.

Goal. We aim to investigate the effect of the use of different custom metrics on HPA's performance.

Figure 8 shows the comparison between performances of HPA using HTTP requests (PCM-H) and HPA using CPU and HTTP requests (PCM-CH). PCM-H provides autoscaling based on the incoming request rate. On the other hand, PCM-CH combines both the rate and the CPU usage. When multiple metrics are specified, HPA scales up if either one of the metrics reaches its threshold. From the CPU usage comparison in Figure 8a, we can see that in general, CPU usage of PCM-H is significantly higher than that of PCM-CH in general. While in Figure 8b, PCM-CH's average request rate is considerably higher than PCM-H's average request rate from 30th to 60th second. After that, it remains lower until the 250th second. This has resulted from the fact that PCM-CH's request rate rises sharply, which triggers the first scaling reactions, as shown in Figure 8c, after which the CPU usage is still higher than its threshold and causes subsequent rises to the maximum number of replicas. Here, because PCM-CH has more replicas, so the average request rate is lower. On the other hand, PCM-H uses only the request rate. After the first scale, the rate decreases and remains below the threshold, which does not trigger any more scaling-up. Moreover, because now it has only 13 replicas, the CPU usage rises and stays at a high level. Furthermore, since there are fewer increases of pods, PCM-H produces only approximately half the number of failed requests of PCM-CH.

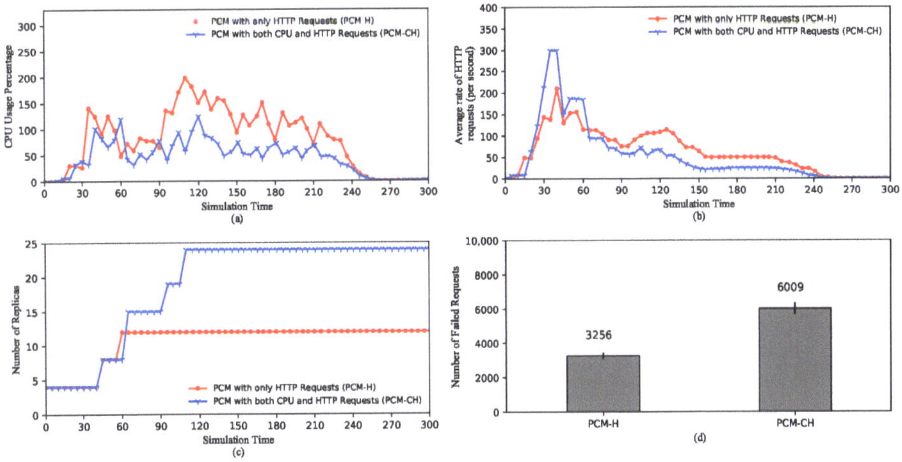

Figure 8. Comparison of HPA Performances with different custom metrics. (**a**) The average CPU usage. (**b**) The average rate of HTTP requests. (**c**) The scaling of the replica set. (**d**) The total number of the failed requests.

5.2.7. Comparison of HPA Performances with and without Readiness Probe

Setup. Regular HPA is set up in one case while it is accompanied by Readiness Probe in the other.

Goal. We aim to investigate the use of Readiness Probe, its effect on the number of failed requests and response time.

Figure 9a shows the comparison between HPA coupled with the use readiness probe (RP) and regular HPA in terms of numbers of failed requests. It is obvious that in the case of using RP, there are no failed requests, because when the additional pods are getting ready, Kubernetes service does not route any traffic to them. Only once they have been deemed ready, they can receive and process incoming requests. Therefore, we can expect that the failed requests observed from the previous experiments can be avoided by using RP. However, this is rather a tradeoff, as the general response time is significantly higher than the case of Not using RP as shown in Figure 9b, since more traffic is routed to the existing pods.

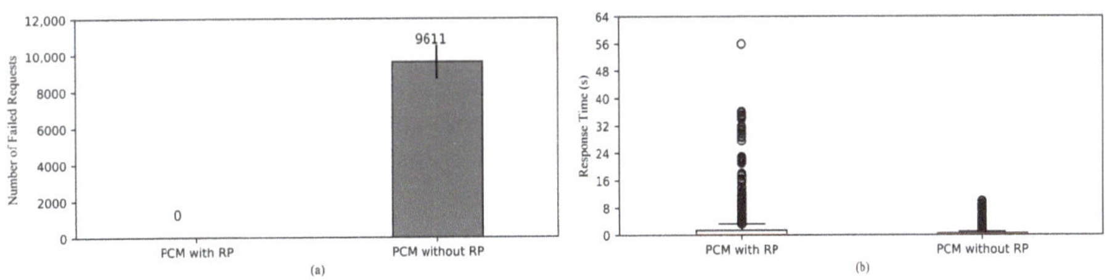

Figure 9. Comparison of HPA performances with and without Readiness Probe. (**a**) The total number of the failed requests. (**b**) Response time of the successful requests.

5.3. Discussion

To summarize the previous experiments and analyses, we list out a few key points on Kubernetes HPA's behaviors.

- On KRM and PCM: KRM only reports values of metrics and is able to change only once every scraping cycle as opposed to PCM, which are able to maintain the trend of metrics' values even during the middle of a scraping cycle or if there are missing data points. As a direct consequence, KRM expands the replica set slower and mostly to a smaller number of replicas compared to PCM. The advantage of this behavior is obviously less resource consumption. On the other hand, when under high load pods could be crashing or becoming unavailable. Therefore, we suggest KRM for applications with more stable loads such as video processing services. In this case, the number of requests from viewers is usually small as it takes at least a few minutes to a few hours for a video. On the contrary, PCM is more suitable for applications with frequent changes in metrics. For instance, e-commerce websites may experience continuous surges in a few hours during a sale event, thus require fast system reactions.

- On the scraping period: Adjustments to the scraping period of PCM do not strongly affect the performance of HPA. Therefore, a longer period can be set to reduce the amount of resource used for pulling the metrics. However, it is worth noting that a overly long period can cause imprecision in calculating the metrics. Regarding KRM, the scraping period has a significant influence on the performance of HPA. A longer period can reduce the amount of resource allocated for new pods, but it can cause decreased quality of service. Therefore, the scraping period should be carefully chosen having considered the type of service and the capability of the cluster.
- On the cluster size: It is obvious that a 4-worker cluster has more computational power, which allows it to perform HPA operations faster, than a 2-worker cluster, assuming workers of the two clusters are identical in terms of computational capabilities. In addition, the communicational capability of the 4-worker cluster is superior to the 2-worker cluster. This results in the difference in request response time of the two clusters. However, even if the 2-worker cluster has equal computational and communicational power, it is safer to spread pods to a wider cluster as half of the pods can become unavailable when a node crashes, compared to a forth of the pods in the case of the 4-worker cluster.
- On HPA with different custom metrics: Prometheus enables the use of custom metrics such as HTTP request rate to meet specific demands. Especially combining multiple metrics together can also increase the effectiveness of HPA as changes in any individual metric will cause scaling reactions. However, as a downside, this may result in waste of resource. Therefore, metrics, or combinations of metrics, should be chosen according to the type of the application. For instance, a gaming application may have various request sizes. Requests to move a character around the map are small in size but their number can be numerous. Thus, the request rate should be considered, so that each request can be quickly served, which reduces the "lagging" effect and improves the overall gaming experience. On the other hand, requests to load new locations' maps are heavy but small in number. Here, computational requirements grow significantly higher, which indicates HPA should scale based on CPU and memory usage. In short, Custom Metrics enable applications to consider various factors such as the number of requests, latency, and bandwidth for efficient horizontal autoscaling.
- On Readiness Probe: It is a powerful feature from Kubernetes to prevent requests from being routed to unready pods, which will reject the requests. However, routing a number of requests to existing pods can cause the rest of the requests to have significantly longer response time. Therefore, between keeping the incoming requests alive or letting them fail and expecting re-requests, one should be chosen carefully based on balancing between system resources and QoS requirements.

6. Conclusions

Kubernetes is a powerful orchestration platform for containerized applications and services, and can be applied into important future technologies including cloud/edge computing and IoT gateways. Its feature HPA provides dynamic and effective scaling for applications without the necessity of human intervention. In this paper, we have given the first comprehensive architecture-level view of both Kubernetes and HPA. How each type of metrics, including Kubernetes Resource Metrics and Prometheus Custom Metrics, are collected, calculated and fetched to HPA was also thoroughly explained. Moreover, we conducted several experiments covering a variety of scenarios and provided clear analysis for the behaviors of Kubernetes HPA.

This paper should serve as a fundamental study for further research and development of Kubernetes and HPA. In the future, we aim to expand our experiments with more HPA scenarios as well as to develop a more efficient scaling algorithm for Kubernetes.

Author Contributions: Conceptualization, T.-T.N., Y.-J.Y. and T.K.; Experiment, T.-T.N. and Y.-J.Y.; Writing—Original Draft Preparation, T.-T.N.; Review & Editing, T.K.; Supervision, D.-H.P. and T.K.; Funding Acquisition, D.-H.P. and S.K. All authors have read and agreed to the published version of the manuscript.

Funding: This work was supported by Institute of Information & communications Technology Planning & Evaluation (IITP) grant funded by the Korea government (MSIT) (No.2018-0-00387, Development of ICT based Intelligent Smart Welfare Housing System for the Prevention and Control of Livestock Disease).

Acknowledgments: The authors would like to show great appreciation for Linh-An Phan's many invaluable comments throughout the research and acknowledge the support from Dinh-Nguyen Nguyen during the early stage.

Conflicts of Interest: The authors declare no conflict of interest.

References

1. Amazon Web Services. Available online: https://aws.amazon.com (accessed on 23 June 2020).
2. Google Cloud Platform. Available online: https://cloud.google.com (accessed on 23 June 2020).
3. Microsoft Azure. Available online: https://azure.microsoft.com (accessed on 23 June 2020).
4. Pahl, C.; Brogi, A.; Soldani, J.; Jamshidi, P. Cloud Container Technologies: A State-of-the-Art Review. *IEEE Trans. Cloud Comput.* **2017**, *7*, 677–692. [CrossRef]
5. He, S.; Guo, L.; Guo, Y.; Wu, C.; Ghanem, M.; Han, R. Elastic application container: A lightweight approach for cloud resource provisioning. In Proceedings of the 2012 IEEE 26th International Conference on Advanced Information Networking and Applications, Fukuoka, Japan, 26–29 March 2012; pp. 15–22. [CrossRef]
6. Dua, R.; Raja, A.R.; Kakadia, D. Virtualization vs containerization to support PaaS. In Proceedings of the 2014 IEEE International Conference on Cloud Engineering, Boston, MA, USA, 11–14 March 2014; pp. 610–614. [CrossRef]
7. Pahl, C. Containerization and the PaaS Cloud. *IEEE Cloud Comput.* **2015**, *2*, 24–31. [CrossRef]
8. Docker. Available online: https://www.docker.com (accessed on 23 June 2020).
9. Amazon Elastic Container Service. Available online: https://aws.amazon.com/ecs (accessed on 23 June 2020).
10. Red Hat OpenShift Container Platform. Available online: https://www.openshift.com/products/container-platform (accessed on 23 June 2020).
11. Kubernetes. Available online: www.kubernetes.io (accessed on 23 June 2020).
12. Chang, C.C.; Yang, S.R.; Yeh, E.H.; Lin, P.; Jeng, J.Y. A Kubernetes-based monitoring platform for dynamic cloud resource provisioning. In Proceedings of the GLOBECOM 2017—2017 IEEE Global Communications Conference, Singapore, 4–8 December 2017; pp. 1–6. [CrossRef]
13. Prometheus. Available online: https://prometheus.io (accessed on 23 June 2020).
14. Cloud Native Computing Foundation. Available online: https://www.cncf.io (accessed on 23 June 2020).
15. Thurgood, B.; Lennon, R.G. Cloud computing with Kubernetes cluster elastic scaling. *ICPS Proc.* **2019**, 1–7, doi:10.1145/3341325.3341995. [CrossRef]
16. Rattihalli, G.; Govindaraju, M.; Lu, H.; Tiwari, D. Exploring potential for non-disruptive vertical auto scaling and resource estimation in kubernetes. In Proceedings of the IEEE International Conference on Cloud Computing (CLOUD), Milan, Italy, 8–13 July 2019; pp. 33–40. [CrossRef]
17. Song, M.; Zhang, C.; Haihong, E. An auto scaling system for API gateway based on Kubernetes. In Proceedings of the 2018 IEEE 9th International Conference on Software Engineering and Service Science (ICSESS), Beijing, China, 23–25 November 2018; pp. 109–112. [CrossRef]
18. Jin-Gang, Y.; Ya-Rong, Z.; Bo, Y.; Shu, L. Research and application of auto-scaling unified communication server based on Docker. In Proceedings of the 2017 10th International Conference on Intelligent Computation Technology and Automation (ICICTA), Changsha, China, 9–10 October 2017; pp. 152–156. [CrossRef]

19. Townend, P.; Clement, S.; Burdett, D.; Yang, R.; Shaw, J.; Slater, B.; Xu, J. Improving data center efficiency through holistic scheduling in kubernetes. In Proceedings of the 2019 IEEE International Conference on Service-Oriented System Engineering (SOSE), Newark, CA, USA, 4–9 April 2019; pp. 156–166. [CrossRef]
20. Rossi, F. Auto-scaling policies to adapt the application deployment in Kubernetes. *CEUR Workshop Proc.* **2020**, *2575*, 30–38.
21. Balla, D.; Simon, C.; Maliosz, M. Adaptive scaling of Kubernetes pods. In Proceedings of the IEEE/IFIP Network Operations and Management Symposium 2020: Management in the Age of Softwarization and Artificial Intelligence, NOMS 2020, Budapest, Hungary, 20–24 April 2020; pp. 8–12. [CrossRef]
22. Casalicchio, E.; Perciballi, V. Auto-scaling of containers: The impact of relative and absolute metrics. In Proceedings of the 2017 IEEE 2nd International Workshops on Foundations and Applications of Self* Systems (FAS*W), Tucson, AZ, USA, 18–22 September 2017; pp. 207–214. [CrossRef]
23. Casalicchio, E. A study on performance measures for auto-scaling CPU-intensive containerized applications. *Clust. Comput.* **2019**, *22*, 995–1006. [CrossRef]
24. Taherizadeh, S.; Grobelnik, M. Key influencing factors of the Kubernetes auto-scaler for computing-intensive microservice-native cloud-based applications. *Adv. Eng. Softw.* **2020**, *140*, 102734. [CrossRef]
25. Santos, J.; Wauters, T.; Volckaert, B.; De Turck, F. Towards network-Aware resource provisioning in kubernetes for fog computing applications. In Proceedings of the 2019 IEEE Conference on Network Softwarization (NetSoft), Paris, France, 24–28 June 2019; pp. 351–359. [CrossRef]
26. Santos, J.; Wauters, T.; Volckaert, B.; Turck, F.D. Resource provisioning in fog computing: From theory to practice. *Sensors* **2019**, *19*, 1–25. [CrossRef] [PubMed]
27. Zheng, W.S.; Yen, L.H. Auto-scaling in Kubernetes-based Fog Computing platform. In *International Computer Symposium*; Springer: Singapore, 2018; pp. 338–345. [CrossRef]
28. Heapster. Available online: https://github.com/kubernetes-retired/heapster (accessed on 23 June 2020).
29. InfluxDB. Available online: https://www.influxdata.com (accessed on 23 June 2020).
30. Grafana. Available online: https://grafana.com (accessed on 23 June 2020).
31. Apache JMeter. Available online: https://jmeter.apache.org (accessed on 23 June 2020).
32. Syu, Y.; Wang, C.M. modeling and forecasting http requests-based Cloud workloads using autoregressive artificial Neural Networks. In Proceedings of the 2018 3rd International Conference on Computer and Communication Systems (ICCCS), Nagoya, Japan, 27–30 April 2018; pp. 21–24. [CrossRef]
33. Dickel, H.; Podolskiy, V.; Gerndt, M. Evaluation of autoscaling metrics for (stateful) IoT gateways. In Proceedings of the 2019 IEEE 12th Conference on Service-Oriented Computing and Applications (SOCA), Kaohsiung, Taiwan, 18–21 November 2019; pp. 17–24. [CrossRef]
34. Prometheus Operator. Available online: https://github.com/coreos/prometheus-operator (accessed on 23 June 2020).
35. CAdvisor. Available online: https://github.com/google/cadvisor (accessed on 23 June 2020).
36. Prometheus Adapter. Available online: https://github.com/DirectXMan12/k8s-prometheus-adapter (accessed on 23 June 2020).
37. Gatling. Available online: https://gatling.io/open-source (accessed on 23 June 2020).

sensors

MDPI

Article

Balanced Leader Distribution Algorithm in Kubernetes Clusters

Nguyen Dinh Nguyen and Taehong Kim *

School of Information and Communication Engineering, Chungbuk National University, Cheongju, Chungbuk 28644, Korea; nguyennd@cbnu.ac.kr
* Correspondence: taehongkim@cbnu.ac.kr; Tel.: +82-43-261-2481

Abstract: Container-based virtualization is becoming a de facto way to build and deploy applications because of its simplicity and convenience. Kubernetes is a well-known open-source project that provides an orchestration platform for containerized applications. An application in Kubernetes can contain multiple replicas to achieve high scalability and availability. Stateless applications have no requirement for persistent storage; however, stateful applications require persistent storage for each replica. Therefore, stateful applications usually require a strong consistency of data among replicas. To achieve this, the application often relies on a leader, which is responsible for maintaining consistency and coordinating tasks among replicas. This leads to a problem that the leader often has heavy loads due to its inherent design. In a Kubernetes cluster, having the leaders of multiple applications concentrated in a specific node may become a bottleneck within the system. In this paper, we propose a leader election algorithm that overcomes the bottleneck problem by evenly distributing the leaders throughout nodes in the cluster. We also conduct experiments to prove the correctness and effectiveness of our leader election algorithm compared with a default algorithm in Kubernetes.

Keywords: containers; Kubernetes; leader election; load balancing; stateful

Citation: Nguyen, N.D.; Kim, T. Balanced Leader Distribution Algorithm in Kubernetes Clusters. *Sensors* **2021**, *21*, 869. https://doi.org/10.3390/s21030869

Academic Editor: Paolo Bellavista
Received: 21 December 2020
Accepted: 25 January 2021
Published: 28 January 2021

1. Introduction

Recently, container-based virtualization has emerged as a key technology to deploy applications in Cloud computing [1]. Unlike traditional virtual machines [2], a container runs at the software level within a host machine and shares the kernel with the host operating system [3]. It consumes fewer resources than the traditional virtualization method because it does not consist of an entire operating system; only the application and its dependencies are bundled into a single package. These features make containers more efficient in the deployment and scalability of applications.

In a large-scale system, it is important to have an orchestration platform to manage the container deployment. Kubernetes [4] is the most popular orchestration platform for container-based applications. It provides several powerful functions, such as automated application deployment, resource management, scaling, and load balancing. In a Kubernetes cluster, the application, which contains several replicas, is generally categorized as either stateless or stateful [5]. A stateless application has no persistent storage associated with it, whereas a stateful application requires a persistent datastore. This means that each replica in the stateful application should have its own persistent datastore. Therefore, it is important to maintain consistency among these distributed data stores of the stateful application. This consistency problem can be handled using a leader-based consistency maintenance mechanism in which an elected leader is responsible for maintaining consistency and coordinating tasks among replicas. Kubernetes provides a leader election algorithm, which is implemented by leveraging existing components in Kubernetes, to facilitate the process of using leader election in a Kubernetes cluster [6]. In the leader-based mechanism, the leader solely handles all the data update requests; therefore, the leader replica consumes a load that is heavier than that consumed by other replicas owing to the inherent design of the mechanism.

In this paper, we target a scenario in which the Kubernetes cluster acts as a Fog computing infrastructure, and several stateful applications that use a leader-based mechanism are deployed in the infrastructure. If the leaders of these applications are concentrated on a specific node, it can cause a problem wherein user requests also concentrate on a single node, possibly resulting in a bottleneck within the system. Please note that the preliminary version of this paper [7] proved that the default leader algorithm in Kubernetes does not consider such leader concentration problem that results in a significant performance reduction. Therefore, to improve the performance of the system, it is highly required to balance the number of leaders among nodes in the cluster. In this paper, we propose a balanced leader distribution (BLD) algorithm for the leader election process of stateful applications in Kubernetes clusters. The BLD algorithm improves the default leader election algorithm in Kubernetes by evenly distributing leaders throughout the nodes, thereby balancing the workload among nodes in the cluster. Through experimental evaluations, we verified the correctness and effectiveness of the BLD algorithm in Kubernetes clusters. Consequently, the main contributions of the BLD algorithm can be summarized as follows:

- It facilitates the use of the leader election process, so users can easily deploy the leader election for application in the Kubernetes cluster.
- It balances the number of leaders throughout all the nodes in the cluster, so the system performance can be improved.

The remainder of this paper is organized as follows. Section 2 presents related work. Section 3 provides an overview of the Kubernetes architecture and the leader-based consistency maintenance mechanism. Section 4 describes a default leader election algorithm in Kubernetes, and the leader concentration problem. Section 5 presents the proposed balanced leader distribution algorithm. Sections 6 and 7 present the performance evaluation and discussion, respectively. Finally, Section 8 presents the conclusions.

2. Related Work

Recently, there have been several studies based on Kubernetes. The paper [8] proposed a prediction model to improve the performance of auto-scaling in Kubernetes. This model combines empirical modal decomposition with an autoregressive integrated moving average model to predict the load of the pod. The goal is to expand the capacity of the application before the peak load by adjusting the number of pods in advance according to the prediction result. In [9], the authors investigate the horizontal pod autoscaler (HPA), which is one of the most important features in Kubernetes. They conduct various experiments to deeply analyze HPA based on several metrics, such as Kubernetes resource metrics (e.g., CPU and memory usage) and Prometheus custom metrics (e.g., the average arrival rate of HTTP requests) [10]. Based on the analysis, researchers and developers can gain a deep insight on optimizing the performance of HPA in Kubernetes. In [11], a monitoring platform was presented for dynamic resource provisioning based on Kubernetes. It collects the system resource use (CPU, RAM) and application quality-of-service metrics (response time) by using several open-source applications, such as Heapster and Apache JMeter [12]. Based on these data, it automatically analyzes and scales the application according to the resource provisioning strategy. The paper [13] presented a Reference Net-based performance and management model for Kubernetes. The goal is to identify the effects of the different interference sources (e.g., CPU usage and network usage) on the applications; therefore, the developer can consider such interference sources and improve the application's performance. In [14], a component in Kubernetes was used to build a protocol, named DORADO (orDering OveR shAreD memOry), which coordinates requests in Kubernetes. A leader is elected from among the replicas of an application. To coordinate the requests accessing the application, only the leader has the authority to define orders for handling requests, and all the replicas must execute the requests following this order. In [15], the authors proposed a load balancer for Kubernetes. The proposed load balancer can consider the running status of servers and applications (e.g., CPU and network status) to distribute requests. The users can configure the load balancing rules based on several metrics, including CPU, memory usage, and network IO. Subsequently, the server running status is collected, and the real-time load of each server is calculated to find a back-end server that can forward the requests. The paper [16] proposed a solution that allows for automatically redirecting client requests to healthy pods. A state controller was implemented to integrate with Kubernetes, and it can determine the status of a pod and assign the "active" or "standby" label to the pod. The client requests are redirected to the pods that have the "active" label, and the messages containing the state data are replicated to the standby pods. The paper [17,18] proposed a network-aware scheduling approach that is extended from the default scheduling mechanism in Kubernetes. This approach is used to deploy container-based applications in a smart city. In the Fog computing environment, the paper [19] presented a framework that is based on Kubernetes. It collects the network traffic status to provide elastic resource provisioning of the container-based application among geographically distributed Fog nodes in real time. Additionally, a few experiments were conducted to evaluate the efficiency of Kubernetes on NFV management and orchestration [20] or on deploying microservice-based applications [21].

Regarding the consensus problem in distributed systems, many studies have been conducted over several decades. Some well-known algorithms have been applied to ensure a consensus of data among replicas in distributed systems. Paxos [22], proposed by Lamport, is one of the most famous distributed consensus algorithms. One or more proposed values are proposed to Paxos, and the consensus is achieved when a majority of the replicas accept one of the proposed values. Raft [23] is a consensus algorithm that applies specific techniques that make it more understandable than Paxos. It separates the consensus problem into relatively independent subproblems, such as leader election and log replication. In OpenDaylight (ODL), which is an open source project for Software Defined Networking (SDN) controller, the datastore is distributed into shards; and these shards can be located in any node of the cluster [24]. The Raft algorithm was implemented to maintain the consistency in these distributed datastores. Paper [25] integrated Raft consensus algorithm with Kubernetes. They present evaluations of the Raft algorithm running on the physical machine and on containers managed by Kubernetes. The results showed that the throughput when executing the Raft algorithm on Kubernetes approximately was 17.4% lower than that when running directly on a physical machine; however, it is acceptable because of the many powerful features offered by Kubernetes. Paper [26] presents a solution for replica stateful containers management in Kubernetes. A coordination layer that uses Raft as a consensus algorithm was implemented. A leader replica was determined from among the replicas of an application, and the write operations were performed by only the leader replica. However, the proposed approach is complicated because it requires developers to integrate the Raft algorithm into Kubernetes and implement a firewall to redirect the requests to the leader replica.

To simplify the use of the leader election process, a leader election algorithm was implemented by leveraging existing information and components in Kubernetes [6]. However, as demonstrated in [7], this leader election algorithm does not consider the leader concentration problem in a specific node of the cluster, which results in the bottleneck problem and decreases the system performance. In this paper, a new leader election algorithm is proposed to solve this problem by attempting to balance the number of leaders among nodes in the Kubernetes cluster.

3. Overview of Kubernetes

This section presents an overview of the Kubernetes architecture and a leader-based consistency maintenance mechanism for stateful applications in the Kubernetes cluster.

3.1. Kubernetes Architecture

Kubernetes is a well-known open-source platform for automating the deployment, scaling, and management of container-based applications. A pod, which is the smallest unit in Kubernetes, represents a single instance of an application. Each pod contains a set of one or more containers, and these containers are tightly coupled, use the same I and data storage. The architecture of Kubernetes is shown in Figure 1. A Kubernete cluster usually has one or more master nodes and several worker nodes. The mast node is the control plane that is responsible for managing and controlling the cluster. contains four main components: *etcd*, *kube-apiserver*, *kube-scheduler*, and *kube-controller* [The *etcd* is a datastore that is used to store the configuration and the state of the clust The *kube-apiserver* is the front end of the Kubernetes control plane. In other words, th user or management request needs to communicate with the *kube-apiserver* to intera with the Kubernetes cluster. The *kube-scheduler* watches unscheduled pods and assign them to a node to run on based on multiple factors, such as resource constraints, affinit and anti-affinity rules. The *kube-controller* continuously watches the state of the cluster maintain the desired state. For example, it ensures the correct number of running replica of an application according to the desired configuration.

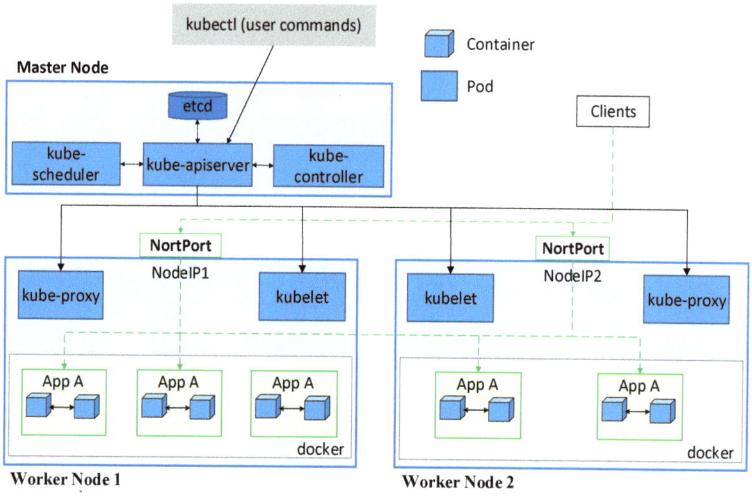

Figure 1. Kubernetes architecture.

The pods are scheduled and orchestrated to run on the worker nodes that consist o three main components: *kubelet*, *kube-proxy*, and *container runtime* [4]. *Kubelet* ensures tha the pods are running and healthy (e.g., by restarting failed pods). It reports the status o pods and node to the api-server and receives commands from the control plane. *Kube-prox* is responsible for maintaining the network rules, which allow communication with the pods from inside and outside of the cluster. *Container runtime* (e.g., Docker [27] or containerd [28] pulls the container image from a container registry and deploys the container based on that image. In Kubernetes, a pod can be created and destroyed frequently, and its IP address is updated after a restart; therefore, it is difficult to access an application using the pod's IP address. Kubernetes provides a Service that is an abstract layer enabling network access to a set of pods. The pods are selected based on their label, and all pods belonging to a Service have the same label. The Service is assigned an unchanging IP address (ClusterIP), and the requests accessing the Service are load-balanced among the pods. The load balancing policy depends on the proxy mode of *kube-proxy*. By default the *userspace* mode uses a round-robin algorithm to select the pods, whereas the *iptable* mode selects pods randomly. The *IP Virtual Server* can load balance traffic to the pods in several ways, such as destination hashing, source hashing, and round-robin methods The ClusterIP is reachable only from within the cluster. To expose the application outside the cluster, the NodePort and LoadBalancer Service can be used. The NodePort Service exposes the application on the node's IP address at a static port (NodePort). As shown in

Figure 1, clients from outside the cluster can access to the NodePort Service by using the NodeIP:NodePort address. The traffic accessing the NodePort Service is then forwarded to a backend pod according to the configuration in *kube-proxy*. The LoadBalancer Service exposes the application externally using a cloud provider that provides a public IP address, and the load balancing policy depends on the cloud provider implementation.

3.2. Leader-Based Consistency Maintenance Mechanism

Stateful applications are services that require saving data to persistent data storage, such as a database or key-value store, for use by servers, clients, and other applications [29]. The pods in Kubernetes are ephemeral in nature and do not persist data, so the data in a pod is lost once it is destroyed or restarted. To support persistent data storage, Kubernetes provides a PersistentVolume (PV) and a PersistentVolumeClaim (PVC) object. A PV is a persistent storage that has an independent lifecycle with the pod. A PVC defines several criteria (e.g., capacity and access mode) to choose the persistent storage, so it is used to claim a persistent storage that satisfies the criteria. Therefore, each pod replica of the stateful application can create its own persistent data storage by using the PVC. This PVC binds the pod to a PV that satisfies the criteria defined in the PVC.

In Kubernetes, one application can have multiple replicas to provide high availability and performance. For example, throughput and latency can be improved by using the load balancing feature in Kubernetes, which distributes the incoming requests among replicas of the application. Because each replica of the stateful application has its own data storage, deploying a set of replicas for the stateful application requires an approach to handle the inconsistency problem among these distributed databases. To handle this problem in the Kubernetes cluster, the paper [7] introduced a leader-based consistency mechanism, as shown in Figure 2. In this mechanism, a replica is elected as a leader, and the other replicas run as the followers. Read operations that clients require to read data from the storage is handled by both the leader and follower. However, only the leader is responsible for handling write operations that clients write the data into the storage. Thus, if a request for a write operation comes to a follower, it must be redirected and handled by the leader.

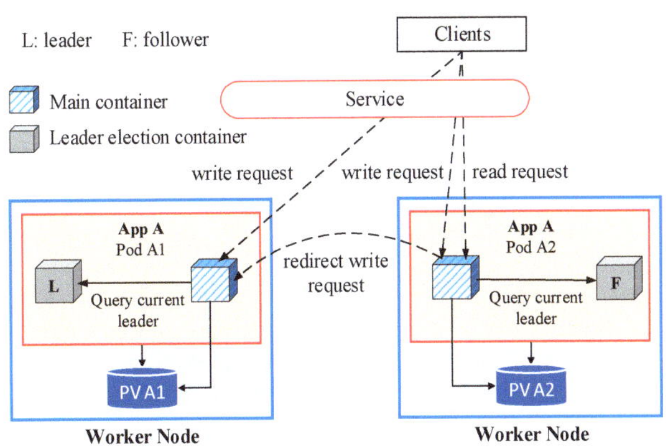

Figure 2. Leader-based consistency model [7].

To determine the leader among the pod replicas, each pod consists of two containers: a main container and a leader election container. The main container is responsible for handling incoming requests from clients, whereas the leader election container is responsible for the leader election process among replicas of the application. The leader election container provides a simple web server that returns the name of the current leader; thus, the main container can easily determine its role (leader or follower) by querying this web server in the leader election container.

4. Default Leader Election Algorithm in Kubernetes

This section presents the default leader election algorithm in Kubernetes and discusses the leader concentration problem of the default algorithm.

4.1. *Default Leader Election Algorithm in Kubernetes*

To use the leader-based consistency maintenance mechanism in a distributed system, an approach to elect a leader among the replicas is essential. Implementing leader election often requires deploying either algorithms or software such as Raft [23], Zookeeper [30], or Consul [31]. However, to avoid high implementation costs and facilitate the use of leader election in the Kubernetes cluster, a simple leader election algorithm was implemented by leveraging existing components in Kubernetes [6].

Typically, in a leader election algorithm, a set of candidates compete to become a leader in several ways. For example, the first one who successfully declares itself as a leader or the candidate who receives a majority of votes from other candidates can become a leader. Once the leader election process is completed, the leader continuously sends "heartbeats" to retain the leadership. If the current leader fails for some reason, the other candidates can be aware of that status and start a new election process to become the leader. The leader election algorithm in Kubernetes uses an annotation in the Endpoint object (EP) to hold a leader record. An example of the leader record in the EP is shown in Figure 3. The leader record includes the name of the leader (*holderIdentity*), the time when the leader renews the leader record in EP (*renewTime*), and the timeout duration (*leaseDurationSeconds*) that the follower has to wait to acquire the leadership.

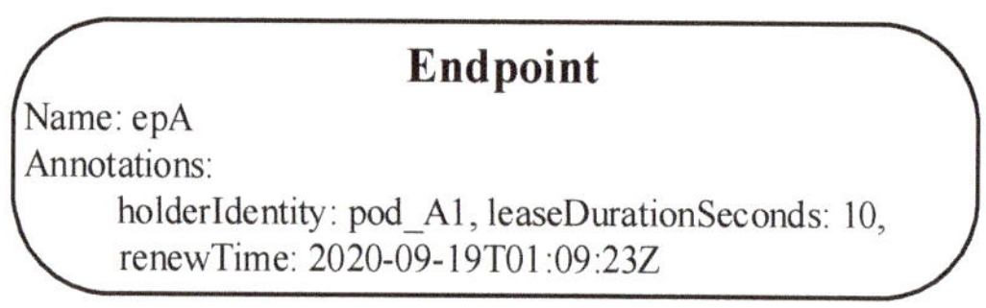

Figure 3. Example of leader record in the EP.

The procedure of the default leader election algorithm is presented in Figure 4. Once a replica starts, it runs as a follower and periodically checks the leader record in the EP to try to acquire the leadership. Please note that each replica maintains an observer record that contains the leader record copied from the EP and the observer time when the observer record was updated. First, the follower checks for the existence of an EP; if an EP has not yet been created, it creates a new EP and updates the leader record in that EP to become a new leader. If the EP did exist, the follower obtains the leader record from the EP and compares it with its own observer record to determine whether the leader record was renewed or not. If the leader record was renewed (the leader record in the EP differs from the leader record in the observer record), it updates the observer record and remains in the follower state. Otherwise, it checks the timeout by calculating the total elapsed time from the latest observation (when the observer record was updated) to the current time. If this value exceeds the predetermined timeout, the candidate updates the leader record in the EP to become a new leader. Otherwise, it remains in the follower state and periodically tries to acquire the leadership by checking the EP. The leader also has to periodically renew the leader record by updating the *renewTime* in the EP to retain its leadership.

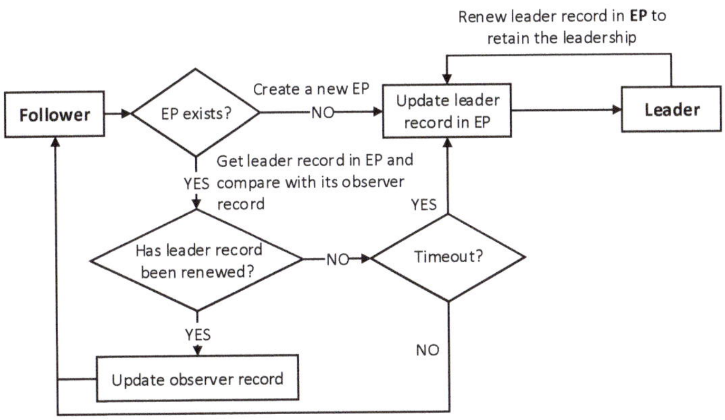

Figure 4. Procedure of the default leader election algorithm.

4.2. Leader Concentration Problem

We consider a scenario in which several stateful applications are deployed in the Kubernetes cluster as a Fog computing infrastructure. Each application has multiple pod replicas, and the application employs the leader-based mechanism for maintaining consistency among data storage of the replicas. It is obvious that the workload of the leader is higher than that of the followers because all requests for the write operation are handled only by the leader. Therefore, if the leaders of these applications are concentrated on a specific node, it may lead to a workload imbalance problem among worker nodes–one node with many leaders has a heavier workload than other nodes do. Eventually, it can cause a bottleneck in the cluster, which results in a significant decrease in the system performance, as already discussed by [7,24]. An example of the leader distribution is shown in Figure 5. There are three worker nodes, and five applications are deployed in this cluster. Figure 5a presents a concentrated leader distribution (concentrated leaders) with five leaders in node 1, while Figure 5b presents a balanced leader distribution (balanced leaders) with 2, 2, and 1 leaders in node 1, node 2, and node 3, respectively. To prove the workload imbalance problem, four clients simultaneously send write requests to each application over a period of time. Figure 6 shows the average CPU use and standard deviation of each worker node in concentrated leaders and balanced leaders. In the case of concentrated leaders, the average of CPU use in node 1 is approximately 70%, while that in node 2 and node 3 is only approximately 20%. In the case of balanced leaders, we can observe a balanced CPU use among nodes, which is approximately 50% in both node 1 and 2 and approximately 40% in node 3. Therefore, it is clear that the workload imbalance problem can occur when five leaders are concentrated in a specific node. This hinders the ability to fully exploit the computational and networking resources of the distributed system. Meanwhile, the workload can be balanced among worker nodes in the cluster when the leader distribution is balanced.

Moreover, it is important to note that although the default leader election algorithm in Kubernetes can facilitate the use of leader election in the Kubernetes cluster, it does not consider where the leader is running. Consequently, it may lead to a leader concentration problem on a specific node, resulting in a significant decrease in the system performance.

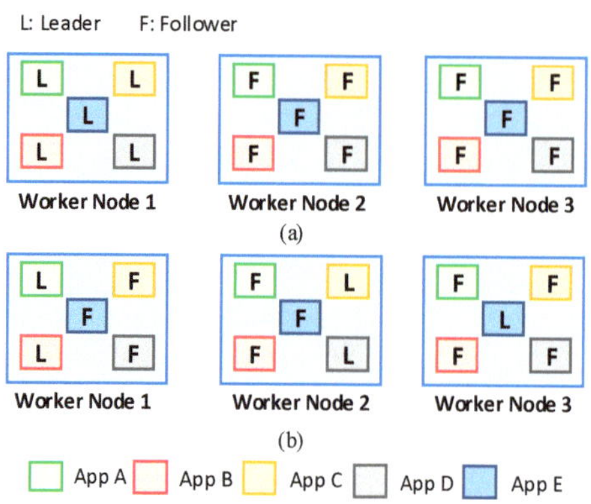

Figure 5. Example of leader distribution: (**a**) concentrated leaders, (**b**) balanced leaders.

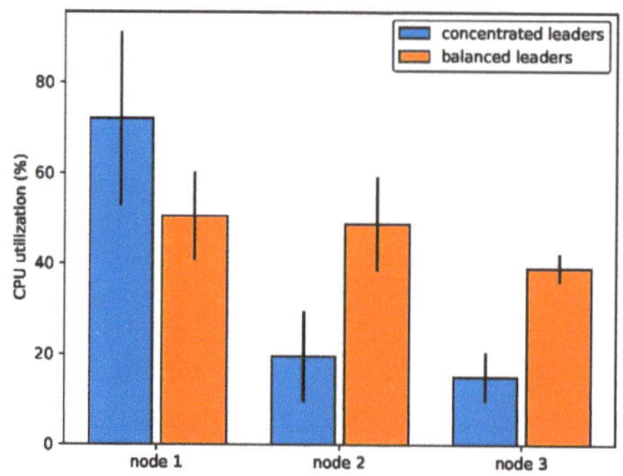

Figure 6. CPU use on each node when clients send write requests.

5. Balanced Leader Distribution Algorithm

In this section, we present the BLD algorithm that overcomes the weakness of the default leader election algorithm in Kubernetes. The proposed algorithm considers the number of leaders in the nodes to achieve a balanced leader distribution among nodes in the Kubernetes cluster. To store the information of the number of leaders on each node, we newly define an Endpoint object named Leader Management EP (LMEP), as shown in Figure 7. The number of leaders on the node is updated by the leader of the application. Once a new leader is determined, it realizes the node where it is running by using Kubernetes API and updates the leader information in LMEP. Using this information, we can calculate the current total number of leaders in the cluster ($L_{cluster}$). The number of worker nodes (N) can also be retrieved by using Kubernetes API. Let us assume that the maximum number of leaders on each node is denoted as M; the value of M can be calculated as $M = (L_{cluster} + 1)/N$, to balance the number of leaders on each node. The balanced leader distribution condition is satisfied if the number of leaders in the node where the candidate is running is smaller than or equal to M. The overview of the algorithm is shown in Figure 8. First, the replica frequently checks the leader record in the EP to try to become a leader (if it is a follower) or to renew the leader record to retain the leadership (if it is a leader). After satisfying the original conditions of the default algorithm, the follower obtains the leader information of the cluster from the LMEP and checks the BLD condition to investigate the status of the leader distribution. It becomes a leader if the BLD condition is satisfied.

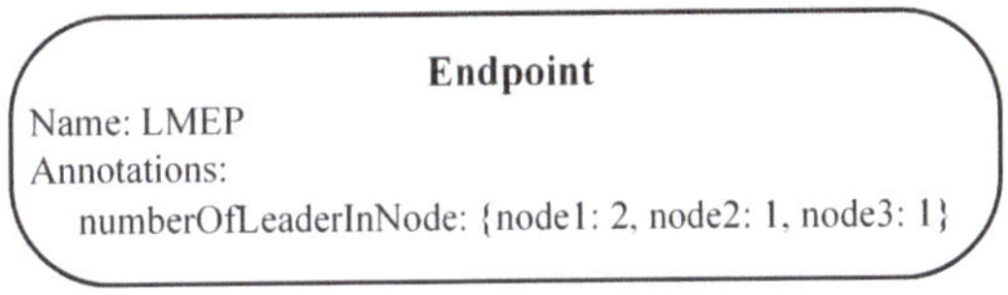

Figure 7. Example of leader information in the LMEP.

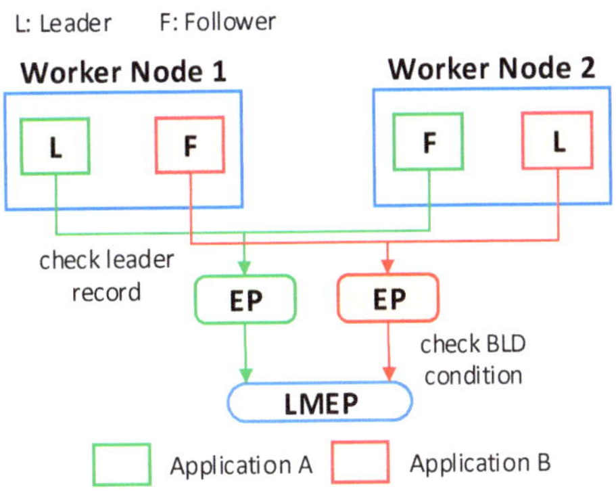

Figure 8. Overview procedure of the BLD algorithm.

The detailed procedure of the BLD algorithm is depicted in Figure 9. The replica pods start as followers, and they race to become a leader by trying to be the first one who successfully declares itself as a leader in the EP. As in the default algorithm, the candidate checks the existence of the EP. If the EP has not yet been created, it creates a new one. Then, the BLD condition is checked. If it satisfies the condition, it updates the leader information in the LMEP and EP to become a leader. Otherwise, if the EP already existed, steps similar to the default algorithm are performed—it checks the leader record in EP, updates the observer record, and checks whether the timeout duration is over. If the timeout has not expired, it remains as a follower. Otherwise, the candidate checks the BLD condition. If it satisfies the BLD condition, it updates the information in the LMEP and EP to become a new leader. If it does not satisfy the BLD condition, it remains as a follower and periodically tries to acquire leadership by repeating the aforementioned procedure. Similarly, the leader periodically renews the leader record in the EP to retain its leadership.

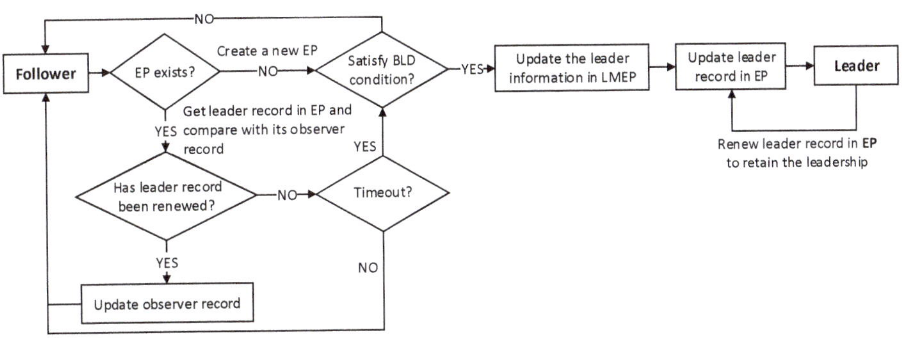

Figure 9. Detailed procedure of the BLD algorithm.

6. Performance Evaluation

In this section, we describe our experimental setup. Then, we compare the evaluation results of the leader distribution and leader election latency between the default algorithm and BLD algorithm. Finally, the effects of the leader distribution is analyzed in terms of throughput.

To evaluate the correctness and effectiveness of the BLD algorithm, we set up a Kubernetes cluster that contains one master node and three worker nodes, using Kubernetes version 1.14.10 and Docker version 18.09.6. The master node has 4 GB of RAM and four CPU cores, and the three worker nodes have 3 GB of RAM and four CPU cores. Several stateful applications are deployed in the cluster. The Hey program [32] is used to create and send requests to the applications.

6.1. Leader Distribution and Leader Election Latency

To evaluate the leader distribution and the leader election latency, we deploy a different number of stateful applications that use the leader-based mechanism to maintain consistency among replicas of the application. Each application is set to have five replicas and the experiment is repeated 100 times. Figure 10 shows a comparison of the leader distribution among the worker nodes between using the default algorithm and using the BLD algorithm. The number of leaders in each node is sorted in a descending order; therefore, highest, medium, and lowest indicate the highest, medium, and lowest number of leaders concentrated in one node. For three applications, the leader distribution is 2.2:0.68:0.1[…] using the default algorithm, whereas it is 1:1:1 using the BLD algorithm. For five and seven applications, the leader distribution in the default algorithm is unbalanced among worker nodes, with 2.94:1.44:0.62 and 4.42:1.88:0.7, respectively. The leader distribution in the BLD algorithm is balanced among worker nodes, with 2:2:1 and 3:2.04:1.96 for five and seven applications, respectively. The standard deviation, minimum and maximum number of leaders in a node are presented in Table 1. We can see that a high number of leaders can be concentrated in a specific node with the default algorithm. For example, all leaders of the applications may be concentrated in one node in case the number of applications are three and five. Therefore, it is clear that the leader distribution among nodes is balanced when the BLD algorithm is applied, whereas it is unpredictable when the default algorithm is used.

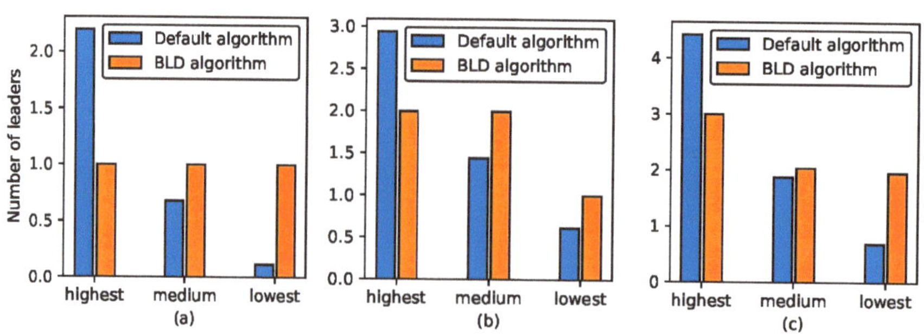

Figure 10. Leader distribution among nodes in the cluster: (**a**) 3 applications; (**b**) 5 applications; (**c**) 7 applications.

Table 1. Statistics of leader distribution.

Algorithm	Default			BLD		
Number of Applications	**3**	**5**	**7**	**3**	**5**	**7**
Std. dev.	1.01	1.2	1.79	0	0.47	0.5
Minimum value	0	0	0	1	1	2
Maximum value	3	5	6	1	2	3

Figure 11 shows an analysis of the latency of the leader election process, measured from when a new election process starts until a candidate becomes a leader. The average leader election latency in the case of the default algorithm is approximately 12 ms for three, five, and seven applications. This is because the default algorithm does not consider where the leader is; therefore, the first replica that starts the election process is highly possible to become a leader. The average latency for the leader election process in the case of the BLD algorithm is slightly higher than in the default algorithm, approximately 34 ms for three, five, and seven applications. Besides, the variation and the maximum value of the leader election latency in the case of the BLD algorithm are also higher than that in the default algorithm. This is because the BLD algorithm requires additional rounds of leader election in case the BLD condition is not satisfied. Table 2 presents the mean, standard deviation of the results, and maximum and minimum obtained values for the leader election process. Clearly, they are higher than the values obtained using the default algorithm; however, it can be considered to be a trade-off to improve throughput, which is shown in the next subsection.

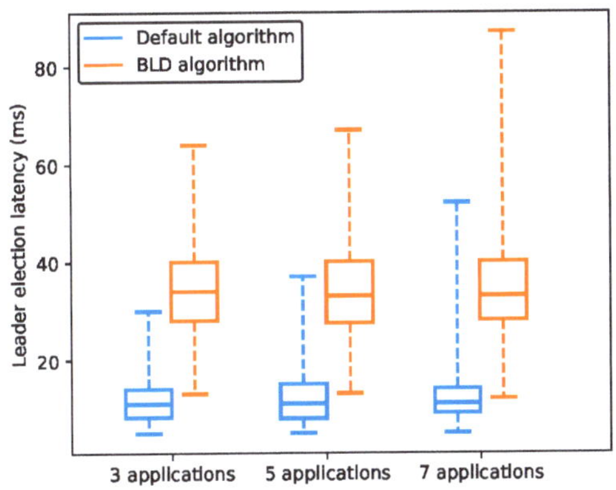

Figure 11. Leader election latency.

Table 2. Statistics of leader election latency.

Algorithm	Default			BLD		
Number of Applications	3	5	7	3	5	7
Mean value (ms)	11.91	12.44	11.92	34.46	33.41	33.77
Std. dev. (ms)	4.65	5.71	5	10.51	10.32	10.93
Minimum value (ms)	5	5	5	13	13	12
Maximum value (ms)	30	37	52	64	67	87

6.2. Effect of Leader Distribution in Kubernetes Cluster

Here, we analyze the throughput according to the replica's role and the leader distribution. The number of concurrent clients accessing each application is increased from 1 to 16. The requests are sent to the applications for 60 s. To evaluate the throughput of read and write operation based on the replica's role, one application with five replicas is deployed. The requests are sent directly to the follower or the leader of the application. The evaluation results are shown in Figure 12. For the read operation, the leader and the follower have a similar trend with increasing concurrent requests, because the read operation can be handled immediately by the leader or follower. The throughput for the write operation handled by a follower is significantly lower than that handled by a leader. For example, the throughput for the write operation in the case of the follower is 161.55

and 1330.7 reqs/s with 1 and 16 concurrent clients, respectively; whereas, in the case the leader, it is 1037.13 and 4136.76 reqs/s, respectively. This is because the write reque are handled only by the leader; if a request comes to a follower, it will be redirected to t leader. Please note that the write operation requires more computational resources a time to handle the requests than the read operation does; this is why the throughput of t write operation is considerably lower than that of the read operation. In the case of o application, we can conclude that the throughput of the write operation can be significan improved if the requests are handled directly by the leader, and the write operation tak more time and resources to handle than the read operation does.

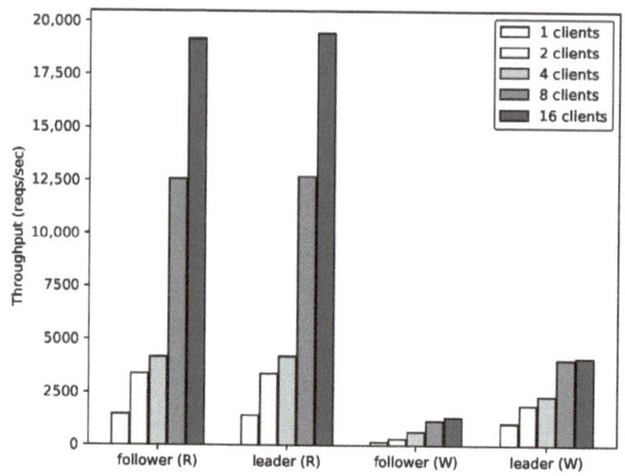

Figure 12. Throughput of one application according to replica's role and read/write operation.

To analyze the importance of balanced leader distribution, we evaluate the throughpu between two leader distribution scenarios: "concentrated leaders" and "balanced leaders which are representation results for the default leader election algorithm and the BL algorithm, respectively. Five applications, each of which contains five replicas, are deploye in the cluster. In case of "concentrated leaders", all the leaders of these applications ar assumed to be concentrated in one specific node. In case of "balanced leaders", the numbe of leaders in each node is balanced, such that three worker nodes have 2:2:1 leaders fc five applications. In this experiment, we use the NodePort service in Kubernetes to expos the service of the application to the outside cluster. The requests accessing the applicatio through the NodePort are distributed among replicas of the application using the *iptabl* proxy mode. Figure 13 shows the cumulative throughput of five applications accordin to the leader distribution scenarios and the request operations (read, write, and smar write operation). For the read and write operation, the requests access the applicatio through a NodePort service, and they can be redirected to either a follower or a leader. Th smart write operation is defined as forwarding the write requests directly to the leader o the application to avoid the overhead of forwarding requests to the leader in the leade based consistency maintenance mechanism. Figure 13a shows that the throughput for th read operation in both "concentrated leaders" and "balanced leaders" has a similar tren because requests are handled immediately by any replica. Meanwhile, the throughput fo the write operation in the case of "balanced leaders" is significantly higher than that in th case of "concentrated leaders", as shown in Figure 13b. The throughput obtained with client in the case of "balanced leaders" is approximately 20.16% higher than that in the cas of "concentrated leaders". It tends to become worse as the number of concurrent client increases. For example, the throughput in the case of "balanced leaders" is approximatel 35%, 52.03% higher than that in the case of "concentrated leaders" with 8 and 16 concurren clients, respectively. In the case of "concentrated leaders", because all the write request are handled at one specific node and that node reached its maximum capacity. By contrast in case of " balanced leaders", the write requests are distributed throughout the nodes in the cluster; therefore, the throughput keeps increasing as the number of concurrent requests increases. Figure 13c shows the throughput for the smart write operation; the "balanced leaders" case shows a significant improvement over "concentrated leaders" in terms of throughput. For example, the throughput in the "balanced leaders" case is approximately 21.93% higher than that in the "concentrated leaders" case with 1 concurrent client, and it is approximately 63.7% with 16 concurrent clients. Notably, the write request is forwarded directly to the leader in the smart write operation, whereas in the normal write operation, it is randomly forwarded to the replicas regardless of their roles. Therefore, the throughput obtained in the smart write operation is significantly higher than that obtained in the normal write operation. In the case of "balanced leaders", the throughput obtained with 1 and 16 concurrent requests in the normal write operation is 1399.76 and 3347.77 reqs/s, respectively; whereas it is 3196.41 and 7640.08 reqs/s, respectively, in the smart write operation. Overall, we can conclude that balancing the leaders of multiple stateful applications among nodes significantly improves the performance, and it can be further enhanced by implementing a smart network service that can forward requests to an appropriate replica according to the replica's role.

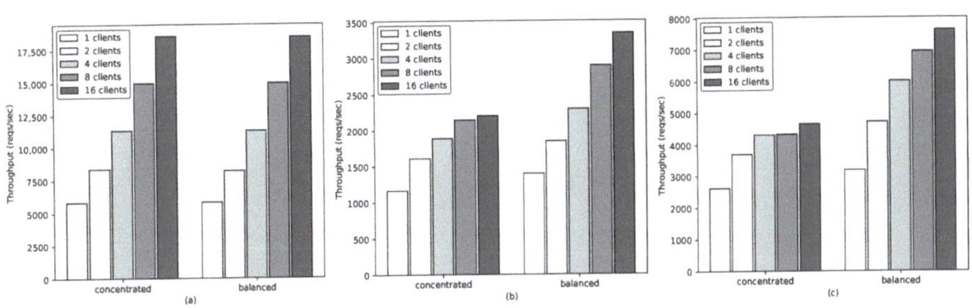

Figure 13. Cumulative throughput of multiple applications according to leader distribution: (**a**) Read operation. (**b**) Write operation. (**c**) Smart write operation.

7. Discussion

Throughout the performance evaluations, we have proved that the proposed BLD algorithm evenly distributes multiple leaders to nodes in the cluster and enhances the throughput of the cluster by balancing the workload of nodes. However, it is worth discussing the limitation of the BLD algorithm. As we have discussed in Section 6.1, the BLD algorithm causes a relatively high leader election latency compared to the default algorithm due to the design of the BLD condition check. The absence of a leader can lead to temporal service interruption. Moreover, although the leader election latency of the BLD algorithm in our experimental environment is not significantly high, there is a possibility that it may increase as the number of replicas and applications increase. Since the absence of a leader may lead to temporal service interruption, in the future works, we will investigate the effect of the leader election latency in large-scale infrastructures to improve both throughput and availability of the service.

It is also interesting to note that the throughput can be improved significantly in case the smart write operation is applied. Hence, implementing a network service in Kubernetes that is aware of the role of a replica and can forward requests to an appropriate replica according to its role is worth considering in future works.

8. Conclusions

In this paper, we described the Kubernetes architecture and the leader-based mechanism for maintaining consistent data storage among replicas of a stateful application in the Kubernetes cluster. Because the leader concentration problem can cause unbalanced resource usage among nodes, the full exploitation of the computational resources of the cluster is hindered. Therefore, we proposed a leader election algorithm that not only facilitates the use of leader election in Kubernetes but also evenly distributes the lead throughout all the nodes in the cluster. The evaluation results showed that the propos BLD algorithm can effectively balance the number of leaders among all nodes in the clus The effectiveness of the BLD algorithm was proved through a performance evaluation wi multiple applications, demonstrating that the throughput can be significantly improved distributing the number of leaders evenly throughout the nodes. There have been mo and more systems using the leader-based mechanism, and we expect that the idea of balanced leader distribution throughout the nodes is widely applied to leader electic algorithm design, to maximize the performance of the cluster.

Author Contributions: Conceptualization, N.D.N. and T.K.; methodology, N.D.N. and T.K.; softwa N.D.N.; investigation, N.D.N. and T.K.; writing—original draft preparation, N.D.N.; writing—revie and editing, T.K.; visualization, N.D.N.; supervision, T.K.; funding acquisition, T.K. Both autho have read and agreed to the published version of the manuscript.

Funding: This work was supported by the National Research Foundation of Korea (NRF) gra funded by the Korea government (MSIT) (No. NRF-2019R1F1A1059408).

Institutional Review Board Statement: Not applicable.

Informed Consent Statement: Not applicable.

Conflicts of Interest: The authors declare no conflict of interest.

References

1. Bernstein, D. Containers and cloud: From lxc to docker to kubernetes. *IEEE Cloud Comput.* **2014**, *1*, 81–84. [CrossRef]
2. Felter, W.; Ferreira, A.; Rajamony, R.; Rubio, J. An updated performance comparison of virtual machines and linux container In Proceedings of the 2015 IEEE international symposium on performance analysis of systems and software (ISPASS), Philadelphi PA, USA, 29–31 March 2015; pp. 171–172.
3. Soltesz, S.; Pötzl, H.; Fiuczynski, M.E.; Bavier, A.; Peterson, L. Container-based operating system virtualization: A scalabl high-performance alternative to hypervisors. In Proceedings of the 2nd ACM SIGOPS/EuroSys European Conference o Computer Systems 2007, Lisboa, Portugal, 21–23 March 2007; pp. 275–287.
4. Kubernetes, Production-Grade Container Orchestration. Available online: https://kubernetes.io/ (accessed on 19 October 2020
5. Deshpande, U. Caravel: Burst tolerant scheduling for containerized stateful applications. In Proceedings of the 2019 IEEE 39t International Conference on Distributed Computing Systems (ICDCS), Dallas, TX, USA, 7–10 July 2019; pp. 1432–1442.
6. Simple Leader Election with Kubernetes and Docker. Available online: https://kubernetes.io/blog/2016/01/simple-leade election-with-kubernetes/ (accessed on 19 October 2020).
7. Nguyen, N.; Kim, T. Toward Highly Scalable Load Balancing in Kubernetes Clusters. *IEEE Commun. Mag.* **2020**, *58*, 78–8 [CrossRef]
8. Zhao, A.; Huang, Q.; Huang, Y.; Zou, L.; Chen, Z.; Song, J. Research on resource prediction model based on kubernetes containe auto-scaling technology. In *IOP Conference Series: Materials Science and Engineering*; IOP Publishing: Bristol, UK, 2019; Volume 56 p. 052092.
9. Nguyen, T.T.; Yeom, Y.J.; Kim, T.; Park, D.H.; Kim, S. Horizontal Pod Autoscaling in Kubernetes for Elastic Container Orchestratio *Sensors* **2020**, *20*, 4621. [CrossRef] [PubMed]
10. Prometheus. Available online: https://prometheus.io/ (accessed on 19 October 2020).
11. Chang, C.C.; Yang, S.R.; Yeh, E.H.; Lin, P.; Jeng, J.Y. A kubernetes-based monitoring platform for dynamic cloud resourc provisioning. In Proceedings of the 2017 IEEE Global Communications Conference, Singapore, 4–8 December 2017; pp. 1–6.
12. Apache JMeter. Available online: https://jmeter.apache.org/ (accessed on 19 October 2020).
13. Medel, V.; Rana, O.; Bañares, J.Á.; Arronategui, U. Adaptive application scheduling under interference in kubernetes. In Pro ceedings of the 2016 IEEE/ACM 9th International Conference on Utility and Cloud Computing (UCC), Shanghai, China, 6–9 December 2016; pp. 426–427.
14. Netto, H.V.; Lung, L.C.; Correia, M.; Luiz, A.F.; de Souza, L.M.S. State machine replication in containers managed by Kubernetes *J. Syst. Archit.* **2017**, *73*, 53–59. [CrossRef]
15. Liu, Q.; Haihong, E.; Song, M. The Design of Multi-Metric Load Balancer for Kubernetes. In Proceedings of the 2020 International Conference on Inventive Computation Technologies (ICICT), Coimbatore, India, 26–28 February 2020; pp. 1114–1117.
16. Vayghan, L.A.; Saied, M.A.; Toeroe, M.; Khendek, F. Microservice Based Architecture: Towards High-Availability for Stateful Applications with Kubernetes. In Proceedings of the 2019 IEEE 19th International Conference on Software Quality, Reliability and Security (QRS), Sofia, Bulgaria, 22–26 July 2019; pp. 176–185.

Santos, J.; Wauters, T.; Volckaert, B.; De Turck, F. Towards network-aware resource provisioning in kubernetes for fog computing applications. In Proceedings of the 2019 IEEE Conference on Network Softwarization (NetSoft), Paris, France, 24–28 June 2019; pp. 351–359.

Santos, J.; Wauters, T.; Volckaert, B.; De Turck, F. Resource provisioning in Fog computing: From theory to practice. *Sensors* **2019**, *19*, 2238. [CrossRef] [PubMed]

Nguyen, N.D.; Phan, L.A.; Park, D.H.; Kim, S.; Kim, T. ElasticFog: Elastic Resource Provisioning in Container-Based Fog Computing. *IEEE Access* **2020**, *8*, 183879–183890. [CrossRef]

Gawel, M.; Zielinski, K. Analysis and Evaluation of Kubernetes based NFV management and orchestration. In Proceedings of the 2019 IEEE 12th International Conference on Cloud Computing (CLOUD), Milan, Italy, 8–13 July 2019; pp. 511–513.

Vayghan, L.A.; Saied, M.A.; Toeroe, M.; Khendek, F. Deploying microservice based applications with Kubernetes: Experiments and lessons learned. In Proceedings of the 2018 IEEE 11th international conference on cloud computing (CLOUD), San Francisco, CA, USA, 2–7 July 2018; pp. 970–973.

Lamport, L. The part-time parliament. *ACM Trans. Comput. Syst.* **1998**, *16*, 133–169. [CrossRef]

Ongaro, D.; Ousterhout, J. In search of an understandable consensus algorithm. In Proceedings of the USENIX Annual Technical Conference, Philadelphia, PA, USA, 19–20 June 2014; pp. 305–319.

Kim, T.; Myung, J.; Yoo, S.E. Load balancing of distributed datastore in opendaylight controller cluster. *IEEE Trans. Netw. Serv. Manag.* **2019**, *16*, 72–83. [CrossRef]

Oliveira, C.; Lung, L.C.; Netto, H.; Rech, L. Evaluating raft in docker on kubernetes. In *International Conference on Systems Science*; Springer: Berlin/Heidelberg, Germany, 2016; pp. 123–130.

Netto, H.V.; Luiz, A.F.; Correia, M.; de Oliveira Rech, L.; Oliveira, C.P. Koordinator: A service approach for replicating Docker containers in Kubernetes. In Proceedings of the 2018 IEEE Symposium on Computers and Communications (ISCC), Natal, Brazil, 25–28 June 2018; pp. 00058–00063.

Docker. Available online: https://www.docker.com/ (accessed on 19 October 2020).

Containerd. Available online: https://containerd.io/ (accessed on 19 October 2020).

Deploying a Stateful Application. Available online: https://cloud.google.com/ (accessed on 19 October 2020).

Hunt, P.; Konar, M.; Junqueira, F.P.; Reed, B. ZooKeeper: Wait-free Coordination for Internet-scale Systems. In Proceedings of the USENIX Annual Technical Conference, Boston, MA, USA, 23–25 June 2010; Volume 8.

Consul. Available online: https://www.consul.io/ (accessed on 19 October 2020).

Hey. Available online: https://github.com/rakyll/hey (accessed on 19 October 2020).

MDPI
St. Alban-Anlage 66
4052 Basel
Switzerland
Tel. +41 61 683 77 34
Fax +41 61 302 89 18
www.mdpi.com

Sensors Editorial Office
E-mail: sensors@mdpi.com
www.mdpi.com/journal/sensors

www.ingramcontent.com/pod-product-compliance
Lightning Source LLC
Chambersburg PA
CBHW041406210926
51850CB00013B/3190
9783036514567